MINIMUM RISK STRATEGY

for

Acquiring

Communications

Equipment and Service

The Artech House Telecommunication Library

Vinton G. Cerf, *Series Editor*

MINIMUM RISK STRATEGY

for

Acquiring

Communications

Equipment and Service

Nathan J. Muller

General DataComm, Inc.
Middlebury, Ct.

Artech House

Library of Congress Cataloging-in-Publication Data

Muller, Nathan J.
    Minimum risk strategy for acquiring communications equipment and services
/ Nathan J. Muller.
        p. cm.
    Bibliograhy: p.
    Includes index.
    ISBN 0-89006-304-4 : $66.00
        1. Management — Data processing.   2. Mangement — Communication
    systems.   3. Computers — Purchasing.   4. Telecommunication —
    Apparatus and supplies — Purchasing.   5. Risk management.   I. Title.
    HD30.2.M85 1988
    658'.05 — dc19   88-8097
                                                                              CIP

Copyright © 1989

ARTECH HOUSE, INC.
685 Canton Street
Norwood, MA 02062

International Standard Book Number: 0-89006-304-4
Library of Congress Catalog Card Number: 88-8097

    10   9   8   7   6   5   4   3   2   1

To Linda, Matthew, Robin, Barbara, Marie, and James. And my parents, Helen and Adler. In memory of Peter, Karen, and Theresa.

Contents

Introduction

The emerging consensus in corporate America is that today's communication network is a strategic resource that can be utilized for competitive advantage or, at the least, that can create the opportunity for competitive advantage. A properly planned and maintained network can help differentiate a company from its competitors. This is especially true in the service sector of the economy, particularly with regard to those companies which operate in highly regulated markets: commercial banks, financial institutions, brokerage houses, and insurance companies. Although such entities cannot easily increase profitability through product diversification, they can compete for customers through service differentiation; that is, by providing better, faster, more economical service.

Unregulated companies have more latitude in their quest for profits. Not only can they introduce new products, they can use technology to improve employee productivity, reach new customers, and better their responsiveness to existing customers — all of which can translate into increased market share over the long term.

Considering the cost of acquiring, maintaining, and expanding communication resources, a risk avoidance strategy is an absolute necessity. Risk avoidance permeates consumer buying decisions; there are dozens of books available that provide advice about the purchase of real estate, cars, and household appliances like stereos, TVs, and VCRs. Until now, however, there has been no single-source reference for corporate managers on how to buy expensive high-technology products, let alone a reference that is both a technical tutorial and a systematic strategy for assuring vendor stability and product reliability.

In a five-year study released in 1988 by Digital Equipment Corporation (Maynard, Mass.) [1], it was found that personnel costs accounted for 29.3 percent of the total cost of network operations among 17 major corporations. At the same time, the costs for hardware and communication facilities combined amounted to 50 percent of network costs, which would suggest that continued capital investments in communication hardware and services are deemed necessary to keep pace with organizational change and to respond effectively to changing market conditions. At the same time, the heavy investment in specialized personnel required to operate and maintain corporate networks points to the need to provide that personnel with the proper decision-making tools for expanding and upgrading those networks. Without such tools, there can be no leadership in discovering new applications of technology, or in making more economical use of network resources. Without that leadership, organizations cannot be sure of an adequate return on their personnel and capital investments.

Companies can incur other penalties as a result of poor purchasing decisions:

- Missed corporate objectives;
- Damage to competitive position;
- Wasted capital resources.

It is not the purpose of this book to compare the specific features and capabilities of the multitude of products currently available through numerous vendors. Rather, the purpose of this book is to introduce you to an innovative approach to making purchase decisions — an approach that is based on asking the right questions to minimize the risk inherent in evaluating vendors and selecting the products that are most appropriate for your applications.

In eliminating risk, the organization will more quickly realize its objectives and enhance its competitive position, thereby getting the most out of its strategic resources. All this is not possible, however, unless managers are equipped with the means to make more informed purchasing decisions. A sound purchasing decision that results from a systematic evaluation process can help managers maintain their credibility with subordinates, peers, and top management.

This book is intended to be used as a blueprint with which managers can navigate their way through the vendor's organizational structure to pinpoint areas of weakness that can affect future performance in supporting the product line. This evaluation strategy will greatly reduce the risks of doing business with every kind of high-technology vendor, be it start-up, venture capital financed, privately owned, or established industry giant.

The reader will typically have the responsibility for approving, recommending, or specifying the purchase of computer or telecommunication products. Some readers may occupy an executive or managerial position and have oversight responsibility for implementing and maintaining a corporate or government agency network. Other readers may have planning responsibilities, determining organizational needs and preparing proposals for top management review.

Individuals who anticipate moving into management positions in the future will find this book useful in preparing for such responsibilities. The book will also be of value to consultants, who typically perform such evaluations for their clients, and who frequently find themselves in the unenviable position of having to provide those clients with detailed justifications for their decisions.

Venture capital firms and institutions may derive insights that will prove useful in determining the investment potential of promising high-technology companies, thereby minimizing the risk of committing investment funds on behalf of their shareholders.

To any students who may be reading this book as a part of their course work in business administration, management information systems, computer science, or telecommunication studies, I would like to be the first to welcome you to the real world of high technology. This book will give you a leg up on the competition you will shortly experience in getting your first job in one of these exciting fields. I hope you will hang on to this book both as an immediately useful tool and as a long-term reference.

Specific companies and products are mentioned in this book for illustrative purposes only. Such references do not constitute an endorsement of any kind, expressed or implied. I have made every effort to verify the accuracy of information regarding specific products and companies. Some facts and figures may have changed slightly during the interim period between final manuscript preparation and the book's publication. In a dynamic industry marked by hyperpaced innovation, that is to be expected. Nevertheless, any errors that may have escaped notice are regrettable and will, I hope, not detract from the book's overall value.

Nathan J. Muller
Southbury, Connecticut

REFERENCE

[1]   Digital Equipment Corporation (Maynard, Mass.), "The Cost of Network Ownership," Rev. 3.0, February 1988.

PART ONE

You have to be very careful of vendors. If you don't ask the right questions, they aren't going to tell you the answers. [1]

Chuck Garrison
Assistant Vice President
Telecommunications Trading Operations
Chicago Board Options Exchange
Chicago, Illinois

Chapter 1
Risk Avoidance: A Twelve-Point Overview

In days past, if you wanted to upgrade your network or changeout an outmoded mainframe entirely for something that promised more horsepower, features, and configuration flexibility, you issued a Request for Proposal (RFP) outlining your requirements. You were a shrewd network manager, so your RFP asked vendors for recommendations on specific products and their application to the planned requirements.

Vendors, eager for the chance to begin a long-term relationship by demonstrating their understanding of your unique needs, would respond with product recommendations, pricing, and availability information. They poured forth their advice on the best way to implement the upgrade or changeout. They pinpointed perceived weaknesses in your plan and tactfully proposed alternatives that would deliver the performance you needed, while bolstering their credibility by identifying additional cost savings.

More often than not, the buying decision was made based on the reputation of the vendor, the demonstrated reliability of its products, trusting words from its technical support staff, and the flawless reasoning that jumped off the pages of its impeccably drafted proposal. Yes, those were the good old days . . . .

Building or upgrading today's communications network has become a pretty risky business. No matter how impressive vendor financial statements look, it is not uncommon for companies inadvertently to buy nonexistent products, to find themselves boxed into proprietary technologies, or to become stranded without ongoing support. Some of the variables that have combined to intensify the risk of purchasing network products include:

- The deregulated communications industry, now crowded with start-up companies compelled to compete on price to break into high-end markets (see Table 1.1 for typical profile).

Table 1.1 Profile of Typical Start-Up Company.

Research and Development:	Technically-oriented founder or founders finalize prototype of the first product. Disorganized administratively, but entirely product focused. One or two other products are in the discussion stage. Self-financed at the start; limited bank credit; overly optimistic about market demand and revenue projections. Will probably need infusion of capital to finish development, market the product, and start production.
Applications Engineering:	Founder or founders install the product to suit customer's application and configuration. They work closely with a small number of key customers who "trial" the product and provide feedback for possible design modifications. Otherwise, no dedicated field-service staff.
Quality Assurance:	Quality-assurance procedures evolve with the product, mostly through trial and error.
Repair and Return:	Founder or founders test and repair faulty units returned by key customers, making design changes as necessary. Until production begins, there will be no dedicated test and repair facilities.
Customer Service:	Founders work with key customers to monitor product performance and to provide advice on feature implementation.
Documentation:	Little or no documentation available; founders will "hand hold" customers until they can afford to develop documentation.
Training:	Nonexistent; founders provide introduction to the product during initial sales presentation and then follow up with assistance over the phone as necessary.
Line of Business:	Single-product oriented until sales revenues justify the development of new products.
References:	Will provide the names of key customers who are satisfied with the product; but may have to rely on résumé of founder or founders to demonstrate industry experience and performance record.
Escrow Protection:	Founders may be persuaded to open an escrow account, if it will result in a sizable sale; otherwise, will resist the concept as not worth bothering with.
Quality of Salespeople:	As the principal salesmen, founders are very knowledgeable about their product and its applications. Initially, they will be reluctant to trust others with approaching potential key customers. They will give up sales activities only when other operations threaten to overwhelm them.
Human Resources:	Founders have high emotional stake in the success of their fledgling enterprise. They tend to be overly critical of others' performance and insist on doing things their own way, ignoring advice from new hires, consultants, and potential investors in the belief that nobody knows their business as well as they do. At the same time, founders do not communicate their objectives, find it difficult to delegate responsibility, and have weak management skills.

Risk Factor: Very high.

Table 1.2 Profile of Typical Venture Capital–Financed Company.

Research and Development:	Technically-oriented founder or founders finalize prototype of the first product, but have run out of capital and credit to bring it to market. First-round venture-capital financing provides infusion of capital to finish development, market the product, and start production. Founders may have to give up majority stake in their company and accept professional management to oversee operations.
Applications Engineering:	Under supervision of the founders, new employees assume responsibility for installing the product to suit customer's application and configuration.
Quality Assurance:	Quality-assurance procedures that have evolved with the product are refined by professional management brought into the company by the venture-capital firm.
Repair and Return:	Faulty units are tested and repaired by production personnel as required until increased workload justifies dedicated staff and facilities.
Customer Service:	Experienced technical help replaces the founders as the principal point of contact for customers experiencing problems.
Documentation:	Crude documentation is put together to satisfy customers until the comprehensive reference package is ready for release.
Training:	Customer training is performed by technicians until product demand justifies experienced trainers who are dedicated to that responsibility.
Line of Business:	First product into production; second product in prototype; third product in the design stage.
References:	Company routinely provides the names of key customers who are satisfied with the product.
Escrow Protection:	Company is receptive to the concept of escrow accounts.
Quality of Salespeople:	Professional salespeople expand customer base; founders go after key accounts.
Human Resources:	Founders retain high emotional stake in the success of their company, but, as the company grows, increasingly feel the loss of control. This may become a source of frustration and eventual disillusionment.

Risk Factor: High.

- The emergence of small, aggressive start-up companies funded by equally aggressive venture capital firms, the overriding concern of which is to bail out of the toddler company with windfall profits within three to five years (see Table 1.2 for typical profile).
- The so-called "yuppie" philosophy, which has become entrenched in the minds of young marketing and sales support professionals. In their quest for instant success and recognition, the yuppies see nothing wrong with promoting the reliability and performance of products that are still in the design stage.
- The ease with which bankruptcy statutes can be invoked by financially strapped vendors to protect themselves from creditors — and customers who have put up money to insure timely customization or product delivery.

Managers who recommend, specify, or approve product purchases cannot afford to ignore the potentially harmful effect these variables can have on their networks, organizations, or careers. For example, it is taking more time and becoming more difficult to reconcile the differing claims of vendors. New innovations in technology, quantum leaps in the rate of product obsolescence, and the ever-increasing number of suppliers not only make easy comparisons impossible, but also increase the chance of making the wrong decision; and if all that is not enough, there is an emerging trend, particularly among software developers, toward "dis-improvement," a word coined by Stephen Manes to describe changes to products that provide only marginal increases in performance, but at the cost of requiring substantially more memory, increasing the product's complexity, and offering less support for peripheral devices than previous releases [2].

The wrong purchasing decision can have a ripple effect throughout the organization. In addition to risking loss of capital, expending the time and effort to remedy adverse situations diverts valuable human resources from other productive pursuits. Beyond that, degraded network performance may be severe enough to jeopardize the company's competitive position. An uninformed purchasing decision can create an irreversible credibility gap that may limit a manager's advancement opportunities, and, if the mistake is serious enough, it can cost a network manager his or her job.

Recognizing these risks, many network managers routinely ask for a financial statement and credit references in the Request for Proposal (RFP) as a condition for considering submitted vendor proposals. As a quick and dirty screening technique, this "snapshot" approach may or may not be effective in identifying and disqualifying undercapitalized firms. At any rate, this kind of information is not an indicator of a vendor's organizational stability, nor can it be used to determine the reliability of its products.

A vendor with a promising but unproven product can make itself look good on paper. For example, loans may be secured against order backlog instead of capital assets. If you have no experience in accounting, you might easily conclude that the vendor is solidly in the "black." At the other end of the spectrum are the vendors who started out genuinely cash rich, but ended up product deficient. The well-publicized experience of digital private branch exchange (PBX) manufacturer Ztel, Inc. (Cambridge, Mass.), in recent years amply demonstrates the pitfalls of dealing with such firms.

Ztel attracted a lot of investment capital in the early 1980s by being among the first companies to pursue development of an integrated voice-data PBX. Eventually marketed as the Private Network Exchange (PNX), its development hit snag after snag, production stalled, and the first units proved to be quite unreliable. A probe by the investors ended with accusations of gross mismanagement and lavish spending. The new management team could not overcome the damage that had already been done. One by one, the original investors pulled out. By late 1987, the company existed in name only, leaving its 40 customers to fend for themselves without a source of spare parts or service.

As companies try to compete effectively in their respective markets, network managers will increasingly be drawn into the fray. In trying to update and reconfigure their networks to achieve ever-greater efficiencies, they must not only become more market-oriented, they must keep pace with changes in technology to protect their company's competitive position, and, as more companies come to realize the value of their communications networks as strategic resources, the network manager will have to develop finely honed evaluation skills for properly selecting vendors and their products. Thus, network management promises only to become a more vendor-intensive profession. For network managers, the question that begs an answer is: "How do I protect myself and my company from high-risk vendors in today's fiercely competitive marketplace?"

To improve the quality of purchase decisions, a systematic evaluation of a vendor's organizational structure, operating philosophy, and performance record in a variety of categories is required. This twelve-point plan is offered as the core for such a risk avoidance strategy. Also offered is a decision-making model highlighting the risk avoidance strategy within the overall decision-making process (Figure 1.1).

1.1 RESEARCH AND DEVELOPMENT

Your objective in probing the vendor's research and development efforts is to determine its track record in translating research and innovative thinking into production prototypes, then into finished, marketable goods. You can start by looking for a formal, budgeted program with dedicated staff. If product development looks like it is a haphazard effort under a master tinkerer, you will not have a basis for predicting the product's long-term reliability, nor will you be reasonably assured that enhancements will be developed to extend the useful life of your investment.

To validate vendor claims of offering "innovative" products, find out for which unique features the product development group is responsible. Does the company hold any patents or software copyrights? Does it license proprietary technology to other manufacturers? Does it supply key subsystems for any products already on the market? Does it design and build key componentry based on Large Scale Integration (LSI) or Very Large Scale Integration (VLSI) to give its products price or performance advantages? If the answer to these questions is no, find out what the vendor means when it describes its products as "advanced," "state-of-the-art," "leading edge," "innovative," or "unique." Many times, these words and phrases are just marketing jargon designed to get your attention.

After establishing the definition of terms — and quite possibly finding out that the vendor cannot live up to them — inquire about the performance record

THE ROLE OF RISK AVOIDANCE IN THE PURCHASING PROCESS

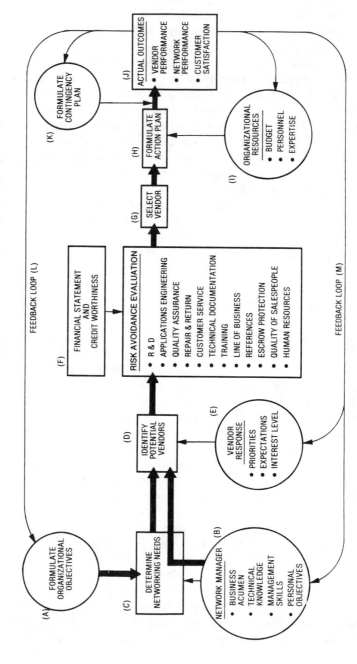

Figure 1.1 The role of risk avoidance in the purchasing process.

Key to Figure 1.1

(A)   Building, upgrading, or reconfiguring a network starts with the articulation of organizational objectives. A simple objective might be to "serve customers faster and more efficiently." A more complicated objective might be to reorganize the company to permit a faster response to changing market conditions. Whatever the objective, the network can play a pivotal role in the organization's ability to achieve it.

(B)   With the organization's objective in mind, the network manager will draw upon his or her business acumen, technical knowledge, and management skills to determine what is required to build, upgrade, or reconfigure the network. The network manager's personal agenda also comes into play. This agenda might include the hope of a promotion or raise for a job well done, the expectation of obtaining additional staff, or increasing visibility among fellow employees or peers in the industry.

(C)   From a hardware-software standpoint, the requirements of the network have been at least tentatively determined. With input from other departments which will be affected by the project, a plan is developed with all milestones and time frames for completion identified. At this point, an RFP may be developed.

(D)   The RFP is issued to appropriate potential vendors who will respond with specific information on products, pricing, and availability.

(E)   The vendor's response will reflect its priorities, expectations, and interest level at the time the RFP is received. For example, the vendor may respond to the RFP with a "no bid" because it cannot possibly deliver the product within the specified time frame, due to its order backlog. The vendor's expectations would include the price and terms it can exact from the customer, and the resulting publicity and marketing value that would come from a job well done. The vendor's decision to respond to the RFP may also be influenced by its willingness to be evaluated according to the criteria listed in **(F)**.

(F)   The network manager or evaluation team will select the best proposals and evaluate the vendors according to financial statements and credit references, in addition to the twelve risk factors.

(G)   The vendor is selected on the basis of the strength of its proposal and the outcome of the risk-avoidance evaluation.

(H)   Together, the network manager and the vendor formulate an action plan that includes delivery, installation, and cutover dates, as well as training schedule. A contingency plan is devised, which becomes effective upon missed deadlines or poor vendor performance.

(I)   The action plan takes into account the organization's budget for the project, the availability of personnel to work with the vendor, the cumulative technical expertise available to support the plan, and the organization's level of commitment to achieving its objective.

(J)   Upon completion of the project, vendor performance is evaluated. The network is monitored for proper operation in relation to performance expectations. Users would provide informal evaluations to the network manager to make known their level of satisfaction. If any of these outcomes does not result in the achievement of the organizational objective, several options are available to remedy the situation, including **(I)** and **(K)**.

(K)   If an adverse outcome cannot be corrected by simply revising the action plan, the contingency plan may be invoked. For example, the vendor misses the cutover date and incurs a financial penalty.

(L)   This feedback loop has two possible directions. The project is evaluated against the organizational objective **(A)**. The entire process is repeated, with appropriate modifications, for each successive upgrade or reconfiguration. Alternatively, the contingency plan **(K)** may be invoked if the outcomes do not result in the achievement of the organizational objective.

(M)   This feedback loop has three possible directions. The outcome of the project may require additional organizational resources **(I)** and a revision of the action plan **(H)**. For example, the organization may decide at this point to implement a product enhancement earlier than anticipated, in which case it must be factored into the decision-making process at points **(I)** and **(H)**. The outcome of the project may reveal the need for additional products or services to achieve the organizational objective. In this case, the feedback loop brings the network manager back to point **(E)** in the decision-making process for a second look at other vendor responses to the original RFP, or to point **(B)**, where the network manager uses his or her combined skills and experience after determining another network need **(C)**.

Note: This decision-making model may take a year or more to implement, depending on the nature and scope of the project. Also, the network manager may find this model useful for projects already in progress. For example, if RFPs have already been issued, it is a simple matter to perform risk assessment **(F)** to select the vendor from among the two or three strongest proposals. If the vendor has already been selected, this model can be implemented at point **(H)**.

of products already deployed in networks similar to yours. If the product has undergone design changes since its introduction, find out how many changes were made and the vendor's rationale for each change. The answers may help you determine product quality, levels of customer satisfaction, and the vendor's responsiveness to changing customer requirements.

To determine the vendor's commitment to the product, check its research and development budget, as well as the percentage of product development costs to annual revenues. Make sure the staffing level for the product is adequate to accomplish its objectives within a reasonable time frame. Determine what kind of enhancements are planned or being developed, their time frames for release, how they will be offered to customers when they become available, and at what cost. Find out to what extent future enhancements will impinge on the product's basic design. For example, will the enhancement require factory modification of the existing product, or will it be accomplished on customer premises with a changeout of chips, board, or software? The answers may reveal the extent to which the vendor has thought through the life cycle and growth path of its product.

Determine how effective the vendor is in planning new products and enhancements by checking into the number of completed *versus* canceled projects. If the vendor's record of canceled projects strikes you as high, it may indicate that money is being thrown at projects to establish instant market position. If so, keep in mind that this strategy is very risky, even for industry giants.

Compare the present product development budget with those of previous years. If the current budget is significantly lower than past budgets, find out the reason. If the vendor has recently instituted layoffs and other cost-cutting measures, find out to what extent product development is affected. You must walk away convinced that the vendor is not shortchanging long-term development efforts in the hope of improving short-term revenues. If this appears to be the case, you may not be getting the enhancements that were promised at the time of purchase or the level of technical support you might require in the future.

Another way to check the vendor's record in product planning is to compare product announcement dates with product release dates. If there has been a discrepancy between the two dates, ask for an explanation. Such questions may uncover serious problems that should be factored into your decision making. Some red flags include software problems, personnel turnover, and budget cuts. In these circumstances, you should investigate further before you buy.

A variety of problems can sidetrack a product's release, as the experience of Avant-Garde Computing, Inc. (Mt. Laurel, N.J.), clearly shows. The company had announced the release date of its highly publicized Net/Advisor, an expert system designed to manage multivendor networks. Then, in third quarter 1987, the company postponed the product's release date indefinitely, citing the difficulty of designing a system that can monitor the network and make recommendations on alarms and vendor-variable data.

In response to market pressures, many companies routinely announce availability dates for products before the completion of development. This is strictly a marketing ploy intended to stir industry speculation and confuse customers just enough so that they will stop purchasing the products of competitors. The danger in waiting for such products, of course, is that their availability may be delayed beyond your ability to wait — worse, the product may never see the light of day — and if the company has been bootstrapping product development with deposits from customers, there may be no money left in company coffers to refund.

Some companies make it very difficult for buyers to determine whether a product is available now, or is still under development. Performance numbers are touted as if the product exists, but they may be based on projections. Sales literature even refers to the product in the present tense, and demonstrations of the product at trade shows are only representations of intended functionality, rather than present levels of functionality.

It should be pointed out, however, that the failure of many companies to deliver what they promise is not always the result of deliberate deception. More often than not, the failure is the result of good intentions gone awry; typically, the result of underestimating the nature and scope of the product development effort. Upon agreeing on the next product or enhancement to be developed, determining that its market potential is very high, and that there is no real competition, the mood within the company is understandably very upbeat. Unfortunately, this exuberance results in unrealistic goal setting; everybody from hardware engineers and software developers to systems integrators thinks they can accomplish their part of the job sooner than their abilities or work environment will permit. Only when the various projects are well underway does a sense of reality take hold. Suddenly people need more time, modifications are made, deadlines lapse, and the release date for the product is postponed, leaving an ugly black spot on the company's image.

Despite the risks, software can be a very lucrative business. The bulk of a software company's costs involves spending for research and development and, to a lesser extent, marketing. After a certain sales volume, costs are covered and subsequent sales are largely pure profit. This sounds like a surefire formula for success, and a few companies like Microsoft Corporation (Redmond, Wash.) have even made it look easy with DOS-based products — generating growth and profits that have catapulted them into positions of formidable strength. Start-up companies, however, are finding it difficult to achieve even modest levels of financial success with products based on UNIX. Part of the problem is that UNIX standards are still evolving, and every company uses its own version. For customers, the risk in dealing with such firms is twofold: a UNIX package purchased from one vendor may not be compatible with the UNIX package purchased from another vendor; and UNIX-based products purchased now may not be supported for very long because the industry is on the verge of consolidation, which may be necessary to establish standards and insure the long-term viability of UNIX.

Another way to evaluate a vendor's product development plans is to discuss what the product will look like in one to two years. You should ask about the levels of functional control the product will have, what features it will provide, and what technology might be used to achieve design requirements. With Integrated Services Digital Networks (ISDN) already in the formative stages of implementation, it pays to find out if the products you are buying today will be upgradable or compatible with ISDN tomorrow. Detailed "futures" information will be viewed as proprietary, of course, but your immediate objective is to get a general idea of how the vendor intends to protect your investment, and to satisfy yourself that some kind of plan does indeed exist.

Find out if the vendor is taking advantage of any new components in its prototypes. If so, it indicates that the product will be endowed with operating efficiencies that substantially improve the price-performance ratio of the existing line, but be on the alert if the vendor is sole sourcing critical componentry to leapfrog the competition. To minimize risk, check into the availability of production quantities of these components to insure that delivery of your purchase will not be delayed by their initial scarcity. All too often a vendor will base its product development efforts on trial quantities of components with the idea of committing to bulk quantities once the product reaches the manufacturing stage. This kind of arrangement is no guarantee that the vendor will get the production quantities needed, when they are needed. Bigger customers with the strategic relationships already established will have priority. For some components, it may take six months or more for the vendor's supplier to fulfill pent-up demand. Of course, the vendor can compensate for shortages of critical components by engineering around them and using off-the-shelf parts that are available. The disadvantage of doing this, however, is that the price-performance ratio of the product may not be quite the same.

Sometimes, however, initial component scarcity cannot be overcome easily even by industry giants. One extreme and widely publicized example of how component scarcity can affect product introduction and delivery comes from Intergraph Corporation (Huntsville, Ala.). The company had designed a line of high-end interactive graphics and mapping workstations and servers that used Fairchild Semiconductor Corporation's (Cupertino, Calif.) 32-bit Clipper chip set as the core processor. The Clipper was chosen by Intergraph for its enormous performance rate, approximately seven times the performance rate of DEC's venerable VAX 11/780, but the scarcity of the chip sets was so severe that Intergraph's order backlog rapidly reached 2,000 by mid-1987, threatening its continued expansion into various computer aided design (CAD) markets. To satisfy demand, shipments of the workstations were diverted from software developers who needed them for designing applications programs. The lack of applications programs for specific markets, in turn, threatened continued demand for Intergraph's products in those markets. To resolve this dilemma, Intergraph had considered buying a 20% stake in Fairchild just to insure adequate supplies of the chips. In the end, Intergraph was able to buy Fairchild's Advanced Processor Division — which developed the Clipper — from National Semiconductor Corporation (Santa Clara, Calif.) when it purchased Fairchild.

Intergraph's experience is not an isolated case; component shortages are really quite common. By the mid-1980s, U.S. companies like Motorola abandoned the production of dynamic random access memory (DRAM) chips to Japanese conglomerates, which had demonstrated their ability to produce higher capacity chips far more economically than their American counterparts. With the introduction of 80386-based microcomputers and the success of high-powered RISC-based (Reduced Instruction Set Computing) workstations, however, the demand for DRAM chips in first quarter 1988 quickly overtook the supply, and prices soared accordingly. The situation was exacerbated by some companies which double and triple ordered the chips to insure adequate supplies for themselves. On the open market, however, one-megabit chips that were initially priced at $15 apiece soared to around $40 by March of 1988. Small board manufacturers and system developers were forced into raising the prices of their products, or shipping their products with minimal memory. Other companies had to stop taking orders and discontinue manufacturing their products because of the unavailability of memory. At this writing, the DRAM drought is expected to continue into first quarter 1989.

Do not be surprised to find vendors single-sourcing their componentry and other piece parts. Twenty years ago, alternate sources were popular as a means to keep production lines going despite strikes, natural disasters, and the vagaries of the marketplace. In doing business with a multitude of suppliers, however, purchasers could not possibly keep close ties with every one of them, and having too many suppliers diluted purchasers' buying power. With the cost of maintaining alternate suppliers rising, purchasers turned to material requirements planning (MRP) and the "just in time" (JIT) method of staging materials for the production line. By developing strategic alliances and arm's-length partnering arrangements with a single supplier, purchasers not only eliminated the necessity of stockpiling an inventory, but enhanced their buying power to reduce the cost of materials.

Single-sourcing is still a risky way of doing business, but today there is no getting around it; the safety net of having multiple suppliers is just too expensive for most firms to maintain. Supplier reduction programs will continue throughout the 1990s, which means that customers will have to factor that risk element into their decision making. To minimize the risk associated with doing business with a vendor who single-sources key componentry, find out why the vendor chose a particular supplier for its componentry, how long it will take for the pipeline to dry up in case of disruption, and what contingency plan the vendor has for obtaining components from another source on a temporary basis.

If the vendor is using a start-up supplier as the source for key componentry, the risk of not getting your purchase delivered on time increases exponentially. The reason ties back to the previous point, the industry trend toward supplier reduction. It is the small, start-up firms that fall victim to this practice first, and this hinders their ability to survive. After all, small suppliers have limited product lines and less flexibility in pricing. They count on doing business with a very broad customer base so they can bootstrap their development efforts and, they hope, become leading components suppliers. You can minimize your risk by finding out what makes this small firm so special to your vendor. It may be that the small firm offers your vendor a migration path to an emerging technology. It is quite possible that the small firm is more willing to customize its offerings than the broad-line components makers, who are more interested in batch selling whatever comes off the production lines. The small firm may be supplying your vendor with components that have distinct advantages, such as greater speed, smaller packaging, and less power consumption — all of which contribute to the performance of the end product, which is probably the reason you are interested in buying it. All of these factors contribute to the survivability of small suppliers in an environment that promises only to become more competitive, but you will not appreciate all of this unless you ask the right questions.

Pay attention to rumors and follow them up if they pertain to your situation. Sometimes industry rumblings can yield valuable information about problems with componentry, as happened when word leaked out of Intel Corporation (Santa Clara, Calif.) that there were design flaws in its newly released 80386 microprocessor chip. As it turned out, as many as 60% of the units shipped during first quarter 1987 were afflicted with a bug that caused problems during execution of 32-bit instructions to multiply. Purging the pipeline of defective units and redesigning the chip caused a severe shortage of high-grade silicon, resulting in the postponement of high-volume deliveries until fourth quarter 1987. The unanticipated redesign work delayed release of the 20-MHz version of the 80386 microprocessor, which had already been announced.

These cases point to the need to keep abreast of developments in key componentry because they may affect equipment prices and delivery schedules, as well as product performance. Failure to stay informed and take such factors under consideration before you buy could consume limited organizational resources as you seek problem resolution.

You must also insure that the vendor's product development effort includes compliance with industry standards. You want to avoid one-of-a-kind products that are not compatible with the offerings of other manufacturers. Do not become so imbued with a vendor's technology that you overlook compliance with industry standards, or you may have to invest more time and money just to get your network operating as smoothly as before. Regaining lost credibility among top management, however, may not be quite so simple.

Unless your company has agreed to beta test a new product at no obligation to buy, or has a budget that allows you to experiment with new products and technologies, steer clear of start-up firms using proprietary technology. Over the years, relatively few entrepreneurs have been lucky enough to achieve success with proprietary technology, mostly with products like T1 multiplexers and modems where, until recently, features and performance were more highly valued than standards, but implementing a proprietary local area network (LAN) operating system, for example, may be unwise, if only because the industry is rapidly coalescing around standards that promise greater connectivity among competing devices.

With modems and other communications products, the Consultative Committee for International Telegraph and Telephone (CCITT) has established global standards to insure the compatibility of like devices. Choosing modems, for example, without regard for standards could result in the accumulation of devices that could not communicate with each other on the same network, and, in this go-go era of megabuck mergers and acquisitions, adherence to industry standards can smooth the integration of previously separate corporate entities. When it comes to trimming communications staff to eliminate redundancy, the network manager who did have the foresight to base procurement decisions on industry standards is more likely to prevail over the manager who did not.

It is through standards that the network manager can truly avoid unneccessary risks. Should the vendor go out of business, products can be more readily maintained through third-party maintenance vendors, if only because such vendors find it more profitable to service the products with the broadest installed bases.

If you are considering products for the international segment of your network, it is imperative that you get written assurances from the vendor that its products have been "homologized." This term refers to the various certification procedures used by telephone administrations and PTTs worldwide to insure that products do not interfere with the proper operation of the public switched network or pose safety hazards to technicians and maintenance personnel who may have to service such products. Even U.S. companies like MCI must undergo rigorous trials before their long-distance services are approved to compete with the monopoly carrier which, in many cases, is AT&T. Sometimes sensitive political considerations enter into these certification processes, as when American companies attempt to compete head-on with the dominant carriers or vendors, which may be government owned or government sponsored. Sometimes approval hinges on a vendor having permanent plants or company offices in the country in which certification is sought. All of these considerations go unstated, of course, but may affect your ability to operate your network at international locations. To avoid potential service disruptions or product installation delays, make sure you select a vendor whose products have already been "homologized."

The ultimate in risk avoidance would be to use compatible products from different vendors on the same network, thereby not putting all your eggs in one basket, but this is only possible with products that adhere to industry standards. The single-vendor approach is only viable if the vendor has a family of integrated networking products with a common architectural base. Further, its product line must be able to grow and expand with changing network applications and emerging technologies, while providing a migration path from the lowest to the highest product level. Before tying your company's future to a single vendor, however, you must probe for any weak links in its product line. Only a handful of large firms like IBM and Digital Equipment Corporation come close to meeting this criteria when it comes to computers, but even they must forge alliances with other vendors to make up for deficiencies in their networking products.

There is another reason to be concerned about a vendor's commitment to standards: since the communications industry is showing support for standards by participating with users in their adaption and implementation, it enhances the vendor's ability to compete successfully in the marketplace, thereby insuring that your purchases will continue to be supported by the vendor throughout their useful lives.

Of course, for every rule there is an exception, and such occurrences are usually to be found in new types of products catering to emerging markets. In the area of engineering workstations, for example, some companies that once touted their adherence to standards have become spectacularly successful by deviating from the norm. Companies like Apollo Computer Inc. (Chelmsford, Mass.), Hewlett-Packard Company (Palo Alto, Calif.), and Sun Microsystems Company (Mountain View, Calif.) are building their products around proprietary RISC (Reduced Instruction Set Computing) architectures, a technology designed to achieve faster performance in CAD-CAM applications. At this writing, the performance of these workstations counts more than standards, or even the ability to port software across incompatible RISC architectures. The only way to minimize the risk associated with such products is to buy from an established company until the standards issue sorts itself out. If current trends continue, the products of one or two companies capturing the greatest market share in the shortest time frame will become *de facto* standards, similar to the way Novell's (Provo, Ut.) NetWare became a *de facto* LAN operating system and the way IBM's Structured Query Language (SQL) — its many derivatives notwithstanding — has emerged as the standard language for accessing and developing reports from mainframe data bases.

The merits of the products notwithstanding, these examples illustrate the practical realities of standards setting: it is easier for vendors to line up behind — and standards committees to adopt — what already enjoys widespread usage than to go against the grain by forcing the world to implement something entirely different, even if it is demonstrably better.

The race for market share can make standards quite difficult to set, especially in such fast-moving segments of the industry as modems. Despite virtually universal agreement on the need for modem standards, and years of study by members of the U.S. Modem Working Party, it seems that a consensus on various standards is more elusive than ever. Even when vendors are on the verge of agreeing among themselves what standards to support, something happens to negate what little progress has been made. For example, the Microcom Inc. (Norwood, Mass.) Network Protocol (MNP) had been on the verge of becoming a standard for data compression in high-speed transmission. To facilitate acceptance, Microcom even granted its competitors access to various layers of its protocol, which could be used to enhance the performance of their own high-speed modems. Microcom, however, eventually came under fire from its competitors for repeatedly releasing modems supporting a more advanced MNP level before making that level available to them.

This allowed Microcom to stay one step ahead of its competition in the race for market share, a stepping stone to CCITT acceptance, which it eventually received. James Dow, president of Microcom, defended his company's actions by saying, "We're sharing our innovation with the industry so the industry can have some standards. Any advantage we get, we deserve, because we developed the technology." [3]

In early 1988, however, Racal-Vadic (Milpitas, Calif.), Concord Data Systems (Marlborough, Mass.), and Universal Data Systems (Huntsville, Ala.) announced plans to support an alternative compression algorithm developed by Adaptive Compression Technology (Santa Clara, Calif.). The compression technique is no better than the one offered by Microcom's MNP, just different enough to make a unified standard for data compression even more elusive. At this writing, ACT's compression scheme gives Microcom's competitors a way to differentiate their products, but that may not be enough incentive to draw users away from the more established protocols used by Microcom or Hayes Microcomputer Products (Atlanta, Ga.).

In time, these types of struggles may work themselves out. Meanwhile, *de facto* standards will continue to be set by users through their purchasing decisions, and their choices will hinge on the protocols that are most likely to be supported by those to whom they are trying to connect.

On balance, proprietary products and standards-based products have their advantages. Standards will ultimately allow you to interconnect with more types of equipment, in turn providing more configuration flexibility and more choices in equipment selection, but standards setting can take more years than you can afford to wait, whereas proprietary systems are available now. The choice is yours, but it should be an informed choice.

1.2 APPLICATIONS ENGINEERING

The complexity and technical sophistication of today's networks usually require that a vendor take into account the customer's applications so that the most appropriate products, interfaces, cabling, and environmental controls may be selected. The lackluster performance of many vendors, however, both in the quality and timeliness of such services, suggests caution as the best plan of action. It is not adequate to inquire about the number of installers and field service people the vendor employs; this information will not reveal anything about the vendor's ability to help you select the right products and plan their integration into your network. If your network is a hodgepodge of products from numerous manufacturers, find out if the vendor has a comprehensive applications engineering capability. This capability entails a thorough assessment of your current network configuration, design services that match products to your specific requirements, and the development of a maintenance program tailored to your needs.

Find out if the vendor is willing to perform a preinstallation site survey to help you determine space availability and utility interfacing requirements. Do not wait until the equipment arrives to discover that you do not have room for it, or cannot accommodate its power requirements, or that it throws off more heat than you can handle with your present environmental controls. Beware of vendors who try to push their products on you before even inquiring about the physical characteristics of your site; they may be putting quick sales over your long-term interests.

Also check into whether the vendor has the capability of integrating its products with previous generations of products from other manufacturers. Ask about the vendor's specific experiences in bridging generation gaps. If the vendor does not support the older protocols still in use by equipment on your network, it may mean that you will have to give up some of the products you already use to accommodate the products you want to buy. Because this adds to the cost of doing business with particular vendors, it must be factored into your purchasing decision.

Do not rely on vague statements in the vendor's proposal or accept a sales representative's verbal assurances that these kinds of integration services will be provided after the sale. Look for specific commitments from the vendor. Such commitments should not come from the vendor reluctantly as the result of strong-arming on your part; they should be offered freely by the vendor as standard operating procedure and then spelled out on paper. Make sure you include a "weasel" clause that gets you out of the purchase agreement with no penalty if the vendor cannot deliver the product on time or make it work on your network at an agreed-upon performance level by the cutover date.

Whenever possible, pay only a small fraction of the product's purchase price in advance — your objective is to provide the vendor with maximum incentive to follow through on all provisions of the contract. If you have to put up a substantial amount of the purchase price to insure timely delivery or to have the vendor start work on a customized version of its product, insist on a clause that penalizes the vendor for missing the cutover date. The penalty may consist of discounting the outstanding balance a percentage amount for every day that the cutover is delayed.

Specify penalties for delays, even when you are dealing with the most reputable vendor who has been servicing your company for years — these days you have to protect yourself against potentially disruptive corporate mergers, hostile takeovers, or industry shakeouts.

Risk can also be averted by talking directly with the vendor's operations staff to get an accurate picture of what can and cannot be done within a specified time frame. You cannot always count on sales representatives or marketing people for this kind of information. Many times they are not up to date on the schedules of the operations people, the whereabouts of technical crews, or what is actually involved in a product's installation. Yet, the accuracy of such information is very important, especially during a major network upgrade that involves multiple carriers and vendors. The failure of a hardware vendor to deliver channel banks on time, for instance, can have a ripple effect on the cutover schedules of T1 circuits, which may entail penalties being imposed by the carrier. The last thing you want to hear later from the hardware vendor is, "Sorry, our sales person wasn't authorized to offer that delivery date." Also, as a network manager, you do not want to get into the position of blaming everyone else for a never-ending series of snafus. It does not take very long for excuses to wear thin; top management may begin to doubt your ability to handle the job. It is best to do some investigation up front rather than having to settle for compromise solutions later on.

The problem of integrating products into your existing network is especially pronounced when dealing with vendors who have only suddenly realized the marketing wisdom of selling "connectivity," whereas previously they were only peddling proprietary products. Whether mere marketing savvy or a genuine reversal in operating philosophy, the connectivity orientation brings with it new responsibilities that the vendor must be able to handle. LAN vendors, for example, must extend service and support to an entirely new level because there are now more products that must be pulled together, each with its own communications protocol, some of which may be shrouded in vendor secrecy. When dealing with a vendor who offers "connectivity solutions," thoroughly explore its track record to separate reality from marketing hype.

1.3 QUALITY ASSURANCE

A program of quality control is generally a good indicator that the product is as reliable as the vendor says it is. A good quality assurance (QA) program includes multiple inspections, starting with the receipt of batch componentry and other raw materials at the receiving dock. While touring the vendor's manufacturing facilities, look for multiple inspection points at various stages of the assembly process, including automated testing stations and visual inspections by operators.

Inadequate commitment to and organization for product development testing and system integration testing lies at the root of many problems vendors are experiencing today with product reliability. How much testing is enough? Generally, the answer depends on the product and its intended application. If the item is a plug-on option module that inserts into a product's motherboard and is designed for on-site replacement and disposal after being diagnosed as faulty, simple economics dictates that a final functional test is all that is required. If the product is a digital cross-connect, however, extensive multi-level system integration testing under conditions that approximate the customer's environment is required — the more, the better. You would think that economics would be the determining factor: it is much easier and less costly for the manufacturer to test during the various assembly stages than continually to rip apart finished products to identify and correct problems, or to try to fix problems in front of angry customers, but in the face of order backlogs, even the largest companies may not be able to resist the temptation to skimp on testing to keep up production levels. In such cases, delivering the product on time to avoid penalties may take precedence over final testing. Of course, the company hopes to complete final testing at the customer site before the cutover date. If the customer encounters serious problems later, an unscrupulous vendor might try to attribute them to something else, like the customer's application or inexperience with the product, rather than to its own shoddy workmanship, hasty production, or inadequate testing.

You can minimize risk by inquiring about a product's MTBF, or Mean Time Between Failure. This measurement, usually predictive, is expressed in hours, and is derived using statistical methods. While this indicator may be useful in evaluating new products with no performance history, a better indicator of reliability for existing products is their MTTR, or Mean Time to Repair. Instead of predicting reliability, MTTR is derived from the performance record of the product in the operating environment. The MTTR is very important from the vendor's perspective because it may point to the need to replace weak componentry.

In questioning plant personnel, do not be surprised to discover printed circuit board failure rates as high as 25% or more. There may be a variety of reasons for this, such as shorts due to solder splashes, the carelessness of workers in implementing electrostatic safeguards, or mistakes in the placement of oversized components that do not lend themselves to automatic insertion. High failure rates may also be attributable to the vendor purchasing components in bulk quantities that have undergone little or no testing by the manufacturer. Such componentry may be purchased at a substantial discount. If this is the case, make sure the vendor uses automated tests and multiple inspections to assure proper performance of all board componentry. Further, make sure that all boards are tested at the finished product level. With high failure rates, statistical sampling methods at this juncture should be viewed as unacceptable.

In your review of QA procedures, try to determine whether the vendor's administrative support for the program is automated, or mired in paperwork. In a manual support operation, technicians typically record various test results on a dozen different forms. All this data is keyed into the company's central computer and batch edited. The results are routed back to the technician and his or her supervisor for correction. The corrections are entered into the computer and the test data stored on magnetic tapes for retrieval. Since data retrieval must be scheduled with MIS-dp (Management Information System–data processing), so other corporate operations are not disrupted, it may be two or three days before a retrieval request is fulfilled. Under such circumstances, it may be faster for the QA supervisor to pull historical information from the original forms — if they are readily available.

In an automated QA support operation, technicians enter test results directly into the computer using their own terminals, eliminating the bulk of the paperwork and cutting down the data-entry error rate. Test equipment also records test results directly into the computer, totally eliminating the human interface. The computer uses an application program to calculate and display differences between test specifications and actual measurements instantly. The results may be immediately reviewed by the technician and supervisor, who may either release the data to the historical file or flag it for further analysis. This level of automation is important because it has a direct bearing on the performance and reliability of the vendor's products. By cutting down the overhead associated with quality control, automated processes play a key role in producing the best products at the least cost.

In your tour of the vendor's production facilities, take note of the cleanliness and orderliness of work areas, provisions for electrostatic control, and potential safety hazards. Granted, these are elementary points, but ones which may go unnoticed by years of management complacency.

When considering the purchase of new products with no performance history, obtain the location and key contact of the beta site — and make sure the results of those tests come from a customer site and not from the vendor's own laboratory. Find out what benchmark tests are being used and why they were chosen.

The concept behind benchmarks is pretty simple: a performance parameter is selected and then a test is designed to simulate a real-world application of that parameter. Ideally, performance benchmarks should help network managers know what to expect from a product when it is installed and operating in an environment similar to their own, but performance benchmarks have tended to oversimplify the purchasing decision. One reason is that vendors tout the benchmark result that favors their own products. Another reason is that customers do not really know what the benchmarks are designed to measure. For example, it is not always clear whether the tests compare the machines, the compilers, or the operating systems. Some customers do not know enough about their own operations to know what loads to expect, or which performance parameters are most important to them.

When a variety of benchmarks are available, you must insure that you are basing your purchase decision on the right one for your application. Bigger computers like the VAX 11/780* from Digital Equipment Corporation (Maynard, Mass.) or the Sun 4/260 from Sun Microsystems Inc. (Mountain View, Calif.) measure raw CPU (central processing unit) power in terms of "millions of instructions per second" (MIPS). With these or similar mid-range computers, make sure the vendor is using commonly accepted benchmarks like the Dhrystone, Whetstone, and Linpack tests. If your MIS-dp group uses the high-level language C to develop its applications programs, for example, then you should find out how various mid-range computers fare in the Dhrystone benchmark, which simulates

---

* Although the VAX 11/780 has been discontinued by DEC in favor of the 8000 Series, the 11/780 continues as the *de facto* standard against which most mid-range computers are compared. Some vendors even use the term "VAX MIPS," the assumption being that the VAX 11/780 is a 1-MIPS computer.

the C-programming environment. The Dhrystone benchmark contains 100 statements of which 53 are assignments, 32 are control statements, and 15 are function calls. As it turns out, the Sun computer comes out the winner by executing 13 times as many Dhrystones per second as the VAX.

The Dhrystone benchmark is not adequate for comparing the performance of computers that will be used primarily for scientific applications like data collection, statistical analysis, and analytical graphics. In this case, the Linpack benchmark is more appropriate. It solves very dense sets of linear equations to arrive at a measure of floating-point performance. As measured against the single-precision Linpack benchmark, the Sun 4/260 is only six times faster than the VAX 11/780.

While comparisons of raw CPU performance are useful, choosing one product over another based only on its MIPS number makes for a poor purchasing decision because it trivializes the need to evaluate products based on their fit with specific applications, the availability of appropriate software, features that facilitate rather than complicate management, and the commitment of the vendor to a unified product line.

Some products like modems do not lend themselves to easy benchmark comparisons because there are no standard tests. Many of the product comparisons issued by vendors are biased in that test results are released on only those features that beat the competition. To add "credibility," they might list a relatively trivial feature or two that do not perform so well — but only by a narrow margin.

The independent benchmark testing conducted by consumer and trade magazines — although well intentioned — further confuses buyers because of inadequate test designs. For example, when devising a benchmark test to compare the performance of error-controlled modems, it is a common mistake to simulate line impairments on the receive side of the data path, whereas a valid test must induce impairments on the send side as well. Also, most tests do not include the most common transient line impairments like impulse, phase, and gain hits, which tend to degrade the throughput of high-speed modems.

When a vendor offers you benchmark tests to convince you of a product's performance, you must satisfy yourself that the test was conducted in real-world conditions, or under conditions that adequately simulated real-world conditions. You must also satisfy yourself that the benchmark test is relevant for your application.

Be aware that many vendors use performance benchmarks as marketing tools rather than consumer aids. Privately, many vendors admit that benchmark tests can be designed to say anything they want them to say. Only when all variables are uniform from product to product can benchmarks can be useful in head-to-head comparisons.

With information systems, for example, the processors, software, memory, and application must be the same. If only one item differs even slightly, the results can be entirely different. In addition, a variety of unknowns may contribute to skewed benchmarks. In a comparison of similarly equipped microcomputers that use Intel's 32-bit 80386 processor, the same spreadsheet program may run differently because one model has the intelligence to avoid the bad sectors of a hard disk, whereas the other model continues to perform number crunching, but unless you checked for errors in the finished spreadsheet, you would have selected the wrong microcomputer as the one that performed best.

An emerging trend among software vendors is to sell partially finished products, or "betaware." Under this arrangement, the user gets to debug the program and recommend changes to the vendor, who will incorporate the suggestions into higher-priced releases of the product. Users can benefit by the opportunity to shape the final version of the product, which may make it more relevant to their needs. This kind of arrangement is perfectly all right, as long as the vendor lets you know up front that its product is only a beta version. You should have a written agreement governing this arrangement, which should include provisions guaranteeing you free upgrades, or a substantial discount on the finished product. Problems with this scheme arise when the vendor "forgets" to tell the customer that the product he is paying for is not quite finished.

Quality assurance is just as applicable to software as it is to hardware. Traditionally, software QA consisted merely of a check of the completed product to see if it performed as intended by the developer, whose responsibility was limited to making the product work, no matter how circuitous the route. Only recently have software companies invited QA people to become involved during the product's design stage. Among the larger vendors, it is now the responsibility of QA staff to help review specifications and establish a clear understanding of how the product is put together. Armed with this knowledge, the QA people put themselves in the shoes of customers to uncover every conceivable way the product can fail, and it is not only new products that should pass muster with the QA people, but user interfaces and product enhancements as well. This ensures that software products meet customer expectations regarding ease of use, as well as functionality. Beyond that, the early involvement of QA people enhances the vendor's ability to solve customer problems. You can minimize the risk inherent in major software purchases by making sure the vendor has a formal QA program, appropriately staffed and budgeted.

When a vendor uses continuous feedback from the real-time environment, as in beta test sites or follow-up evaluations at customer sites, coupled with extensive testing throughout the manufacturing process, you can be reasonably assured that all the design flaws have been discovered and weak componentry replaced. This should make for reliable product performance down the road and a safe purchasing decision — without unknowingly becoming the vendor's test site!

1.4 REPAIR AND RETURN

In most cases, technicians dispatched to your location are not trained or equipped to perform board-level repairs. Even if your own technicians perform the maintenance, they are usually only trained to isolate the faulty board and swap it out with a spare. The faulty board is then sent to the vendor's depot test and repair facility. To minimize your exposure to risk, you should make sure that the vendor has the facilities with which to fulfill the product warranty. The warranty is worthless if the vendor is not adequately staffed and equipped to service products after the sale.

In-warranty items sent in for repair should be returned within ten working days. Emergency repair service should take only three days. Critical items like control logic boards should be same day shipped as "loaners" until the faulty unit can be repaired. Items that test positive should be returned to the customer without charge.

Be suspicious of vendors who appear to be doing you a favor by "throwing in" repair and return services to get your business because they probably do not have the resources to follow through when your "chips are down."

It is pretty easy to evaluate the quality and timeliness of repair and return services. Find out how past and present customers rate the vendor in these areas. If the vendor has not invited you to visit its facilities by the time you have made the purchase decision, invite yourself. While you are there, find out if items returned for repair go through the same quality assurance procedures as newly manufactured products.

1.5 CUSTOMER SERVICE

A vendor's commitment to customer service should go far beyond just having a 24-hour hot line to technicians who can resolve problems over the phone; the customer support unit should be staffed with people who will "own" the problem until it is resolved. Find out the budget for this operation and inquire about the qualifications of the customer support people, including the number of years they have been in the industry. The qualifications of customer support people are very important when you have remote network nodes that are staffed entirely with nontechnical professionals. It takes special interpersonal skills as well as in-depth technical knowledge to guide such people through diagnostic routines and restoral actions.

A vendor who continually stresses the quality of customer support will have measured response times to service calls and performed surveys to assess the level of customer satisfaction — all this with the goal of maintaining a continuing relationship with its customers. Any reputable vendor will be happy to share this information with you. In fact, some companies are so proud of their performance in this area that the results are used as a marketing tool. This kind of information is especially important if you have a multinode network that is dispersed over a large geographical area. The vendor's ability to provide prompt service to any node may be a critical factor in your purchasing decision.

You must also check into the availability of local support. The shrinking demand for some products has forced many companies to pull back local field service staff into larger regional centers. You may be paying a premium price for local support, which is bundled into the product price. If you are paying a premium price for the product because it includes local support with a response time of one or two hours, you do not want to be at the mercy of vendors who decide later to centralize support operations and delay response times to three or four hours. To insure the maintenance response times you contract for, specify a system of response time credits and component downtime credits. With response time credits, the vendor discounts maintenance charges for every hour that maintenance personnel fail to arrive within the agreed-upon time. With component downtime credits, the vendor discounts maintenance charges for every hour that equipment is out of service.

The question of local support is particularly important when your network has international locations. Many American companies must rely on third parties to provide maintenance and support services in other countries. Find out the response times of these vendors under worst case conditions. Find out the nearest locations of depot repair centers, the sources for spare parts and componentry, and replacement equipment.

In times of economic crisis or market uncertainty, corporate budget trimmers usually do not think twice about cutting back service staff; on paper the function looks like a liability rather than an asset. Find out the philosophy of the vendor regarding customer service, then compare it with what customers say in your check of references. Depending on what you find, it may be desirable to seek a third-party maintenance vendor with a national service presence, or consider purchasing products from an alternative source.

If you want to survive in the multivendor environment, check the vendor's commitment to standards as well as its track record in supporting the products of other manufacturers. Check references to insure that the vendor's performance record in this area has not been exaggerated to get your business.

Ask the vendor to justify the price of customer support. You will typically find that prices bear an inverse relationship to product reliability or performance — the higher the quality, the lower the price for support; the lower the quality, the higher the price for support. In fact, some vendors confide that pricing customer support is merely a question of finding out what the market will bear. With this in mind, you may be able to negotiate these costs downward quite easily.

If you are buying computers from retail dealers, the question of support is more important than you may realize at the time of the purchase. Not only is technology becoming increasingly complex, but dealers must contend with multiple hardware architectures, as well as multiple operating systems. As if all this were not enough, dealers must now provide seamless file access across different operating systems, being forced to play the role of systems integrator to keep customers happy. Dealers feel pressured into carrying the full range of products or risk not meeting sales quotas, which qualify them as authorized dealers and for higher discount levels. Not having the resources to support this range of offerings, they are having to rely more heavily than ever before on manufacturers for technical assistance in solving customer problems. The manufacturers claim that the dealers are ultimately responsible for support because they are the interface to the customer.

The support issue will not go away, and, ultimately, it will separate the strong players in the dealer channel from the also-rans. In any case, it is always a good idea to compare the service agreements of the product vendor and a third-party maintenance provider before entering into any purchase agreement. Third-party service vendors are becoming more competitive in both terms and pricing, offering two-hour response times and a 15-20% cost savings, and, since they typically service the equipment of multiple manufacturers, they may provide you with additional flexibility in your future choice of equipment.

1.6 TECHNICAL DOCUMENTATION

Until recently, technical documentation typically received scant attention from vendors. As products moved from the design stage to the production stage, rudimentary documentation was hastily thrown together in the hope of placating customers who were not really accustomed to expecting anything more. Although this is changing, many vendors still do not appreciate the customer's need for quality documentation. Too many vendors still try to smooth things over by delivering production drawings, circuit schematics, photocopied internal memoranda, and parts lists that are of little or no use to customers.

Today's complex computer and telecommunications technologies require that vendors view documentation as an integral part of the product, inseparable from the hardware or software. Without a comprehensive documentation package from the vendor, you could be leaving yourself — and your network — vulnerable to the whims of the vendor, especially if you have a problem with frequent staff turnover.

Review the vendor's product documentation to validate the claims of salespeople. Check it for comprehensive installation procedures, initialization or setup instructions, and a complete explanation of the product's features. In addition to appendices that amplify aspects of the product's operation, look for a detailed index. A good documentation package will also include a troubleshooting guide that will help the network manager to determine the nature and scope of problems before calling the vendor's customer support people.

Products typically evolve over time as a result of enhancements. Unfortunately, many vendors do not keep their documentation up to date. Find out how the product documentation will be maintained and distributed to customers.

If you are buying products from an OEM (original equipment manufacturer) or VAR (value-added reseller), find out to what extent the original product's technical references have been substituted with the OEM-VAR's own manuals. Sometimes the original reference materials are deliberately trashed with the idea of increasing the level of customer dependence on the OEM-VAR. With this ploy, the OEM-VAR can recover costs or even turn a profit over the life of the product when its "free" short-term maintenance agreement expires, or by charging for services not specifically covered in the service agreement.

Quality documentation benefits the vendor as well as the customer. It provides the vendor with a marketing edge against competitors, while reducing the cost of customer support activities that involve mere "hand-holding." From the customer's perspective, quality documentation shortens the lead time required to bring new staff members up to speed on the use of software or hardware. It also substantially reduces periods of inactivity and minimizes disruption to other company operations while waiting for the vendor to respond with corrective action. In essence, quality documentation reduces the hidden cost of purchasing the vendor's products.

Keep in mind that if you let vendors get away with inadequate documentation, you not only hurt yourself in the near term, but also encourage poor performance which may spill over into other areas of customer support.

1.7 TRAINING

A reputable vendor will offer a full line of instruction about its products and technology; dedicated training staff and facilities are a pretty good indication that the vendor appreciates long-term customer relationships and not just "sales."

Many times it is not adequate for the vendor to provide only formal classroom training at its own facilities. Check into the availability of training at your location or the locations of your various network nodes. Find out about the cost, if any, of additional training as you purchase enhancements or add new staff. Depending on your circumstances, you may want to negotiate with the vendor to provide training as needed — and to noncustomer personnel as well.

When Hewlett-Packard Company (Palo Alto, Calif.) upgraded its northwest area network in 1986, Tellabs (Lisle, Ill.) was selected to provide the equipment, which included cross-connect systems, channel service units, and ADPCM-based transcoders (circuit multiplication devices that use the adaptive differential pulse code modulation technique for doubling T1 channel capacity from 24 channels to 48 channels). Hewlett-Packard had many reasons for choosing Tellabs, not the least of which was the company's willingness to provide extensive training on the use of its products.

The key to Hewlett-Packard's upgrade was the location of its private network hub off a Tellabs cross-connect system inside a Pacific Northwest Bell (PNB) exchange in Portland, Oregon. The Tellabs cross-connect system was installed at the digital centrex exchange to improve T1 fill and groom traffic from several satellite locations. It also provided alarms and management reports with which Hewlett-Packard could insure the continued availability and efficiency of T1 facilities. Not only did Tellabs provide training in the use of the cross-connect to Hewlett-Packard personnel, but to the personnel of PNB as well.

Upon cutover, Tellabs engineers were on hand at Hewlett-Packard's network control center in Palo Alto and in Portland, Oregon, to insure the proper operation of transcoders with the cross-connect system, which involved a tightly coupled integration. With the close involvement of Tellabs, the entire network upgrade went smoothly.

Ask about the experience and qualifications of the trainers. Make sure the vendor does not just send out technicians to provide training; they have other priorities and they generally do not make good instructors unless they have been appropriately trained for that responsibility.

Review the training materials for scope and clarity with the idea of determining if they will meet your training needs. If the vendor looks like it is skimping in this area, you may be getting less out of your capital investment than anticipated at the time of purchase.

1.8 LINE OF BUSINESS

Find out if the product you are considering is a major or minor part of the vendor's business. If it is only a sideline or if the product is viewed only as the means to gain entrée into more lucrative markets, you may not get the attention you deserve when you have a problem. An effective way to determine long-term commitment to the product is to look into the vendor's product development efforts, described earlier in this chapter.

Be skeptical of small companies who want to become your "single-source supplier" for all of your networking needs. In today's competitive environment even large vendors are pulling back to become more responsive to customer needs and to serve specific territories and market niches better. The very idea that a small vendor can step in as your single-source supplier is ludicrous.

One way that companies try to become more responsive to customer needs is by repackaging an existing product to make it more appealing. Some vendors of matrix switches, like Data Switch Corporation (Shelton, Conn.) and Dynatech Data Systems (Springfield, Vir.), for example, started in 1987 to offer scaled-down versions of their products, enabling customers to install them at the departmental or work-group level rather than in the computer room for company-wide access. In addition to saving valuable floor space in the computer room, these distributed matrix switches permit more efficient configurations, which also result in substantial savings in cabling costs. The distributed approach also enhances departmental control over network operations, relieving the MIS-dp group of day-to-day management responsibilities.

As customer needs become more diverse and sophisticated, it is quite natural and understandable that vendors look for ways to deliver the products and services their customers demand. Micom Systems, Inc. (Simi Valley, Calif.), provides a good illustration of how acquisitions are used to plug holes in the product line to stay abreast of customer needs.

Micom started out selling modems, data switches, and low-end statistical multiplexers, but by the time it grew into a $200 million a year business, sales of these products declined as its customer base of mid-range minicomputer users stepped up to more sophisticated products. The company acquired Interlan, Inc. (Boxborough, Mass.), in 1987 to broaden its product line to include an Ethernet LAN. The same year, Micom also acquired Spectrum Digital Corporation (Herndon, Vir.) for its expertise in advanced T1 multiplexer technology. Packet switching technology was obtained from Amnet (Framingham, Mass.) under an OEM arrangement. With a customer base that is ready to move up to hybrid communications systems providing a mixture of T1 and packet on both private and public networks, Micom wants to be positioned to take them there. To do so, however, Micom must shed its image as a mere components company in favor of one that is systems-oriented, but in rounding out its product line from so many different sources, it remains to be seen whether Micom will be successful in unifying them under a single management scheme.

Big companies, too, use acquisitions as the means to plug holes in their product lines. In 1987 Unisys Corporation (Blue Bell, Penn.) decided to do something about its deficiency in the area of networking. Appreciating the increasing importance of networking in the mainframe environment, in early 1988 it completed its acquisition of Timeplex Inc. (Woodcliff Lake, N.J.), a leading supplier of T1 multiplexers. With the networking expertise of Timeplex, Unisys hoped to acquire the knowledge base with which it could develop products that can compete and perhaps link up with IBM's through NetView, a host-based network management product. For Timeplex, the infusion of big money into its research and development efforts means that it can fine-tune its network management products and better position itself for future implementations of ISDN. With Unisys, it will also have the opportunity to expand and enhance its maintenance and support capabilities, the lack of which was already emerging as a concern among customers. Interestingly, right before it was acquired by Unisys, Timeplex acquired Cygnus Computer (Dallas, Tex.), a designer and manufacturer of packet switches. With Cygnus, Timeplex hopes to enhance its already formidable position in the T1 networking arena.

Noting this and other partnering activity in 1987 and 1988, industry analysts speculated that it would become increasingly difficult for other communications companies to survive without a large partner. The vendors involved have promoted these efforts as beneficial to customers, and the media have gone so far as to envision the coming of a new competitive climate in which nonaligned companies will consistently lose out to monolithic vendors who, it is claimed, will be positioned to provide one-stop shopping reminiscent of the predivestiture era. This scenario, however, ignores some basic truths about social and organizational behavior that have demonstrated their validity many times over and which heighten the risk of doing business with such firms.

In addition to the wholesale elimination of redundant staff, an acquisition brings with it internal upheavals within the target firm, replete with cultural shock and political maneuvering. Since highly creative people tend to gravitate to smaller firms, they are among the first to leave when their employer is gobbled up by a much larger firm. As product support and product development priorities change, key engineering and management personnel perceive things getting out of their control and start looking for greener pastures. During the transition stage, the remaining staff members of the target firm typically focus their attention inward, during which time customer support suffers. Employees, who are quite naturally concerned about their futures, may be too preoccupied to give customers the level of service they require.

In such situations, salespeople may quite easily lapse into a crisis of confidence. Uncertain of the situation at headquarters, they may not know how to respond to customer questions about the ramifications of the merger or acquisition. Customers want to know how they will be affected by the new scheme of things:

- How will response times to service calls be affected?
- Will the promised hardware upgrade be delivered as scheduled?
- Will future product development directions be compromised to meet the parent company's requirements at the expense of customers?
- Will support for discontinued products be eliminated?
- Will the new network node be cutover as planned?
- Will the present discount structure remain in place?

The inability of salespeople to provide definitive answers to straightforward questions erodes confidence in the company. Sometimes, the rumors become so rampant that the company feels compelled to place advertisements to reassure customers that everything will actually be better under the new regime.

All of this turmoil may take years to resolve. By then, the smaller firm will have been fully absorbed into the parent. This situation carries with it a different set of consequences that are of no less importance to customers.

Total absorption has the potential of strangling the life out of the acquired firm. Often, dealing with the sluggish bureaucracy of the parent is enough to sap the energy and creativity from the smaller firm — qualities that made it successful in the first place. This, in turn, may result in personnel withdrawing from intimate daily involvement with customers. In losing touch with customers, the firm fails to respond quickly enough to changing market conditions. The failure to adapt to changing market conditions signals the loss of vision required to position itself for the future.

Buying products from acquired firms may actually become more complicated, not because there are more products to choose from,but because customers must consider the possible hidden agenda behind the vendor's recommendation of one type of equipment over another:

- Is this product what you really need, or is the acquired vendor merely trying to migrate you prematurely to the high-end offerings of its parent? Worse still, is the parent trying to lock you into a proprietary solution?
- Is that enhancement really going to reduce your operating costs, or is the vendor merely trying to unload the excess inventory of its parent before the end of the fiscal year?
- Is the vendor really offering you a good deal on that component, or is the low price merely the means to move the weak link in the parent's product line?
- Is the vendor dissuading you from purchasing products you want because it knows the other supplier has a poor service record, or merely because its parent will not integrate the products of its competitors?
- Can this configuration be implemented more efficiently with less equipment, or is the vendor merely trying to meet the sales quota of its parent?

There also is always the risk that the parent company will put its acquisition into the "harvest" mode — taking the best of what the smaller firm has to offer and spewing out the rest before finally hanging it out to dry.

As customer needs become more diverse and sophisticated, it is understandable — even commendable — that vendors look for ways to deliver the products and services their customers demand, but simple economics drives most partnering and consolidation activity and, quite often, the parent does not understand the business or the technology of the acquired firm, or lacks the management skills to leverage its diverse assets into a cohesive whole. For customers, the result is often paradoxical: instead of being able to buy a hodgepodge of products from many vendors, they can now buy a hodgepodge of products from a single vendor.

One possible solution for corporate users is to seek out nonaligned vendors who stand out as consistently innovative, financially stable, and strategically positioned. Contrary to popular opinion, there are still a few firms out there which provide complete networks and the capability to manage them. Such firms, by virtue of not being aligned, can actually simplify purchase decisions and eliminate the risk inherent in upgrading or expanding today's complex networks.

- Because nonaligned firms are typically smaller in size, they are not burdened with supporting large infrastructures. The resulting economies mean that customers have more latitude in negotiating volume discounts, as well as contract terms and conditions.

- Nonaligned vendors can be more responsive to individual customer needs, especially when it comes to customization and applications engineering. For the most part, they are more attentive to what customers are trying to accomplish with their networks, and are better organized to stay in step with changes in their strategic direction. Monolithic corporations, on the other hand, prefer to provide customers with cookie-cutter solutions to keep assembly lines moving, and to minimize the inventory of finished goods. As customers learned from predivestiture AT&T, industry giants tend to be quite paternalistic, which is more a hindrance than a help.

- Smaller, nonaligned firms can react faster to changing market conditions than their elephantine counterparts. This enhances their ability to survive, which means they will be around long into the future to support their products. Large firms, as the experience of ITT (New York, N.Y.) demonstrated not too long ago, do not hesitate to abandon products, dislocate thousands of employees at a time, and sell off whole subsidiaries when profits do not meet expectations.

- Nonaligned firms offer continuity of staff, whereas vendors who have been acquired or merged into larger entities undergo many changes of personnel as redundancy is eliminated and as various functions are integrated. As the players change, the natural rift that exists between vendor and customer widens, if only because obligations and expectations are interpreted differently. When that happens, promises have a way of receding into fuzzy oblivion.

Sure, there is pressure on vendors to provide more pieces of the network puzzle and then to tie them together with a comprehensive management package. The continuing trend toward mergers and acquisitions only demonstrates how unprepared most vendors are to meet this emerging customer need. In the final analysis, many users might be better served by the independent firms which have already demonstrated their success in meeting that challenge.

Beyond these considerations, the industry trend toward consolidation portends a slower pace of innovation, if only because such activity stifles fledgling enterprises by hogging the most lucrative markets. Even if a smaller firm experiences high growth by serving very narrow markets, it is not long before its success attracts the attention of industry giants, a development that usually mitigates against the possibility of the small firm surviving to maturity. Another consequence of mergers and acquisitions is that they work against the adoption of standards. A conglomerate that is even moderately successful in luring customers into proprietary technology has little incentive to adopt standards unless, of course, it tries to market its technology as the *de facto* industry standard, but the success of this strategy hinges on the ability of the conglomerate to convince other vendors to abandon their technologies in favor of new R&D expenditures that will make their respective products compatible.

Although mergers and acquisitions may pose unforeseen problems for unwary buyers, risk can be minimized by looking at the reasons behind such arrangements. Often, the smaller firm is ripe for a takeover or buyout because it is in financial difficulty. In such cases, it is easy for the larger firm to exploit the smaller firm for whatever purpose is deemed necessary, which may not be in the best interests of its customers. If two financially healthy industry leaders get together, however, the marriage has the potential of not only dominating the market, but also of stemming the rising tide of competition, as illustrated by the 1987 marriage of two LAN vendors: 3Com Corporation (Santa Clara, Calif.) and Bridge Communications Inc. (Mountain View, Calif.).

3Com had become a leading supplier of work-group networks for workstations, while Bridge had become a leading supplier of connectivity solutions for multivendor host computer environments. They both supported Ethernet and token ring LANs, while developing products based on the Transmission Control Protocol–Internet Protocol (TCP-IP) and the Open Systems Interconnection (OSI) reference model. Together, the two companies have the potential capability of offering a seamless network, instead of only bits and pieces of a network. So far, this merger looks quite promising.

Not even strategic alliances of the kind between IBM and Network Equipment Technologies (Redwood City, Calif.) are without risk. First, it is not likely that either party will base its long-term product development, marketing, and distribution strategies on a company that it does not control. Second, there will always be the temptation on the part of both parties to strengthen their own products, rather than to devote resources to helping the partner. Third, once each partner has achieved its hidden agenda, there is always the chance that the alliance will dissolve, to the detriment of customers who may have made long-term commitments based on the promise behind that alliance.

Still another approach to meeting customer needs is through OEM or VAR arrangements whereby a bare-bones product is improved, repackaged, or modified for a particular application for resale in select market niches. If the product you want to buy is essentially the result of a VAR-OEM arrangement, look into the vendor's relationship with its main supplier. Find out if the supplier views the vendor as a "preferred" account. If not, components or products in critical demand will go to those who are viewed as such. In fact, manufacturer support of VARs and OEMs has never been consistently good because these types of vendors are viewed as direct competitors of the manufacturer's own sales force. It is not uncommon for a VAR to work with a potential customer for six months or more, for example, only to have the sale snatched away by its supplier's own direct sales force who can cut a better deal. Other times, manufacturers neglect their VARs or OEMs because they are more concerned about what they can supply through direct sales than what VARs and OEMs need to fulfill customer requirements.

Matters of expediency also muddy the distribution picture. IBM (Armonk, N.Y.), for example, in mid-1987 started offering bonus incentives to its VARs, but only because it was losing market share to archrival Digital Equipment Corporation and to upstarts like Hewlett-Packard. Data General, too, had decided to put more emphasis on VAR-OEM accounts, but only because it had suffered a $42.6 million second-quarter loss in 1987. Had IBM and Data General reported record earnings and maintained their competitive positions in the marketplace, support for VAR-OEM accounts might still be languishing, and who is to say that support will not be weakened when they do report record earnings in the future and must bow to internal pressure from their own direct sales forces?

In the case of Data General Corporation, there is an added concern; in refocusing its efforts toward low-volume VARs, it severed ties with its five U.S. distributors who formerly supplied the VARs. In doing so, Data General may have put itself in the awkward position of not being able to respond fast enough to small VARs who typically live from sale to sale in narrowly defined market niches. Customers of those VARs may have to wait unreasonably long for the products they have ordered.

In buying from VARs, look carefully into the relationships with their main suppliers. Here is how to determine the health of such relationships:

- Find out if the VAR has the source code and engineering documentation for the products its sells so it can readily customize, modify, or enhance the products you buy.
- Determine how much support the supplier gives to the VAR's maintenance program. The supplier should offer an ongoing training program for VARs, consisting of preventive maintenance, diagnostics and repair, and program management. There should also be substantial discounts on replacement parts as further incentive to the VAR to establish a comprehensive maintenance program.
- The supplier should not lock the VAR into rigid margins by dictating the prices for equipment or systems. The VAR should be able to price its offerings according to the market it serves and the extra value it builds into the products.
- If the VAR offers only the peripherals of the supplier, this may indicate that its product line employs a closed architecture, which will limit configuration flexibility as well as connectivity potential to other manufacturers' systems.

Merger and acquisition activities can also have devastating effects on VARs. If you are considering a purchase from a VAR whose supplier has just been acquired or is currently embroiled in a takeover bid, be aware that distribution channels are typically the last item to be addressed after such marriages. Even if the trade media quote the glowing reports of industry observers that the consolidation process is progressing smoothly, there is usually a six-month period of turmoil brewing just beneath the surface. If the supplier is lucky enough to remain in existence, management issues must be sorted out, support infrastructures rebuilt, and lines of communication reestablished with the VARs.

Not every merger or acquisition heightens the risk of purchasing products from firms engaged in such activities, but you will never know if you do not look behind the scenes and try to project their effect on your organization's network, resources, or competitive position. During your probe of these and other relationships, you must satisfy yourself that the vendor is engaged in or pursuing demonstrably productive associations or ventures with other firms. If so, it indicates that they are serious about staying close to customers for a long time to come. Just keep in mind that there are risks associated with each type of arrangement and that you must look beyond all the media hype typically surrounding their formation. Of course, do not discount vendors whose product lines are so complete as to make these types of arrangements unnecessary. It may be the case that some low-profile firms are just not as skilled in blowing their own horns as they are in developing a comprehensive product line.

1.9 REFERENCES

Ask the vendor for a list of references. Do not accept two or three; ask for ten or twelve so you can get a more representative cross section of opinions about the vendor. Even if the vendor has carefully prescreened the list of references, which any vendor with a modicum of common sense should do, you can unearth the facts. Go beyond the references supplied by the vendor. Ask those references for the names of other users. Check user groups and cooperative purchasing organizations. If you have selected specific features, ask for references who also use those features.

When calling references, be sure to inquire about the timeliness with which installation, integration, or customization was completed. Ask about the cooperativeness of the vendor in solving elusive problems in the operation of hardware or software, and whether the product's performance matched the buyer's expectations. If you stumble upon a reference who no longer uses the vendor's products, find out why!

If your application of a product is unique and the manufacturer is trying to persuade you that some up-front money is necessary to complete product development, customize, or redesign an existing product to fit your application, be extra careful about entering into any agreement that does not include a detailed description of the nature and scope of such activities and a precise list of performance milestones. Even before matters reach this stage, you should have asked the vendor about its experience with such arrangements and obtained appropriate references. If this is the first such transaction for the vendor, it is a good idea to check the local media and the national trade press for any adverse publicity about the firm, its officers, or its products. Be especially alert for published evaluations of the company's products or its marketing efforts written by industry consultants or financial experts, and if you have not yet looked into the company's product development effort, especially to find out the cancelation rate of projects, now is certainly the time to do so. Remember, forking over money to help a company customize existing products or develop new ones is the riskiest purchase of all.

1.10 ESCROW PROTECTION

If you are buying a software product, find out if you are entitled to the program's source code in the event the vendor goes out of business or closes out the product. The source code, which reveals details of the software's architecture, should be deliverable automatically from an escrow account or from a third party specializing in such services. These arrangements require the assistance of an attorney who is experienced in matters of software protection — it is too easy to overturn these agreements in court. Also, make certain that whenever the product is updated that the source code in escrow is also updated.

For hardware, find out how much of the product's technology is proprietary and what provisions have been made to provide customers with continuing support if the vendor should go out of business or discontinue the product. Find out if the circuit schematics, parts lists, and production drawings can be deposited in escrow. You will be amazed at how many vendors balk at such notions, but still expect you to fork over big bucks for products that are not even off the drawing boards! When it comes to sizeable investments to install new systems or upgrade your network with expensive add-ons, do not settle for vendor smiles and the "trust me" send-off.

1.11 QUALITY OF SALESPEOPLE

Examine the quality of the vendor's salespeople. Find out how knowledgeable they really are about their products and technology, and about networking in general.

The current buzz words in the communications industry include "systems integration" and "solutions selling." Many salespeople mouth these words quite authoritatively, but without really understanding the implications of what they are saying or without comprehending the promises they are making. When considering integrated solutions, make sure the performance of each element is up to par and that there are no weak links in the package. If you are evaluating high-speed computers that must support remote access from several satellite locations, for example, find out how the vendor supports T1 products. If the vendor's expertise is really in computers, chances are it cannot support T1, in which case the sales representative should not be trying to peddle "integrated" solutions.

Another red flag to watch for is salespeople who seem to disregard details about your network while attempting to sell you their favorite "solution." It may appear that a particular salesperson is genuinely deficient in technical knowledge; more often than not, however, the details are deliberately ignored because the vendor cannot address your needs with its current product line.

When meeting a salesperson for the first time, it is a good idea to strike up a conversation about industry developments and about networking in general. Some tactful probing on your part will flush out what salespeople know and do not know, and what the vendor has or does not have. If you determine that the salesperson lacks a basic understanding of networking or has no appreciation for your company's objectives, voice your concerns to the vendor, and suggest that someone else handle your account. Any vendor worth doing business with will appreciate the feedback and remedy the situation quickly, and if you can determine the real problem is that the vendor has no real solutions, make that known to the vendor, too. It will prevent you from wasting time listening to generic presentations, and it will enable the vendor to target its customers more effectively. Both parties stand to gain through constructive dialogue.

1.12 HUMAN RESOURCES

Always plan to visit the vendor's headquarters when you are considering a major purchase. Your objectives include verifying the claims of the vendor's representatives and separating fact from fiction in product literature and company brochures. If you discover discrepancies, you have to wonder what else has been exaggerated and whether it is serious enough to disqualify the vendor from further consideration.

There is another reason to visit the vendor's facilities — you want to get a feel for the management styles of the key executives. Many vendors are privately held companies headed by founders with colossal egos which manifest themselves in potentially harmful ways. Some executives have the overwhelming need to control others, so much so that it clouds their judgment. Some exhibit autocratic tendencies, sometimes to the point of deriving great satisfaction from humiliating employees and browbeating them into submission. Some executives actually believe that their unpredictable behavior keeps employees on their toes! In making others grovel to keep their jobs, they experience an intensely euphoric sensation of power and superiority. They will not give up control, even on trivial matters. The fact that they are successful businessmen, possibly millionaires, only reinforces in their own minds the validity of their behavior. Obsequious employees have to put up with it, or move on to other companies — after all, there is no arguing with "success."*

There are sound business reasons why you should avoid such companies when you encounter them. Employees who have nowhere else to turn are too angered or frustrated to give you peak performance when you need it. In extreme cases, it is not inconceivable that an angry technician, for example, may deliberately botch an installation as a way of "getting even" with management for the latest in a long string of inflicted wrongs.

When employees decide to leave harsh environments, the resulting turnover can have devastating effects on small companies, if only because they can least afford it. In addition to inexperienced or insensitive management, there may be other causes for high turnover: benefit plans that are out of step with the industry, jobs that do not motivate or challenge, and the absence of growth potential within the company. It is important to look at attrition because it may have a direct bearing on the timeliness with which problems get resolved and, consequently, affect the continued stability of your network.

Beware of vendors who defend high attrition, claiming that it is natural to employ a large number of people during the product development stage and then cut back once production begins. This rationale is merely a lame excuse for poor planning and management incompetence. Beyond that, it demonstrates the grossest form of insensitivity, which may be turned against you after you become a customer.

If the company has instituted significant layoffs, find out how the situation was handled by management. Were affected employees given assistance in applying for unemployment compensation? Were they given the opportunity to attend coun- seling sessions on the techniques for writing résumés and obtaining interviews? Did management notify other companies in the area about the availability of engineers, technicians, and other professionals? Did the company provide sever- ence pay and pick up the cost for extending medical coverage to ease the transition to new employment? The purpose for asking such questions is to ascertain how you might be treated by the company when the going gets rough. Beyond that, you are trying to assess the level of damage done to the company before you become one of its customers. For example, if the company has recently terminated 10% of its workforce, for whatever reason, find out how the financial community and industry analysts view the situation. They may interpret the layoff as mere

---

* For an excellent account of how a founder's idiosyncratic personality can bring a company to the brink of disaster, I highly recommend reading Jeffrey S. Young's insightful account of Apple Computer, Inc., *Steve Jobs — The Journey Is the Reward,* published in 1988 by Scott, Foreman and Company (Glenview, Illinois).

"fat trimming," a positive step, long overdue, or the layoff might confirm their long-held suspicions that the company is not reacting fast enough to changes in the marketplace and is allowing young, aggressive competitors to overtake it with innovative products. You do not want to get into long-term relationships with companies that are backpedaling to stay alive. Leave these high-risk opportunities to investors and arbitrageurs.

When you tour the vendor's facilities, find out how accessible the vendor's top executives are to customers. Determine if the executive staff members are predominantly academic, technical, marketing, legal, or accounting oriented — there should be a balance. If they are oriented toward engineering or accounting, can they communicate effectively with customers? More importantly, do they have the ability to listen to problems and act on the advice of customers? (The existence of a users group may answer that question.) Have any of the key executives been installed by a venture capital firm to keep an eye on its investment? Answers to these questions may alert you to the need to probe further.

Be alert to executives who are "one-man shows," who demonstrate overly aggressive behavior, or who prefer to operate from gut instinct. Typically, staff development is ignored, turnover is high, projects are poorly executed, company performance is portrayed in overly optimistic terms, and technological advantage is allowed to lapse after a brief bout with success.

Also, if you are dealing with a new company that is touting innovative technology, do not succumb to the founder's vision of where the market for the product is headed. Typically, founders are too emotionally involved with their creation to be objective about its market potential. Assuming that the product you want to purchase offers a promising solution, check out the members of the firm's financial management team. Find out what experience they have in budgeting and forecasting, and how closely they work with the company's founder. These considerations are important for the simple reason that new companies most often fail out of a lack of cash, or the inability to handle growth because the right financial structure was not in place. All too often, founders try to be their own chief financial officers, despite their apparent lack of experience in analyzing numbers.

Before committing yourself to products that have only been announced, but not yet released, you might monitor progress by looking into the stock transactions of key executives. If the product's release date has been eagerly anticipated by the media, but looks as if it will slip, company executives will be the first to know. Sudden liquidations of stock by top executives in anticipation of a lower price per share should serve as a red flag that warrants further investigation. There are several newsletters that monitor the trading activities of corporate executives, who are required to disclose their stock transactions to the Securities and Exchange Commission when holdings in any single company reach five percent of total stock. Such information is believed to be a key indicator of company performance by some stock market analysts.

After applying some due diligence to the vendor's executive ranks, do some probing of the vendor's technicians in an effort to flush out potential problems related to installation and acceptance testing. Because the installation of new equipment or systems may involve interconnection with presently installed utilities, communications facilities, and office systems, you should satisfy yourself that the vendor will be using only the most qualified personnel to work on your premises. It is often a good idea to request the résumés of all known personnel who may be assigned to the installation of the equipment or system you plan to buy. Request that the vendor clearly mark the résumés of those who may be available for backup. For each individual assigned to the project, have the vendor designate the individual's status as full-time, part-time, or consultant. Have the vendor specify the number of years each person has been with the company. You should also insure that the vendor plans to have a supervisor on site during work hours. Ask for an organizational chart of project personnel, including an explanation of the delegation of responsibility. Your objective in obtaining all this information is to avoid having your installation used as the vendor's training camp for new and inexperienced technicians, to minimize installation delays, and to avoid shoddy workmanship that may cause problems long after the installation is complete. Include a request for this kind of information in your RFP.

Because installation problems are frequently attributable to vendor scheduling, you should find out what other installation projects the vendor may have during the same time frame. Ask for a statement of the vendor's other contractual obligations, the dollar value of each project, duration of each project, as well as a description of the vendor's role in each project. This information will give you an idea of the vendor's priorities in case there is an unexpected disruption somewhere along the line, whether it be in the form of a labor strike, natural disaster, the nonavailability of equipment or materials, or the refusal of a customer to accept the product for performance reasons.

1.13 THE BIG PICTURE

When you are evaluating products for your network, it is not enough to look at features and performance — you must also find out everything you can about the vendor's organization to determine its future survivability. Only when you have assured yourself about the vendor's present and future position in the market can you be reasonably certain of continued support throughout the product's useful life.

Do not be fooled, however, by a company's impressive sales revenues. Despite big increases from one year to the next, it is not uncommon for companies in the computer and communications industries to post big losses that can leave their futures in doubt — and you holding one-of-a-kind products with limited connectivity to other vendors' products. Among the many factors that contribute to such situations are:

- The vendor's rush to compete in too many markets at once out of the need to diversify its customer base. With this strategy, the vendor hopes to cushion itself against the decline of one or more volatile market segments. Unfortunately, this strategy tends to consume more cash than it generates.
- The sales cycles for computer and communications systems are very long. While buyers are developing RFPs and evaluating vendor proposals, the company still has to make the payroll and support the organizational infrastructure.
- Customer buying habits have changed in the last few years. Instead of bits and pieces of a network, customers prefer to purchase everything from a single source. This has put more pressure on vendors to broaden their offerings. As a result, profits from increased sales revenues may be squeezed by skyrocketing research and development costs.
- Installing and servicing complex systems and networks is time-consuming and resource-intensive. Even if customers agree to pay 30 days after delivery, they tend to wait until the product is installed and operating properly. Then they take another 30 days!
- Building distribution channels requires a lot of up-front cash that may not be fully recovered through sales for another one to three years.
- Sometimes the company grows too fast — beyond its capacity to handle its growth. A company that grows more than 30% in a single year usually has trouble financing and managing its growth. This results in a loss of confidence among investors, which explains why firms that experience unbridled growth sometimes have difficulty obtaining additional financing.

Obviously, not every purchase decision requires implementation of this rigorous risk avoidance strategy. Sometimes there is no significant risk. In the case of satellite-based long-distance service, for example, the demise of Argo Communications (New Rochelle, N.Y.) in 1987 caused no hardship among subscribers because there were numerous service providers to step in and take its place. Furthermore, there were no significant hardware commitments for Argo's services that could not be used with alternative providers of satellite-based services. The same cannot be said of crippled Ztel, which has essentially left customers stuck with an expensive white elephant, notwithstanding wimpy promises to continue to provide support and enhancements for its digital PBX with fewer than ten staff members.

Even when the due diligence approach is warranted, few vendors will score an acceptable grade on every one of these reliability indicators. You will have to weigh each factor in terms of its relevance to your situation. For example, if you are purchasing a telecom management system that requires extensive support from your MIS-dp department, and that department is plagued with high personnel turnover, you should assign a higher weight to technical documentation and customer service. Training is also important, but the vendor is not likely to provide it with every change of personnel without slapping you with extra charges.

Effective purchasing decisions require business acumen as well as technical knowledge. In today's competitive arena, network managers cannot become so imbued with technology that they forsake logic and practical business considerations. The overriding objective in any major purchase should be to minimize uncertainty, while endeavoring to increase planning flexibility and reduce costs over the long term. There is really no room for wishful thinking or the attitude that everything will somehow fall into place. Nor can you afford to delegate purchase decisions to others; top management has a low tolerance for the kind of finger-pointing that blames subordinates for costly mistakes.

Performing this kind of in-depth analysis on even one vendor is going to be a time-consuming process. All other factors being fairly equal among two competing vendors, performing such analysis to select one over the other may not be possible at all, unless you can afford to dedicate at least one staff member to the task.

If you are shorthanded, it may be advantageous to use an independent consultant who has had no previous relationships with the vendors you are considering. In this case, get a written report that addresses all twelve of the risk factors discussed in this chapter. For financial statements and credit information on the vendor, your best resource may be your organization's internal legal counsel or chief financial officer.

Recapping, the twelve-point risk avoidance plan should take into account the following elements. Together they constitute a systematic strategy for assuring vendor stability and product reliability.

- Research and Development
- Applications Engineering
- Quality Assurance
- Repair and Return
- Customer Service
- Technical Documentation
- Training
- Line of Business
- References
- Escrow Protection
- Quality of Salespeople
- Human Resources

While a careful analysis of vendors' financial statements and competitive offerings are indispensible elements of the purchase decision, they are by no means the only ones. The most innovative product can drain your organization's resources if the vendor lacks the wherewithal to provide support when you need it.

The rest of this book focuses on risk avoidance as applied to specific categories of products.

REFERENCES

[1]   *Network World,* November 30, 1987, p. 19.
[2]   Manes, Stephen, "Disimprovement: One Step Forward, One Stumble Back," *PC Magazine,* Vol. 6, No. 16, September 29, 1987, pp. 85–86.
[3]   *Management Information Systems Week,* April 4, 1988, p. 22.

PART TWO

Current products and services should be evaluated not only in the light of the present conditions but also of future technology, products and competition. [1]

John J. Daly
Principal
Westminster Management Group
San Francisco, California

Chapter 2
PBX, STS, and CENTREX

2.1 PBX

The Private Branch Exchange (PBX) evolved from the operator-controlled switchboards that were used on the public telephone network, the first of which was installed in 1878 by the Bell Telephone Company to serve 21 subscribers in New Haven, Connecticut. The operator had full responsibility for answering a call request, setting up the appropriate connection, and tearing down the path when the call was completed. Interconnectivity among subscribers was accomplished *via* cable connections at a patch panel.

Today's PBXs are much more complicated, of course, but they provide the same basic functionality as the first generation of switching devices. The difference is that the process of receiving call requests, setting up the appropriate connections, and tearing down the paths upon call completion is now entirely automated, and because it resides on the user's premises, the PBX allows organizations to exercise more control over business operations and incorporate communications planning into their business strategies.

In a little over one hundred years, PBXs have evolved from simple patch panels into sophisticated systems capable of integrating voice and data. The four generations characterizing the development of PBXs may be summarized as follows:

- First-generation PBXs included operator-controlled devices and electromechanical switches.
- Second-generation PBXs include programmable, computer-controlled switches. Data may be handled with add-on components.
- Third-generation PBXs include the attributes of second-generation PBXs, plus they are nonblocking and may use a distributed architecture. Data-handling capabilities are programmed into the design.

- Fourth-generation devices combine PBXs and LANs into a single device that can deliver voice and data over the same pair of wires and offer gateways to other networks.

Stored program control, advanced processing power, large-scale integration, and high-capacity memory have not only reduced the cost of PBX ownership, they have also endowed the PBX with an incredible array of features. In fact, among the dozens of products to choose from, hundreds of different features are offered. Whether they are standard or optional depends on the vendor, and as PBX makers integrate new technologies into their products to provide more capabilities, the list of features continues to grow, further complicating the selection process.

There are a lot of decisions to make regarding the selection of a PBX. You will have to wade through numerous features and capabilities, choosing only those that your organization really needs, while identifying those that may be required later as the work environment and business activities change. You will have to delve into design issues to determine how the PBX guards against component or power failures, how lines and trunks can be added incrementally and at what cost, and how the PBX can be upgraded to take advantage of new services. Once you have narrowed your list of vendors to a handful of viable contenders, your work has really just begun. You must now evaluate service and support, negotiate price, and work out the terms of the contract. Even after all this you are not finished. You still have to monitor the performance of the contract.

Given all of the variables that go into the selection of a PBX, the risk of making a wrong purchase decision is quite high, and if you are buying several units for a multinode network, the risk is compounded. Think about what it would be like to live with a wrong purchase decision for seven years — that is when the average PBX gets replaced. Considering, however, that the phone system is among the largest capital investments your organization is likely to make, do not be surprised if you are not around to help choose the next one.

If you are not up to date on the latest products and technologies, or not staffed to do all of the necessary research, you are not alone. About 80% of PBX sales involve the use of consultants to perform needs assessment, develop the request for proposal (RFP), evaluate vendor bids, negotiate the contract, and monitor vendor performance after the sale.

2.1.1 Feature Selection

PBX makers pride themselves on the number of sophisticated features their products can implement. The interesting thing about all of these features is that they can be implemented right from the telephone keypad. Here is a sampling of basic features that most PBXs implement:

- Add-on conference allows the user to establish another connection while

having a call already in progress.
- Call forwarding allows a station to forward incoming calls to another station. This includes forwarding calls when the station is busy or unattended, or as needed.
- Call pickup allows incoming calls made to an unattended station to be picked up by any other station in the same trunk group.
- Speed dialing allows the user to complete calls by dialing an abbreviated number. This feature also allows users to enter a specified number of speed-dial numbers into the main database. These numbers may be private or shared among all users. Entering speed-dial numbers is accomplished *via* the telephone keypad.
- Last number redial allows users to press one or two buttons on the keypad to activate dialing of the previously dialed number.
- Call waiting lets the user know that an incoming call is waiting. While a call is in progress, the user will hear a special tone that indicates another call has come through.
- Call hold allows the user to put the first party on hold so that an incoming call can be answered.
- Camp-on allows the user to wait for a busy line to become idle, at which time a ring signal notifies both parties that the connection has been made.
- Message waiting allows the user to signal an unattended station that a call has been placed. Upon returning to the station, an indicator tells the person that a message is waiting.

Hyperactive sales representatives seem to gush over all the things you will be able to accomplish with their product, but you cannot let yourself get caught up in the latest technical advances and lose sight of basic organizational requirements. Privately, even vendors admit that only a handful of basic features are used on a regular basis, while others are used so infrequently that people eventually forget how to implement them.

Do not trust yourself to know what features are best for your organization. Do some basic internal research to determine the needs of each operating unit in your organization. A busy telemarketing or claims-processing unit, for example, will probably need automatic call distribution (ACD), but you probably would not appreciate the importance of ACD if you solicited input only from the accounting or personnel departments because they typically have no need for it. To make sure you are on target, solicit input from every operating group and compile a matrix chart of features each department will need. You have to find out what the growth plans are for each group so that you can anticipate the need for additional features. Future needs should be projected so that you can determine the cost of potential upgrades. You should also project the time frames in which additional features will be implemented.

Another matrix should list the number of stations required at the time of initial system cutover, the number of stations you would like to have wired for near-term use, and the total number of stations you would like to have installed and wired over the life of the system.

All of this is important information to have handy during price negotiations with the vendor. When vendors see that you have a growth plan and that there is a high probability of future business, they can be more flexible on the price of the initial system. This is because such upgrades may entail sales of software, processors, memory, and circuit cards. You can minimize these costs by selecting a modular system that can accommodate such growth incrementally, and without the need to changeout major components, but more on this later.

When it comes to features, you must also consider that the more you buy, the more training will be required, possibly in multiple sessions so as not to overwhelm users with too much information at once. Even when adequate training has been provided, follow-on training may be required for new employees, as well as "refresher" training for experienced users, and when new features are added, you must consider how best to disseminate instructions. It is very easy to negotiate vendor-provided training — even refresher training — because vendors view this as an opportunity to assess your organization's need for new features. Without adequate training, however, a great many features will go unused. If top management does not support the training effort 100% by setting aside adequate time and facilities, the organization will not be getting an adequate return on its investment.

Because there is so much to learn about using this type of product, you will also have to check the quality of documentation that comes with the PBX, of which there should be at least three categories: the system administrator's manual, the attendant's manual, and the user manual. The administrator's manual contains information of a technical nature, such as how to configure the class of service restrictions that control access to certain functions and external networks on a station-by-station basis. The attendant's manual contains information on features that are typically implemented at the attendant station, such as paging, line status monitoring, recall identification, and night service to redirect calls to an alternative attendant station. The user manual contains very concise instructions on how to implement station features. This level of documentation should include a template that fits over the keypad to provide users with a quick reference on how to implement commonly used features. Alternatively, a "cheat sheet" may be provided that slides out from the bottom of the telephone. You should plan on replenishing these materials from time to time because they get lost, worn out, or trashed. This leaves new employees and transferees without the means to use their telephones to full advantage. Beyond that, the image of the organization may suffer if too many people lack the ability to implement even the most rudimentary features. An executive can talk authoritatively about the intricacies of his new product and boast about his company's fabulous growth rate, but immediately lose credibility because he does not know how to transfer the call!

Matters of training and documentation may seem trivial at first, but they can help an organization protect its investment in technology by insuring that all the features it paid for are fully utilized to achieve the efficiency and productivity gains that motivated the initial decision to purchase the PBX.

2.1.2 Capabilities

There are numerous capabilities that promote the efficiency and administration of the PBX. They are distinguished from features in that they are typically transparent to most users. Some of the most common include:

- Direct inward dialing (DID) allows incoming calls to bypass the attendant and ring directly on a specific station.
- Direct outward dialing (DOD) allows outgoing calls to bypass the attendant for completion anywhere over the public telephone network.
- Hunting is a capability that routes calls automatically to an alternate station when the called station is busy.
- Automatic least-cost routing insures that calls are completed over the most economic route available. This feature may be programmed so that mailroom staff always gets the cheapest carrier, while executives get to choose whatever carrier they want.
- Class of service restrictions control access to certain services or shared resources. Access to long-distance services, for example, may be restricted by area code or exchange. Access to the modem pool may be similarly controlled.
- Call detail recording (CDR) is the capability of the PBX to record information about selected types of calls. This includes the collection of detailed traffic statistics. The special hardware and software required to arrange these data into meaningful management reports can be obtained from numerous third-party vendors.
- System redundancy enables two processors to share the switching load, so that in the event one of them fails, the other can take over all system functions.
- Related to system redundancy is data-base redundancy, which permits the instructions stored on one circuit card to be dumped to another card as a protection against loss.
- Power-fail transfer permits the continuance of communication paths to the external network during a power failure. This capability works in conjunction with an uninterruptible power supply (UPS), which kicks in within a few milliseconds after detecting a power outage.
- Automatic call distribution (ACD) allows sharing of incoming calls among a number of stations so that the calls are served in the order of their arrival. This is usually an optional capability, but it may be integral to the PBX, or purchased separately as a stand-alone device from numerous non-PBX vendors.

- Automated attendant is the capability of the system to answer incoming calls and prompt the caller to dial an extension or leave a voice message without going through the operator.
- Music on hold might seem like a frivolous capability, but it can serve a useful purpose. During peak hours, for example, incoming calls might have to wait in queue until handed off by the ACD to the next available station operator. Music on hold provides assurance to callers that the connection is still being maintained. In this application, music on hold functions as a progress tone.

When determining what capabilities the PBX should have, do not forget to project future needs. As with feature selection, this can help reduce the initial cost of the PBX. With future needs documented, vendors will be more inclined to lower the price of the switch with the expectation that they will enhance future revenues from the sale of add-on capabilities. These capabilities might include voice mail, a message center, automated attendant station equipment, local area networking, and telecommunications management systems — many of which require the purchase of adjunct computers and other hardware.

PBX vendors have long recognized the value of add-on sales. Some, like Northern Telecom (Richardson, Tex.), claim that it is possible to derive as much as six dollars in add-on sales revenue for every one dollar of PBX sales. Use this knowledge to leverage your best deal.

2.1.3 Centralized *versus* Distributed Architecture

There is an ongoing debate among vendors about the merits of distributed architecture over centralized architecture. Actually, there is a place for both, depending on the user's circumstances.

The common concept of a PBX network involves a centralized switching system to which all stations are connected. Having a central point at which decisions are made improves the level of control and allocation of all network resources. This may prove to be adequate for small organizations where all the stations are in close proximity to each other, but for larger organizations spread over a campus-like environment, the centralized approach may incur prohibitive wiring costs.

Under the distributed approach, a number of smaller sites are integrated into a larger network. These nodes may be interconnected with each other *via* tie trunks (analog) or T1 lines (digital), which offer a more sensible wiring scheme than running individual phone lines to every site, but the key aspect of this architecture is its distributed control; there are no master or slave relationships on this type of network. Each node functions independently and contains routing information for all other nodes. Special link protocols send information about the caller through the network to the receiving end, which lets each node on the network handle the call appropriately, based on such information as type of originating equipment and class of service.

Trunk connections to the local central office may originate from any part of the network. With critical resources evenly distributed, the network is better protected from the possibility of catastrophic failure.

The arguments on behalf of distributed control are quite strong, but even PBXs with a centralized architecture may be adequately protected against the possibility of catastrophic failure, and more economically than moving to a distributed architecture. For example, Northern Telecom's SL-1RT offers redundant common equipment, CPU, and memory for medical, utility, and military applications that need "hot standby" redundancy. On top of that, the system continually runs self-diagnostics of its CPU, memory, and power units, making the SL-1RT a fault-tolerant PBX. The SL-1RT sells for $650 to $800 per line, *versus* a distributed system which can sell for $750 to $1,000 per line. The one drawback to the SL-1RT is that it suffers from the same ailment that afflicts other PBXs utilizing a centralized architecture: susceptibility to blocking.

Furthermore, the software needed to control and coordinate a distributed PBX is quite complex and almost always requires some degree of customization. As a rule, software problems are much harder to fix than hardware problems.

Proponents of distributed architecture also claim that hardware problems are easier to isolate and fix than under the centralized approach, but the diagnostic tools available with most PBXs facilitate fault isolation. Repairs or component replacements may be performed faster under the centralized approach because the technician is already on site, whereas under the distributed approach, nodal downtime may be prolonged until the technician can arrive at the site.

Distributed systems do offer an easier upgrade path and permit more economical migration across product families. The trend across the communications industry is for products that are totally modular. A nonmodular product is a dead-end product. In the case of PBXs, some vendors require that you replace the common control system to upgrade from one product to another. Sometimes you must change processors, cabinets, and linecards — all at substantial expense. Some vendors even require a change of handsets when moving from one system to another.

The distributed architecture, then, allows you to increase port capacity, features, and networking capabilities by bunching together add-on modules. Some PBXs, like the NEAX2400 from NEC America Inc. (Melville, N.Y.), allow graceful expansion from 32 lines to more than 23,000, and Rolm's (Santa Clara, Calif.) 9750 Business Communications System allows expansion from 100 to 20,000 lines, while permitting users to retain close to 100% of the original hardware investment. Moreover, all Rolm products may be networked together. Northern Telecom, on the other hand, does not allow easy migration from its 600-line SL-1ST to its 1,500-line SL-1NT or 7,500-line SL-1XT. The SL-1ST requires a change of processors, memory cards, and signaling control cards. The SL-1NT and SL-1XT, however, are totally modular and expansion requires no changeout of hardware or software.

Mitel Inc. (Boca Raton, Fla.) even prides itself on the fact that the first PBX it ever sold may be upgraded to the latest generation system.

Although both types of architecture have advantages, the PBX industry is clearly moving toward the distributed approach, if only to offer a smoother transition between product lines and an economical upgrade path for adding lines, features, and capabilities.

2.1.4 Analog *versus* Digital PBX

Another buying decision that has to be made is whether to choose an analog or a digital PBX. Conventional wisdom says that analog systems are outmoded technology and that digital switches are required if you want to position your network for future services like ISDN. Understandably, users are confused. They would like to stay with proven analog technology, which costs less than digital. On the other hand, they want to position their networks for future services like ISDN, in which case they would need the more expensive digital PBX.

There are essentially three classes of user. One class is composed of heavy data users whose current requirements necessitate buying a digital PBX. There is another class of user whose present and future requirements are solidly voice oriented and, therefore, require only an analog switch. Still another class consists of users who are unsure about their future data requirements. This is the group most vulnerable to the hype surrounding the need to abandon analog and buy into "digital."

One of the catalysts fueling the move from analog to digital PBXs is the anticipated implementation of ISDN, which promises end-to-end digital connectivity and a greatly simplified way of moving voice, data, and video through the network. Several PBX vendors claim that their products are "ISDN ready," meaning that they will be able to accommodate the special interfaces required to provide primary rate ISDN services when they become available from designated CENTREX exchanges.

Another catalyst driving the move to digital PBXs is the seemingly relentless invasion of microcomputers into the office environment. Some industry analysts are predicting an installed base of as many as 26 million microcomputers by 1990. Many of these devices will be networked over LANs. Some PBX vendors have suggested that microcomputers and peripheral devices can be networked economically through the digital PBX. AT&T (Basking Ridge, N.J.) obviously thinks so. Its low-end System 25 is compatible with microcomputers, asynchronous ASCII terminals, printers, and FAX machines. In addition to functioning as an Electronic Tandem Network (ETN) endpoint, the System 25 even provides a gateway to StarLAN, AT&T's local area network.

These developments are compelling reasons for choosing digital over analog systems, but there are equally compelling reasons for retaining or purchasing analog systems. Analog PBXs have a number of advantages over digital systems. They are substantially lower in price, which means that the dollars saved can be invested in meeting other important needs.

Analog systems have the benefit of many years of software development behind them, software which has demonstrated its reliability in a variety of configurations and operating environments, whereas the software of digital systems is still grappling with software bugs. Many analog systems use English-language, menu-driven procedures for administrative functions, whereas digital systems use high-level programming languages that require learning special abbreviations, mnemonic codes, and arcane procedures that greatly complicate administration.

Despite the continued publicity about ISDN, it may still be five to seven years or more before it is available on a national or international basis. Since the average PBX is replaced every seven years, you can defer the high cost of digital by going with analog now. By the time you need to replace the analog switch, the price for digital will have fallen dramatically, and by then ISDN should be well established. On the other hand, if ISDN fails to take off as predicted, you will have saved your company a bundle, and you will be recognized for your foresight and business acumen.

Even if things do not pan out within the time frames you expect, there is still a fail-safe solution: buy an analog PBX that can be upgraded to digital. Mitel Inc., for example, claims that it can upgrade its analog PBXs to digital in as little as an hour. All that is involved is a field visit to the customer location where an exchange of circuit boards takes place. Analog boards are replaced with new circuit boards containing digital software, thus providing a convenient and economical migration path.

While certain organizations have an immediate or foreseeable need for a digital PBX, others may not have that need for some time to come, in which case investing in digital technology is premature. After all, there is no appreciable difference between digital and analog PBXs in terms of voice quality. Consequently, there is no need to pay a premium price for digital technology when an analog PBX can provide the same basic features, including data communications *via* card-mounted modems.

2.1.5 Service and Support

There are two trends in the PBX industry to be aware of. One is the intense competition in recent years that has forced many PBX vendors to slash prices to make sales. As a result, profit margins are being squeezed dry. This has forced many vendors to make up the difference by cutting corners in other areas, such as service and support. Lately, the pendulum has been swinging in the opposite direction. Some companies now recognize that as PBXs increasingly look alike and act alike, they can differentiate themselves from competitors by offering comprehensive service and support. In your evaluation of vendors, however, you must delve into this issue to make certain that they are not just mouthing the words to make new sales. You have to find out if there is any substance behind these claims. After all, since your organization will be living with the new phone system for quite some time, ongoing service and support is all the more important.

One of your first steps in performing such an evaluation is to contact references who use the same type of PBX that you are considering for purchase. An "excellent" level of service adheres to this sequence of events:

- In reporting a problem, you are treated courteously.
- The vendor's service desk operator should attempt to find out the nature of the problem before dispatching a field technician.
- The service desk operator should give you the approximate time of the field technician's arrival at your site.
- The field technician should arrive within the time frame stated by the service desk operator; if there is a change due to unforeseen circumstances, you should receive a call notifying you of this, and an apology.
- When the field technician arrives, he should be equipped with appropriate test equipment and spare components.
- You should be satisfied that the problem has been fixed before the technician leaves your site.
- Upon his departure, he should have you sign a work order which states the length of time he was on your site and the type of work that was performed. There should also be a space in which you can indicate your level of satisfaction with the work.
- Later in the day, or the next business day, you should receive a follow-up call from the vendor's service desk operator, who should inquire about the quality of the field technician's work and whether the system is working properly.

Do not be lulled into complacency just because you may be purchasing your PBX through a big company. You might think that a big company can afford a huge support staff and, therefore, will be able to deliver superior service, but even AT&T has had to tighten its belt in recent years, and its support staff has not been immune from cutbacks. In fact, AT&T's System 85 customers were among the first to feel the effects of the company's cost-cutting program in 1987.

In an effort to be more competitive, AT&T has been known to quote the System 85 with only one processor to keep the price as low as possible and win the job. In such cases, AT&T has tried to convince potential customers that its processor had an MTBF of about two years and, therefore, dual processors were really unnecessary. Some customers who bought into that line of reasoning ended up regretting it because AT&T's processor was not always as reliable as claimed.

One customer is even known to have experienced three such failures in a single day! Although AT&T fixed such problems whenever they occurred, some customers were without phone service for up to four hours at a time.

Customers felt the effects of AT&T's cost-cutting program in other ways. In addition to eliminating the jobs of good field service people, which prolonged response times, customers found that they had to wait as long as 90 days to have equipment orders fulfilled, even on routine items like station phones. This was because in trying to cut costs, AT&T had cut down on equipment inventories.

One way to get the straight scoop on the quality of a vendor's service and support efforts is to contact its user group. Usually, the members are quite knowledgeable about the product and will even recommend — or demand — that the vendor develop specific features and capabilities. Beyond that, the user group is privy to "futures," information about the continued development of the product, and if a vendor fails to deliver a product enhancement on time, or the product experiences problems upon release, the user group members are among the first to know all the details because most likely one of them was the beta test site. Most vendors are very sensitive to the issues raised by these groups because they know that their members carry a lot of influence. They are quoted in the trade press, even if anonymously, and they are contacted by potential buyers for their opinions about the vendor and its products.

Not all vendors have a user group, however. Some vendors believe that user groups will only end up airing a lot of dirty linen that the company is in no position to deal with anyway. If a user group is not available for the PBX you are interested in buying, you should contact a consultant who is knowledgeable about the vendor's system and its support capabilities.

Another method of checking out the vendor's support capabilities is to talk directly with the vendor's local field-service and customer-service representatives. An informal conversation may reveal such things as personal attitude, motivation, responsiveness, perceived job security, personal feelings toward their employer, level of job satisfaction, and desire to improve their knowledge and job skills. All of these will help determine the quality of support you can expect, once you become a customer.

2.1.6 Evaluating the Bids

The PBX industry is more competitive than most, and for good reason: the market for new PBXs is steadily shrinking. Consequently, it is getting a lot tougher for vendors to eke out a living. This means vendors have to conjure up more creative ways to capture new business to carry them through the lean years to come, but this ground swell of "creativity" may not be in your best interest. To minimize the risk of making a bad purchase decision, be alert to these ploys:

- Watch out for the bid that barely meets your requirements. It may look like a bargain, but if the vendor leaves out too much processing power, you may not have enough room to grow. You might think you are getting off cheap — until you discover one day that you have to buy a new shelf or add another cabinet just to accommodate another telephone station.
- Some manufacturers are speeding up production lines to reduce the cost per unit, thereby giving themselves more room to maneuver in competitive situations. While this strategy may stimulate sales revenues over the short term, it may force cutbacks in research and development, as well as jeopardize the long-term financial health of the company. Remember, a sound R&D program and continued service and support depend on the company having a stable infrastructure. Without it, you may have to do without future enhancements and become burdened with additional costs for service and support that go well beyond your original expectations.
- PBX makers are also buying the distributors that once sold their line of products. The purpose is twofold: to reduce markups by eliminating the middleman and to provide more direct contact with customers, but many PBX makers are hardware oriented rather than service oriented. That was the reason for selling through distributors in the first place. It remains to be seen whether such arrangements will work over the long term. If not, be prepared for major disruptions in service and support until this trend runs its course.
- Watch for the vendor whose pricing is substantially lower than the others. This usually indicates one of two things: either the vendor has bid the system too close to capacity or a key requirement has not been addressed. You might want to consider discarding that bid, rather than waste time trying to discover which game the vendor is trying to play. Put your efforts into evaluating bids that are in the mainstream. Chances are good, anyway, that the mainstream vendors will trim their bids if they think they have a good shot at getting your business. Tossing the low bid will intensify the competition.
- Once you have settled on price, you can expect the vendor to try to make up lost margins by bumping up the price of maintenance. Look for the vendor to offer short-term contracts of no longer than a year, so the door will be open for hefty increases every year thereafter. If the size of your purchase is significant enough, you might be in a position to negotiate more favorable terms.
- In checking all bids, make sure the cost of customized software is already included in the vendor's proposal. If you assume too much, you could find yourself in the position of having committed to the vendor, only to find out too late that the software you need costs quite a bit extra. Vendors know that you will prefer to deal with them direct on matters of software, rather than put up with multivendor hassles later.

One last tip: given these lean times, you may want to think twice about going with the smaller vendor. The product might fit your organizational needs very well, but unless the vendor has a large enough customer base, its prospects for survival are doubtful.

2.1.7 Used PBXs

Although this chapter is focused on the purchase of new PBXs, you may want to consider the purchase of refurbished equipment if you have a small budget. Not only do refurbished PBXs offer immediate cost savings, but they also can provide an interim solution until your needs justify a digital PBX, the price of which should drop substantially by the time you are ready to buy. To test the viability of purchasing a used PBX, ask yourself the following questions:

- Does the PBX have a solid track record of reliability?
- Does the PBX offer all, or most, of the features and capabilities your organization needs?
- Does the manufacturer continue to provide spare parts for the PBX?
- Is the manufacturer committed to maintenance and support for the next three to five years?

Although many vendors have been reluctant to provide support and maintenance for products that have been discontinued because they prefer to push new products, others see the benefits. One of those benefits has to do with increasing their customer bases for new equipment when the time comes for users to upgrade or replace the previously owned systems.

Even though some PBXs, like AT&T's Dimension, have been discontinued for many years, the user groups continue to meet. The meetings still draw the participation of AT&T, which provides guest speakers representing such functions as maintenance, service and support, and product management. Despite its image as a dead-end product, there is still a lot of useful life left in the Dimension PBX, as evidenced by the 25,000 that are still in use.

In stark contrast to the way AT&T has handled its secondary market support program is the way IBM handles this matter for its PBX subsidiary, Rolm Corporation. During first quarter 1984, Rolm started shipping three PBXs: the VSCBX 8000, CBX II 8000, and CBX II 9000. Today, used models may be purchased at 40%, 46%, and 52% of the list price respectively. Because IBM wants to migrate these customers to the Rolm 9751 CBX, which offers more functionality and configuration flexibility in a more compact unit, it has been slow to announce a secondary market support program. As long as IBM can stall such an announcement, CBX users might feel compelled to buy the 9751 because they are not sure how much longer the older equipment will continue to be supported.

You can minimize the risk of purchasing a refurbished PBX by determining if the vendor has established a secondary market support program. Without such a program, you really have no assurance that spare parts and qualified service technicians will be around long enough to make the purchase worthwhile. This helps explain the wide differences in the values of comparably featured PBXs of different vendors. Because IBM has not yet committed to establishing a secondary market support program for Rolm PBXs, the values of Rolm's used equipment are much lower than the comparably featured products of other vendors with such programs. Consequently, what appears to be a bargain, may end up costing you dearly in terms of equipment downtime, user dissatisfaction, and damage to your credibility among top management.

2.1.8 Contract Considerations

All of the time and effort spent in implementing risk avoidance procedures for the selection of the best equipment and vendor will be for naught if you fail to apply the same due diligence with respect to negotiating the contract. As applied to the purchase of a PBX, the purpose of contract negotiation is to spell out your expectations, protect your capital investment, and limit your exposure to chronic PBX problems.

As alluded to earlier in this chapter, you will have a better chance at negotiating the price of the PBX downward if you can document your organization's future needs for features and capabilities. You can reduce the cost of adding on to the PBX by making sure that the initial purchase comes with enough processing power to carry you well into the future, and that product upgrades can be accomplished with minimal service disruptions and without changing out major system components. Before getting to the point of negotiating the contract, you should have already satisfied yourself that the PBX meets your organization's present and future requirements. You should have already validated vendor claims of product performance and reliability with references, user groups, and consultants. You should also know what to expect in terms of service and support. After you have done all of your homework leading to the choice of product and vendor, you cannot afford to drop your guard now by accepting the vendor's standard agreement. To be successful in avoiding risk, you must pay close attention to detail.

Among the important provisions in the PBX agreement that deserve special attention are:

- Functional specifications;
- Change orders;
- Installation and cutover;
- Acceptance test;
- Warranty;
- Software license.

The functional specification document describes what the hardware and software must do and provides guidelines for how it must be done. This is different than a design specification, which merely lists features, capabilities, and componentry. The problem with a design specification is that you are responsible for discovering all the variables that could affect the operation of your network. In effect, a design specification thrusts you into the role of engineer. The vendor can claim that it met your specifications, and will be off the hook if trouble develops later. For this reason, you must be careful about how you want the vendor to respond to your RFP. The response should include a functional specification. Any vendor who responds with only parts lists and cabinet diagrams should be eliminated from further consideration.

Many times the functional specification will be taken care of in the contract by referencing the RFP. If this is not possible, for whatever reason, you will have to cross-check your requirements as listed in the latest version of the RFP with those listed in the functional specifications section of the vendor's proposal and then merge that information into the contract. You cannot always delegate this job to less experienced personnel because of the differences in terminology that may be encountered in a detailed comparison of the two documents.

The vendor's hardware and software manuals should be referenced in the contract, as should the advertisements and sales literature which piqued your initial interest in the product and vendor, but many vendor contracts specify that the agreement supersedes any and all previous representations, statements, or claims about the product's performance. Watch out, though! This may even be interpreted to include the vendor's own response to your RFP. Because the vendor will usually insist on having this clause, your only alternative is to include a functional description of all hardware, especially the performance expectations of the processor or processors. When it comes to software, you have to describe all of its features and capabilities functionally, not just list them. Of course, there is nothing to stop you from using the vendor's own advertising, sales literature, and technical manuals as your primary sources for such descriptions — you can always square off with the vendor at the negotiating table.

The change order is another provision of the contract that requires your attention. In making sure that such a provision is incorporated into the contract, you maintain some flexibility to modify your configuration, add features and capabilities, or upgrade hardware and software between the time the contract is signed and the date of cutover. This provision is intended only for fine-tuning, not for drastic changes. After all, the vendor has a production and installation schedule to maintain, which limits the nature and frequency of changes. The vendor will insist on limits, which may include barring changes beyond a certain date or setting a cap on the dollar value of the changes, which varies according to the purchase price. Nevertheless, you should negotiate for as much flexibility as possible, even if you think you may not need it. Above all, never let a vendor attempt to dictate the nature and scope of permissible changes. In this competitive climate, everything is negotiable. Vendors understand this better than anyone.

Describing acceptance test criteria is another important concern. The acceptance test should provide you with an objective confirmation that the installed PBX operates according to the specification document. It is not enough to perform a visual inspection of the installation. You should not accept the vendor's own diagnostic tests as proof that the system works properly, and you certainly should not accept vendor assurances that start-up failures will be covered by the warranty or maintenance contract. If you let this happen, you may never get the problem resolved. At any rate, you should tie the payment schedule to the vendor's successful completion of installation milestones, shifting as large a percentage as possible to the later milestones. A reasonable schedule of payments may be as follows:

- 10% upon signing a mutually acceptable agreement;
- 15% upon PBX installation;
- 20% upon cutover;
- 55% upon passage of the acceptance test.

Your best course of action is to write up a performance test and specify a reasonable prove-in period. Make sure this is included as part of the contract and set a penalty if the vendor fails the acceptance test. The penalty may consist of withholding final payment and deducting a few percentage points from the final payment for every week that the problems persist. You should also limit the number of retests and specify the maximum duration that such efforts will be allowed to continue, after which time the contract may be terminated and a refund made.

The warranty is also a critical component of the contract. This provision should specify what the warranty covers, how long the warranty will be in effect, and how the vendor will respond to problems covered under the warranty. The items that should be covered by the warranty include hardware, software, and installation. The warranty generally promises that these are free from defects in material, design, and workmanship, but do not settle for phrasing which states that such defects must materially affect your use of the product. That is leaving too much room for interpretation. Tie the warranty back to the product's functional specification document.

The warranty period in the PBX industry is generally one year, but if the dollar amount of the purchase is high enough, you might be able to leverage a two-year warranty from the vendor. Under the contract provisions describing the nature and scope of the warranty, you have to spell out the response times for routine and emergency problems covered under the warranty. You must decide what constitutes an "emergency." The definition of this term differs from one organization to another. A PBX problem that may be viewed as trivial by a small business may not be so trivial to a hospital or E911 service provider, where PBX downtime may impede the delivery of life-saving services. Vendor response times to emergency and routine problems range from one to four hours.

When it comes to changes that correct software defects, insist that the changes be covered by a warranty, even if it means that it will extend beyond the period of the original warranty. The failure of the vendor to agree to such a provision might reveal a serious lack of confidence in the abilities of its staff, or in the product itself. A tactful remark aimed in the vendor's direction is usually enough to put this issue into proper perspective.

The software that comes with the PBX is typically not bought, but licensed by the user. This means that the PBX vendor retains all rights to how the software is used, who uses it, and where it is used. The license agreement binds you to confidentiality, which means that you cannot allow a third party to access it. To guard against the loss of support should the vendor go out of business or discontinue the product line, find out what provisions the vendor has made to protect its customers from such an occurrence. If the program's source code and documentation have not been placed into an escrow account, find out why. Again, the dollar amount of your purchase may have a lot to do with how flexible the vendor is in accommodating you. You may even get away with a provision that allows you to buy the software outright if the vendor goes out of business or discontinues the product line, provided that you agree not to resell the software.

To minimize the risk associated with the purchase of a PBX, you must make every effort to weight the contract in your favor, instead of allowing the vendor to weight the document in its favor. That way, the push-pull process of negotiation will more likely result in an equitable arrangement that will benefit both parties. The negotiation process should result in mutual agreement about product performance, especially the criteria for acceptance testing. The agreement should include incentives for the timely performance of vendor obligations. Finally, the contract should describe the responsibilities of both parties as they apply to the delivery of service and support.

2.2 SHARED TENANT SERVICES (STS)

One of the newest trends related to PBXs is that of shared services. Under this arrangement, tenants of an office building share a central PBX, which is under the management of the property owner or an outside firm specializing in this type of service. Such arrangements have been around in one form or another since the invention of the first switchboard, but the concept has resurfaced in recent years. Fueling all the hype about shared PBXs serving the tenants of "intelligent" office buildings has been the divestiture of AT&T, advances in technology, and competition among PBX manufacturers for new markets.

Promoters of the shared tenant services concept cite its many advantages, which include:

- Access to contemporary technology without the high cost of ownership;
- Access to a system that can be customized to meet your present needs and

grow with you to accommodate future needs;

- Access to advanced features that can increase productivity;
- Access to value-added services like FAX, Telex, electronic mail, and computerized message center;
- Access to cost-saving, long-distance carriers like MCI, U.S. Sprint, Allnet, ITT Long Distance, and others *via* programmable, automatic, least-cost routing;
- Access to detailed management reports that can account for telecommunications expenses by employee, project, work group, or department;
- Access to on-site training, maintenance, and consultation services;
- Access to one-stop shopping for a variety of station equipment and add-ons.

In addition to these benefits, the user does not have to allot space for a PBX or install utilities or environmental controls, and, since all support services are provided as part of the arrangement, there is no time wasted in managing the system, no administrative costs, and no need to hire technical personnel.

These are pretty much the same benefits associated with CENTREX, except that there is no telephone company personnel within shouting distance, so response times to service requests will be slightly longer. The shared arrangement might seem best suited for very small organizations, but their modest needs can be met very well by an inexpensive key system. Large users tend not to use shared arrangements because they would have to give up the control that PBX ownership can give them.

The bottom line is that the shared PBX concept really has not caught on. There are many reasons for this, but the principal reason is that the shared arrangement is a high-risk gamble for both the service provider and the user. For example, a lot of the success of such arrangements hinges on the anchor tenant's willingness to use the system. This gives the service provider the means of recovering much of the overhead associated with maintenance and staffing. Once the anchor tenant is persuaded to use the system, other tenants may be more inclined to follow suit.

If the anchor tenant refuses to support the system, however, or fails to renew the contract for the service, the continued viability of the shared arrangement is seriously jeopardized. Without the anchor tenant, the profit margin for the service provider is not enough to justify continuing the operation, and frequently the margins are so low, anyway, that when the anchor tenant pulls out of the arrangement, the service provider is forced into a loss position. If the arrangement deteriorates to the point where service must be discontinued, each tenant is put into the unenviable position of having to make an unplanned, unbudgeted capital expenditure for their own equipment.

Even when things appear to be going smoothly, the continued success of the arrangement hinges on too many factors beyond each user's control. For example, if only a few tenants become dissatisfied with the service, or new tenants do not subscribe to the service, that may be enough to compromise the entire operation. In such cases, other tenants lose confidence and begin looking for a key system or PBX of their own.

Other times, shared arrangements fail because they are poorly managed, or the services are poorly marketed. Many property owners were lured into the concept because it promised to attract a better class of tenant, reduce tenant turnover, and make their buildings more valuable. They did not have the management skills necessary to run the system themselves, and when they looked to PBX manufacturers, long-distance carriers, or environmental control companies to manage the system for them, the damage had already been done and failure was inevitable.

There are some shared systems that are well managed. The first such system was set up at the headquarters of Planning Research Corporation (McLean, Vir.) in the early 1980s. It was looked upon as the model for future "intelligent " buildings. The primary reason for its continued success, however, is that the principal user of the system is the Planning Research Corporation. Thus, the company is intimately involved with maintaining the highest level of service and has a dedicated staff to perform day-to-day operations. Also, PRC is staffed with engineers and technicians from a variety of disciplines, and these resources can be brought to bear on any problem to insure its immediate resolution. This situation is unique, however, and you should exercise caution before getting your company involved with such arrangements.

If you are persuaded that the shared PBX is a viable concept, there are some things to look into before making a commitment.

- Ask for a copy of the service provider's business plan. The fact that a business plan even exists is a good sign because it reveals that key facets of the operation have been thought through and are considered worthwhile pursuing. Do not let it go at that, however. Have an accountant review it to see if the financial projections appear reasonable.
- The business plan should describe sources of funding. If not, that information may have been deliberately purged from the plan. You must satisfy yourself that the operation is being adequately capitalized. Many operators have gotten themselves into trouble by thinking that they could bootstrap themselves into a profitable position by offering basic services first and then adding on more sophisticated features and functionality later. Unfortunately, things do not work this way. This kind of service is capital-intensive. The operator should anticipate being in a loss position for the first three years and be adequately capitalized to carry him through that period. In addition, a line of credit of at least $200,000 should have been arranged to cover unforeseen contingencies. If an operator can provide you with such documentation, it lends more credibility to the business plan.

- Find out exactly who is managing the system. If it is the property owner, consider that a red flag. These people have a different set of skills, and playing telephone company is not usually one of them. If an outside firm manages the operation, find out what other installations they manage and arrange for a visit.
- If the system is already in operation, call on tenants to find out how happy they are with the arrangement. There is no need to ask for references, just use the lobby directory to come up with your own list. Do the same thing when visiting other buildings managed by an outside firm.
- Be sure to talk with tenants who do not subscribe to the service; perhaps they know something that the other tenants do not.
- No matter who is running the operation, find out the qualifications of the on-site staff, especially the manager and technicians. Ask to see résumés and talk with each person individually. If they appear to be too busy to give you the time of day, they are probably spread too thin. If that is the case, they will also be too busy to help you after you sign up for the service.
- Find out what future services are planned in case your needs change. If you find out that electronic mail, videoconferencing, local area networks, and voice mail are among the services to be added later, find out what experience the staff has in implementing these technologies.
- If the operator does not have plans for expanding the service offerings and does not have people dedicated to planning or construction or to sales or marketing, this may indicate a lack of commitment. If there is not a satisfactory explanation for this, you would be wise not to get involved any further.
- Do some legal homework. Find out if the state Public Utilities Commission (PUC) has ruled affirmatively on the issue of shared PBX services. In some states, the telephone companies have opposed such arrangements as infringements on their right to be the sole provider of local telephone service. Some states have not yet decided on the legality of the concept, and so the telephone companies try to make life hard for the providers of shared services. Lines are not installed properly, circuits are not cutover within a reasonable time, or technicians are slow to correct reported problems. This type of friction with the telephone company can undermine the best laid plans for shared services.
- Finally, ask for a sample contract so you can have your lawyer, and possibly a consultant, review it. Make sure it includes clauses for indemnification. Such clauses specify a penalty to be paid by the property owner or manager of the system should an outage exceed a reasonable period, such as two hours.

Shared PBX arrangements can work well, but you will not be able to predict the likelihood of success without asking the right questions and then taking appropriate steps to verify the answers.

2.3 CENTREX

Unlike the PBX, which provides communication services from an on-premise switch, CENTREX (short for *central exchange*) provides communication services from a switch located at the telephone company's central office. CENTREX has been serving small, medium, and large customers for more than a quarter of a century. In the years prior to divestiture, however, AT&T came to recognize that it might very well lose its Bell Operating Companies (BOCs). Up to that time, AT&T sold CENTREX services through the BOCs, while it pursued sales of PBXs. Faced with the very real threat of losing the BOCs, AT&T started putting its R&D efforts into PBXs at the expense of CENTREX. Under increasing pressure from the Justice Department, AT&T finally threw in the towel, divesting itself of the 22 BOCs and liquidating its holdings in two other major telephone companies, Cincinnati Bell and Southern New England Telephone. With divestiture, many new competitors entered the PBX market, wresting what customers they could from the clutches of AT&T. As the pace of innovation accelerated, industry observers were getting ready to ignite the funeral pyre for CENTREX.

After they were cut loose by AT&T in 1984, the Bell Operating Companies did not just roll over and die; with their newfound freedom, they were free to revamp CENTREX completely, turning it from an outdated, expensive, and restrictive offering into the flagship service it is today. In fact, CENTREX has been so enriched with features and functionality that New York Telephone Company's Intellipath II digital CENTREX service boasts 99 percent feature parity with the most advanced PBXs, lacking only automatic call distribution, which is anticipated for tariff before the end of 1988.

Bell Laboratories developed CENTREX in the mid-1950s. In its infancy, CENTREX provided only a few simple services. As is still true today, the central office was configured to recognize and complete calls among groups of lines that were assigned to specific customers. Telephone numbers associated with CENTREX consisted of the standard seven-digits conforming to the nationwide Direct Distance Dialing (DDD) numbering plan. Although both inward and outward calls were associated with a specific number, DID and DOD were packaged as separate CENTREX features, with all outgoing long-distance calls identified for billing purposes through either Centralized Automatic Message Accounting (CAMA) or Automatic Identified Outward Dialing (AIOD).

Because the major telephone companies have changed out a lot of their analog plant in favor of digital switches, today's CENTREX boasts the features, configuration flexibility, functionality, and control that were once attainable only with the largest, most expensive PBXs. AT&T's 5ESS and Northern Telecom's DMS-100 are the two central office switching systems most widely used to provide digital CENTREX services. Both provide computer-controlled time-division switching and support a distributed architecture that uses a host module and multiple microprocessor-controlled switching modules.

These switches are capable of directly interfacing with T1 carrier systems to provide 24 digitized voice channels over a 1.544-Mb/s digital pipe. With these capabilities, digital CENTREX exchanges can provide interfaces to DACS (Digital Access and Cross-Connect System), which acts as a gateway to a variety of services over the public network.

Many CENTREX offices routinely offer automatic route selection, local area networking, facilities management and control, message center services, and voice mail. In addition, numerous CENTREX exchanges are in the process of being upgraded to provide primary rate ISDN service.

Critics once cited the lack of call detail information as a key weakness of CENTREX. After all, without accurate and timely call records, CENTREX users had no means to monitor usage and control costs. Until recently, this was a legitimate complaint. In the past, CENTREX users who wanted to collect, store, and process call records to obtain management reports had either to install line scanners on site or to obtain a billing tape from the telephone company. Line scanners provided a limited amount of information, while billing tapes were typically slow in arriving and took as many as ten days for service bureau processing. Under this arrangement, it took as long as 25 days to obtain the finished management reports. By then, a lot of the information was too stale to be of any real benefit. Now, however, a variety of options are available to monitor usage and control costs, among them, Station Message Detail Recorder to Premises (SMDR-P), which transmits call records directly to the customer premises from the central office. This arrangement provides virtually immediate access to call record data, in the process providing a solution to one of the most persistent complaints about using CENTREX rather than a PBX.

Another criticism voiced about CENTREX over the years has been the lack of management control, but innovations in CENTREX have provided a solution for this problem, too. With an on-premise terminal and interactive software program, users can control which numbers, features, services, and billing codes are assigned to each line. Not only can users review the status of their current CENTREX configuration, they can also plan ahead to meet future demands by determining the effective date of the changes. All such changes are implemented through a management system located at a central location. Individual CENTREX exchanges poll the system daily for any customer changes, which causes the internal records of the telephone company to be automatically updated.

Not only is the service competitively priced against a similarly featured PBX, CENTREX requires very little up-front capital to get started, and in opting for CENTREX users are not burdened with additional PBX-related costs such as insurance, utilities, environmental controls, ongoing maintenance and repair, and in-house technical expertise. This is not to imply that CENTREX is entirely free of start-up costs. Users may have to buy CENTREX-compatible station and attendant equipment, as well as any equipment and special software required to implement special capabilities like call record polling and feature activation or deactivation.

Another advantage of CENTREX is that monthly costs may be stabilized over many years. For example, Pacific Bell promises its CENTREX customers that monthly costs will remain the same for as long as ten years, even if CENTREX charges increase, and if the basic rate for CENTREX services decreases, current CENTREX users may take advantage of the reductions. Pacific Bell will automatically factor the decrease in costs into its CENTREX customers' monthly charges.

In stark contrast to pay-as-you-go CENTREX payment plans, PBX vendors usually require full payment shortly after the system has passed the acceptance test.

There are many more reasons why CENTREX is becoming increasingly attractive to small and large users alike, among them:

- Companies with multiple locations in the same city are often better off with CENTREX than with several PBXs linked with tie lines, or a single PBX configured with off-premises extension (OPX) lines.

- It is easier to accommodate growth with CENTREX than with your own PBX, which typically has a limit before you start incurring additional costs for major hardware changeouts.

- You do not have to worry about blocked calls during peak traffic hours because the central office provides all of the trunk capacity you will ever need. You can even add or delete CENTREX lines as you need them, and be billed for only what you use.

- Many types of key system may operate behind CENTREX through connections made at the main distribution frame (MDF), just as they do behind the PBX. There are CENTREX-specific consoles that let you choose between a one-to-one relationship between CENTREX lines and station phones, or concentrate the ratio so that fewer CENTREX lines are required, thus reducing the overall cost of the integrated system. In the concentrated configuration, station phones would be tied into a CENTREX-specific console, which provides most of the features available with a PBX console, including the ability to monitor line status, recall calls made to unattended stations, and implement camp-on.

- Another alternative to key system attendant consoles is an integrated telephone-terminal that provides the functionality of both a telephone and an IBM Personal Computer AT in a single unit. The terminal provides windows that display call status, enable you to view and scroll through the telephone directory, and access the message center.

- With CENTREX Station Rearrangement (CSR), you can access the telephone company's central data base from a dumb terminal from which you could do such things as add, delete, or change station lines, change a line's class of service, or adjust call pickup groups.

- The risk of technological obsolescence is minimal with CENTREX because the BOCs have a big stake in continually enhancing their switching plant to retain their customers and to attract new customers. As if further proof were needed, this is demonstrated by the BOCs' headlong rush into ISDN, despite its uncertain market potential. Nevertheless, CENTREX users are well positioned to take advantage of ISDN when it becomes available. Unlike PBX users, CENTREX users will be assured of complete ISDN interoperability.

- With CENTREX, you do not have to waste money building an equipment room and equipping it with utilities and environmental controls. Such costs can reach $100,000 for a 125-line system. That is equivalent to buying another 125-line PBX at $800 per line! The on-premise CENTREX equipment can fit in a four-by-five telephone closet and does not require the installation of environmental controls.

- CENTREX also offers 24-hour maintenance services and various levels of redundancy, providing users with peace of mind.

Despite all the advantages of enriched CENTREX, there are some factors you should consider before subscribing to the service:

- Find out the regulatory climate in your state. Some PUCs tightly regulate CENTREX services and require cost-based tariffs. In other states, the regulatory climate is more relaxed, permitting CENTREX to compete more effectively against PBXs. In still other states, the PUCs require cost-based tariffs, but grant special treatment on a case-by-case basis so that the BOCs can design a package of features that better fits a customer's needs. Because of the different regulatory climates from state to state, you must read the fine print in the CENTREX agreement. Although charges may be locked in for a number of years, for example, payment plans may be subject to review and modification by the PUC. The differences in regulatory climate also mitigate against the possibility of obtaining the same good deal from one state to another — it is not like a PBX buy where discounts are structured according to the number of units you take.

- If you have a multinode network that spans many states, keep in mind that CENTREX offerings differ from place to place, even in the same regional Bell territory. Thus, you may be able to obtain call detail information in near real-time from one CENTREX exchange, but not from another, or you may be able to activate and deactivate features and change service options on a per-line basis at one location, but not at another. Thus, the differences in regional CENTREX offerings may complicate, rather than facilitate, administration.

- If you already have a key system or PBX, find out how much of it is usable with the CENTREX service. You may be able to use the same telephone sets and station wiring, in which case your start-up costs will be significantly lower. You should also think about using already installed customer-premises equipment (CPE) behind the CENTREX service to streamline costs even further.
- Finally, check into the disaster recovery plan of the telephone company. Since you may be relying on the CENTREX exchange even for local calls, you do not want to risk disruption of business operations in the event of a disaster. Pacific Bell, for example, assures customers that its CENTREX exchanges are built to withstand earthquakes, as well as storms and fires.

Whether you choose to buy a PBX, participate in shared tenant services, or subscribe to CENTREX, the risk in making the wrong decision can be minimized by matching each with the needs of your organization, comparing the options on such critical points as growth potential, associated costs, level of redundancy, and service response time. Despite all the media hype about "price wars" between vendors and the increasing competitiveness of the BOCs with regard to CENTREX pricing, you should think twice about buying on price alone. The many cost-cutting schemes used today by PBX vendors to get your business may have a devastating effect on their future health, which may impinge upon their ability to deliver enhancements, upgrade the hardware, and provide timely service and support.

When it comes to CENTREX, just keep in mind that this is a regulated service provided by telephone companies, which are subject at times to the whims of regulatory bodies, pressure from public interest groups, and the moodiness of ratepayers. These factors may throw a wrench into the best laid plans of communications managers.

REFERENCE

[1]   Daly, John J., "Steps in Marketing Plan Development: Reality and Vision," *Procomm Enterprises Magazine,* Vol. 2, Issue 13, May 1988, p. 16.

Chapter 3
Private Network Systems

3.1 MODEMS

Modems convert digital signals to analog signals that are suitable for transmission over voice-grade lines. At remote terminals, other modems perform the process in reverse, converting analog signals to a digital form that can be understood by machines.

The concept behind the modem is not new; early nineteenth-century inventors like Samuel F. B. Morse knew about coding information for transmission over wire lines. Human operators did the coding-decoding, thus performing the functions of modems. As the amount of traffic increased, it threatened to overwhelm human operators. This led to the development of mechanical devices to handle the coding-decoding processes. The first printing telegraph made its appearance in 1846. Invented by Royal E. House, the device was roughly the size of a small piano. Then, in the 1850s, a new company, which eventually became Western Union, consolidated the many independently run telegraph companies and added new lines to link the country coast to coast. The House device was abandoned in favor of human operators, which were the norm until shortly after World War I, when the first teletype machines, developed before the war, were placed into service by Western Union. These machines used a 5-bit Baudot code. An internal modem transmitted information at the rate of 50 b/s. It was not until 1931 that AT&T finally set up a similar service which it called Teletypewriter Exchange service, or TWX. While AT&T also used the Baudot code, it did so at the rate of 75 b/s. Western Union eventually bought AT&T's TWX service.

Modem technology progressed steadily over the years, pushing transmission speeds ever higher. With the introduction of mainframe computers in the early 1950s, dial-up modems were used to support remote batch processing and then time-sharing arrangements at 110 b/s. By the late 1960s, 300-b/s modems were available from telephone companies. There were no alternative sources to speak of, mainly because AT&T prohibited customers from connecting their own devices to the public switched network. Even when AT&T relaxed this rule after the Carterfone decision in 1968, it insisted that telephone company–provided "data access arrangements" (DAAs) be used to "protect" the network, which only added to the cost of communications.

By 1976, the Federal Communications Commission (FCC) had done away with the requirement for so-called "protective" devices on the switched network, replacing it with a registration plan, whereby all equipment from alternative suppliers would be evaluated and approved for direct connection to the telephone network. This marked the beginning of a whole new industry, which went on to produce one innovation after another. In no time at all, modems evolved from simple devices that provided point-to-point communications, to the basic building blocks that are used to create and manage complex multipoint networks, involving multiplexers, packet switches, fiber, and satellite. Within the span of only a decade, the speed of dial-up modems went from a pathetic 300 b/s to a blazing 19.2 kb/s, and with innovative compression techniques, transmission at 54 kb/s is now possible. There is also tremendous innovation in synchronous modems, which reach speeds of 1.8 Gb/s over fiber and 565 Mb/s over microwave links.

3.1.1 Basic Operation

The output of such devices as computers, terminals, and printers can be broken down into ones and zeros, which are represented by the presence or absence of pulses, or even by two distinguishable pulses. In the former case, for example, a pulse indicates a one, while the absence of a pulse indicates a zero. Representing information in this way is called "binary encoding." When strung together in unique combinations of seven bits, these ones and zeros represent all the characters and symbols that appear on a terminal keyboard, plus a parity bit used for error checking, bringing the total number of bits to eight. Special control characters are also included in this coding scheme, such as the "end-of-transmission" (EOT) character, which is represented as 00010111. The American Standard Code for Information Exchange (ASCII) lists 127 different character variations. There are other coding schemes in common use, such as IBM's EBCDIC (Extended Binary Coded Decimal Interchange Code), which uses 8-bit structures to represent up to 256 characters. Regardless of what scheme is used, coding and decoding operations are transparent to the users at both ends; that is, neither user is aware that such operations are even taking place.

The trouble is that these digital signals cannot be transmitted over phone lines without quickly degrading. The "squareness" of the high-frequency signals starts rounding off as the total electrical-energy level drops. The inherent resistance of metallic wire is the primary cause of this energy loss. To surmount this problem, digital signals must be changed to analog form for long-distance transmission over voice-grade phone lines, which offer slightly less than 4 kHz of useable bandwidth. This process is called "modulation." At the distant end, the analog signals must be converted back to digital form for acceptance by the computer or terminal through a process called "demodulation." The device that acts as a modulator-demodulator is called simply a "modem."

There are three modulation techniques, each drawing upon the basic characteristics of an analog signal as the means for expressing the ones and zeros used in the digital format. Analog signals vary continuously by amplitude, frequency, and phase. Amplitude refers to the extreme ranges of a waveform — its highest to lowest point — and is usually expressed in volts. Frequency refers to the number of cycles that the waveform completes in one second, which is expressed in cycles per second (c/s), or Hertz. Phase refers to the angularity of the waveform, which is expressed in degrees.

Unlike digital signals, which are predicated on the presence or absence of energy, a basic linear signal is always present in an analog transmission. This is called the carrier signal. It is a single-frequency signal that does not change until information is put onto it through a modulation technique. The information that is modulated onto the carrier signal is called the baseband signal. Information can be modulated onto a carrier signal in several ways (Figure 3.1).

In amplitude modulation (AM), digital signals are converted to analog form using a single-frequency carrier signal the wave height (amplitude) of which varies according to the presence or absence of information imposed on it. A high amplitude represents a binary one, while a low amplitude represents a binary zero. In its purest form, this modulation scheme is not used in today's modems. By itself, AM is easily disturbed by noise, causing binary ones to be misread as binary zeros, and binary zeros to be misread as binary ones. This results in information arriving at its destination in corrupted form.

Frequency-shift keying (FSK) uses a carrier of constant amplitude, which is varied by frequency to indicate a binary one or a binary zero. This technique is more robust than AM, but still does not lend itself to transmission speeds greater than 2.4 kb/s.

Phase modulation takes advantage of phase-shift transitions in the waveform to represent binary ones and binary zeros. In a technique called "phase-shift keying" (PSK), a 180-degree phase change represents a zero, while no change indicates a one.

If it were possible to represent multiple bits of information with each change in the modulated signal, it would be possible to increase the transmission speed. Vendors have done this with the phase-modulation technique to increase the transmission speeds of modems. "Differential phase-shift keying" (DPSK), for example, uses two bits — a dibit — to represent each of four shifts in the phase of the modulated signal. Each shift is represented by one of four possible dibit combi-

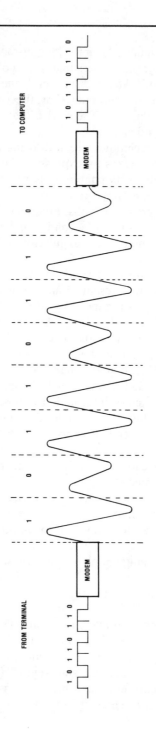

IN AMPLITUDE MODULATION (AM), DIGITAL DATA IS CONVERTED TO
ANALOG SIGNALS USING A SINGLE FREQUENCY CARRIER SIGNAL. A HIGH
AMPLITUDE WAVE REPRESENTS A BINARY 1, WHILE A LOW AMPLITUDE
WAVE REPRESENTS A BINARY 0. THIS MODULATION TECHNIQUE IS RARELY
USED BY ITSELF DUE TO THE SIGNAL'S SUSCEPTIBILITY TO NOISE AND
BECAUSE OF THE TIME REQUIRED TO SAMPLE THE AMPLITUDE OF THE SIGNAL.

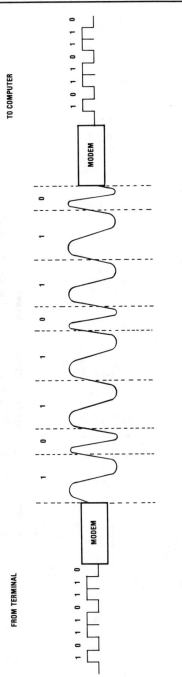

FROM TERMINAL

1 0 1 1 0 1 1 0

1 0 1 1 0 1 0

MODEM

TO COMPUTER

MODEM

1 0 1 1 0 1 1 0

IN FREQUENCY-SHIFT KEYING (FSK), A CONSTANT AMPLITUDE CARRIER SIGNAL AND TWO FREQUENCIES ARE USED TO DISTINGUISH BETWEEN A BINARY 1 AND A BINARY 0. FSK IS A SIMPLE MODULATION TECHNIQUE THAT IS NORMALLY USED IN 2400 bps AND SLOWER MODEMS.

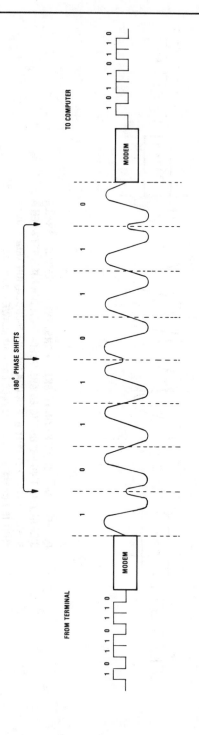

IN PHASE-SHIFT KEYING (PSK), A CHANGE IN PHASE AT TRANSITION POINTS IN THE CARRIER FREQUENCY IS USED TO REPRESENT BINARY ONES AND ZEROS. IN THIS CASE, A 180 DEGREE PHASE SHIFT INDICATES A ZERO. AS MANY AS EIGHT PHASE SHIFTS MAY BE USED IN HIGH-SPEED MODEMS.

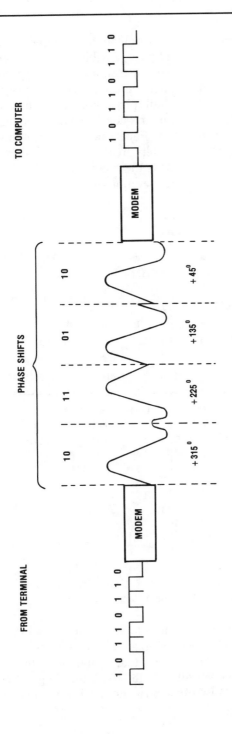

Figure 3.1 Summary of modulation techniques.

IN DIFFERENTIAL PHASE-SHIFT KEYING (DPSK), DIFFERENT PHASE SHIFTS ARE USED TO REPRESENT FOUR "DIBIT" COMBINATIONS. IN THIS EXAMPLE, A 45-DEGREE PHASE SHIFT REPRESENTS 10 , A 135-DEGREE PHASE SHIFT INDICATES 01, A 225-DEGREE PHASE SHIFT INDICATES 11, AND A 315-DEGREE PHASE SHIFT INDICATES 10. THE USE OF DIBITS INCREASES MODEM SPEED. OTHER ENCODING SCHEMES USE THREE OR MORE BITS PER CYCLE TO ACHIEVE MODEM SPEEDS AS HIGH AS 19.2 Kbps.

nations: 00, 01, 10, and 11. Another phase modulation technique called "differential eight-phase shift keying" uses three bits — a tribit — to represent each of eight shifts in the phase of the carrier signal to permit even faster speeds.

The multilevel encoding scheme is carried much further in "quadrature amplitude modulation" (QAM), which uses both phase and amplitude modulation. This encoding scheme entails the use of a matrix consisting of 32 points, which is referred to as a constellation (Figure 3.2). Each change in amplitude may be represented by four bits (quadbits). QAM offers greater immunity from noise than the techniques previously discussed. QAM is the basis of CCITT's V.32 recommendation for full-duplex 9.6-kb/s modem operation over dial-up lines.[*]

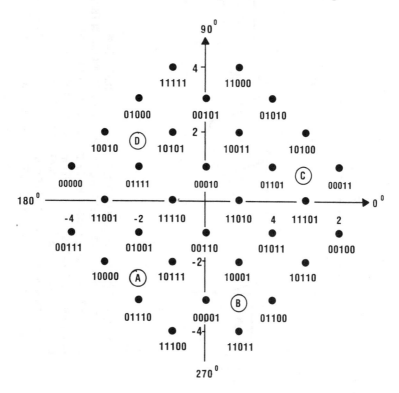

Figure 3.2 Minimum signal constellation for 9.6 kb/s throughput.

The V.32 specification, however, is flexible in that it also allows an error-correction technique called "trellis-coded modulation" (TCM), which is designed to overcome noise problems with dial-up lines that would normally impair data transmission at rates at or above 9.6 kb/s. With trellis encoding, another bit is added to the quadbit under QAM for error correction, bringing the total number

* CCITT, VIIIth Plenary Assembly, 1984.

of bits to five for each point in the constellation. The extra bit is used to identify invalid signal patterns that may have been introduced by line impairments.

Signal patterns that do not conform to the constellation are recognized as errors and corrected before they arrive at the data-terminal equipment (DTE). Forward-error correction, then, is an inherent feature of TCM, providing more than twice the immunity from noise and phase jitter provided by QAM alone.

Some vendors have attempted to improve upon the trellis-encoding technique to obtain higher transmission speeds, in the process creating very dense constellations, which become ever more difficult to demodulate at the receiving end. At this writing, it appears that 64-state, eight-dimensional TCM providing a constellation of 160 signal points is the optimal requirement for insuring the integrity of the data at a speed of 19.2 kb/s. This optimal configuration also permits the encoding of fallback speeds of 16.8, 12, and 9.6 kb/s.

There are other approaches to obtaining high-speed transmission over voice-grade lines, such as that offered by Telebit Corporation's (Mountain View, Calif.) TrailBlazer Plus, which borrows from spread-spectrum communication technology used in secure military systems. As the data are fed to the TrailBlazer Plus they are buffered, so that they may be broken into discrete pieces and spread over the available bandwidth, creating up to 512 subchannels, each of which carries a very slowly modulated signal. Data are transmitted over the link in bit-parallel fashion. The modem tests the signal-to-noise ratio of each subchannel so that those of poor quality are not used. This lets the modem operate at 19.2 kb/s without compression and, more importantly, lets the modem fall back slowly under degrading line conditions. Since the quality of the line is constantly evaluated, the modem can continually adjust to varying line conditions. In addition to supporting the Hayes AT command set, the TrailBlazer features error detection-correction, flow control, and support of Microcom's MNP Classes 1 through 3.

All of these games are necessary to overcome the corrupting influences of noise. Amplifying the signal may seem like a simple solution, but in the process of amplifying the signal, noise also gets amplified. Even if amplification were feasible under certain circumstances, it is not really an option because modems are prohibited from putting more than 0 dB on the line. As a result, modem manufacturers must use alternative means to overcome the effects of noise to increase transmission speed.

Currently, TCM devices are very expensive, sporting a price tag of about $4,000 each. Diagnostic models that can be used in conjunction with a network-management system currently sell for $8,000 to $9,000 each. The high cost is attributed to the powerful processors that are required to perform decoding at the receiving end, which may entail as many as 1,000 processing steps per state. As these processors come into more widespread use, and prices fall, modems that use the trellis technique will become more affordable. Meanwhile, the TrailBlazer Plus, available as a microcomputer plug-in card, costs under $2,000.

No matter how complex these encoding schemes may appear to be, they all have one thing in common: changes in the basic characteristics of the signal are used to represent one or more bits of information. Advanced modulation techniques like QAM and TCM make for more efficient utilization of voice-grade lines. This means, of course, that you will be able to save on communication costs, since you would be able to use existing facilities to carry ever greater amounts of information. Your decision to buy a high-speed modem, however, will not likely be determined by line-cost savings alone. You must weigh the costs for such devices against the likelihood of achieving significantly higher levels of user productivity. The $5,000 cost for a pair of 14.4-kb/s modems to replace two 4.8-kb/s modems may be justified if only because voluminous file transfers may be accomplished three times faster. That means less time is wasted waiting for information, which may not be very impressive until you calculate the loss in man-hours over the course of a month or a year, and come up with a reasonable payback period to justify the expenditure on such devices.

While the type of modulation technique used will provide you with a general indication of a product's performance, it is important to look at other performance indicators as well. This is because vendors use different methods for implementing these techniques. Such variables as the complexity of circuitry, the types of microprocessors and microcontrollers used, the capacity of RAM, and the size of the bus may account for the differing performance levels of products that claim to use the same modulation technique. Thus, even when different vendors claim that their products are V.32 compliant, they may not be compatible with each other in all respects because they may include extra features the others do not have. This can create problems when you try to mix and match the products of two or more vendors.

Other factors influence performance, too, like the interfaces used between the modem and computer operating systems. When a high-speed modem transmission arrives at a computer using an incompatible protocol, data can be lost. Worse yet, the computer may interpret the session as having ended, and drop the connection. Interestingly, current and impending modem standards do not address such concerns. This is not to say that standards are unimportant, but only that interface requirements should enter into your decision-making about modem purchases.

Aside from modulation techniques, design factors, and specific interface requirements, you must give due consideration to compression, equalization, and error-correction techniques, which also have a direct bearing on performance.

3.1.2 Compression

Many modem manufacturers have turned to data-compression schemes to bolster the throughput of their modems. Among the competing schemes are those offered by Microcom Inc. (Norwood, Mass.), Adaptive Compression Technology (Santa Clara, Calif.), and Hayes Microcomputer Products, Inc. (Norcross, Ga.). All three vendors approach compression in a similar manner, using proprietary adaptive algorithms, which are licensed to other modem manufacturers.

The algorithm is adaptive in that it continually analyzes the patterns of user data and adjusts the compression parameters accordingly to maximize data throughput. Some data patterns, however, will benefit more by compression than others. For example, executable .COM and .EXE files are the least compressible, while spreadsheet files are moderately compressible, or even highly compressible, if they happen to have lots of empty space. Word processing and print files are the most compressible. The more pattern that is found in the data, the higher the compression ratio. The more random the pattern, the less likely that there will be a significant increase in throughput. This makes compression more suited for use on dial-up lines carrying asynchronous data than on leased lines, which typically carry more synchronous than asynchronous data.

A modem vendor's claim of 2-to-1 compression, then, should not be taken literally. A 4.8-kb/s modem will exhibit what the user sees as 9.6-kb/s transmission, but only under ideal conditions. This is not to say that you will not benefit at all with compression on leased-line modems, only that the compression ratios will be much lower, which will reduce the productivity gains and prolong the payback period on your investment. Although 2-to-1 compression would be nice to achieve, an overall compression ratio of 1.5-to-1 may be all you need to justify the purchase.

Microcom has pushed compression into a new realm altogether with the introduction of its QX/V.32c modem, which boasted of being the world's fastest commercially produced full-duplex, dial-up modem as of May 1988. The modem operates at 38 kb/s using MNP Level 9, which provides 4-to-1 data compression. This level supports both MNP and non-MNP modems by "negotiating" with them to achieve optimal performance. At the same time, the modem adheres to CCITT standards for 9.6 kb/s and is Hayes compatible. The QX/V.32c sells for less than $1,800. While Microcom was basking in glory over this development, its competitor was already hinting at a soon-to-be-announced 54-kb/s modem. To achieve such throughput, Adaptive Computer Technologies Inc. said that it intended to use 3-to-1 compression on Telebit's 19.2-kb/s T2000 half-duplex modem. The T2000, however, does not adhere to the CCITT's V.32 recommendations, which means that it will not be compatible with other modems, even at the fallback rate of 9.6 kb/s. Nevertheless, such high speeds may be useful in intracompany applications like CAD-CAM, where high-powered workstations are used.

Because you will not really know how well a certain compression technique works with your applications, it is best to obtain demonstration units from the vendors to check the performance claims for yourself. Compare the performance of each product using asynchronous only, synchronous only, and mixed data. Check their performance on dial-up and leased lines. Use a matrix chart to keep track of the results. Even if one modem performs far above all the rest in a specific area, you have to consider whether you want to pay the high cost for a product that limits your flexibility.

If you do not have time to do the testing yourself, you can still minimize the risk of making a wrong purchase decision by ordering test results from independent labs like Telequality Associates (Golden, Colo.), who specialize in equipment testing. Do not take vendor test results too seriously, though. Vendors have a tendency to weight the tests in favor of their own products.

It is a good idea to choose products that allow you to disable compression. If you encounter difficulties with compression because of a line problem, you at least want to be able to try using the product in the normal way until line quality improves.

3.1.3 Equalization

Equalization is a method for compensating for line impairments that distort transmission, thus increasing the error rate and degrading overall performance. The purpose of equalization, then, is to offset debilitating line conditions to take full advantage of the data-rate capability of the line. The faster the modem, the greater the need for equalization and the more complex the equalizer. Many modems intended for use with microcomputers do not have equalization capabilities. Some are only capable of handling the average delay distortion of telephone lines. As such, they may be of no use at all, if the actual connection does not fall within this parameter. Manually adjustable equalizers are set at the time of the modem's installation to match the particular line. They are typically used on leased lines, instead of dial-up lines where line quality may vary widely from connection to connection. A compromise equalizer, mostly used in 212A-compatible modems, compensates only for a fixed distortion on a specific standard telephone line.

These methods of equalization are not adequate for today's modems, which use microprocessors to provide higher speeds, increased reliability, and more functionality. For these modems, better equalization techniques are required to sustain these advantages. The use of microprocessors also enables more sophisticated equalization algorithms to be implemented, allowing equalizations to be performed automatically in response to changing line conditions. With the automatic adaptive-equalization technique, for example, line sampling is performed continuously to determine which parts of the transmitted signal need enhancement to be intelligible.

In the event equalization techniques are not enough to compensate for line impairments, there are modems available that are capable of data-rate fallback or automatic dial backup. With data-rate fallback, the modem automatically switches to a lower transmission speed upon detection of deteriorating line quality, thus allowing the integrity of data to be maintained until line conditions improve.

Some modems automatically switch back to the higher speed upon detecting improved line quality, while other modems require operator intervention to return the modem to the higher speed. Although this technique minimizes the amount of errors transmitted, it can also degrade throughput if fallback and fallforward are too frequent. This feature can also prevent you from getting the most out of your investment dollar. Considering the $4,000 purchase price of a 19.2-kb/s leased-line modem, you expect maximum performance. Continually falling back to 9.6 kb/s means that you are not getting what you paid for, and do not count on line conditioning always to bail you out. Local and interexchange carriers do not make guarantees about line quality; line conditioning is only provided on a "best efforts" basis.

Another equalization technique is forward compensation, which samples the line before transmission, thus enabling the modem to start out at the highest rate permitted by the quality of the line. This is accomplished by analyzing the bit stream before transmission takes place and inserting extra bits into the data blocks. At the other end, the bit stream is decoded. The extra bits that were added are analyzed to determine if the data blocks were received correctly. If not, they are corrected at that point, without the need to retransmit. This minimizes the need to fall back after transmission has already begun. Some vendors consider this type of equalization to be an integral part of their overall error-correction protocol. The addition of extra bits does increase overhead, which adds to delay. This may be a factor to consider, especially if the modems you plan to buy will be used in a multipoint, multidrop configuration where delay must be controlled to meet response time objectives.

Some leased-line modems have a built-in automatic dial backup capability, which establishes a dial-up connection with a distant modem when the line degrades beyond the acceptable error-rate threshold. The modem monitors the leased line at periodic intervals to determine whether line quality has improved. If it detects improvement, it switches over automatically, while keeping the dial backup line open for a few seconds. This eliminates the need to redial in case the transmission over the leased line quickly degrades. The whole operation takes place without operator intervention. Other modems, however, require manual switchover between the leased line and dialup line.

3.1.4 Link-Level Protocols

Other factors that affect a modem's performance are its capabilities for implementing error detection and retransmission (or forward error correction), flow control, and sequencing — all of which determine the integrity of data. These issues are addressed in the handful of very sophisticated link-level protocols, around which many modem manufacturers appear to be coalescing.

Microcom, for example, offers error control through Level 4 of its Microcom Network Protocol (MNP). Through the use of a technique called "adaptive packet assembly" (APA), MNP assembles larger data packets to increase throughput during periods when the data channel is virtually error-free. Should errors increase, MNP adapts by assembling smaller data packets. Although the use of smaller packets increases overhead, the throughput penalty is still much less than would be incurred by retransmissions. Several modem manufacturers have incorporated MNP into their products.

In May 1988, the CCITT issued the V.42 recommendation for the link-level protocol in full-duplex modems, which includes a compromise standard for error correction that provides for both MNP and LAP-M (Link Access Protocol–Modem). LAP-M is important because of its ability to support multiple virtual circuits within an asynchronous data stream, and for its future capabilities, such as data compression, which will provide hooks into ISDN. MNP also has benefits: it is available now and has demonstrated its reliability. Beyond that, the CCITT could not easily ignore the installed base of modems that use MNP Level 4. Microcom reported an installed base of 300,000 MNP modems as of first quarter 1988, and claims a growth rate of 10,000 per month. [1] Therefore, V.42-compliant modems will support a secondary mode of operation that uses MNP.

Although some manufacturers refuse to adhere to Microcom's MNP for competitive reasons, others see no significant difference between MNP and LAP-M and will not spend the extra money to implement the new standard fully. A few modems will appear that boast full compliance with V.42, enabling manufacturers to differentiate themselves further from competitors. Such modems will cost more, but purchasers will have error-correction compatibility with many more modems than if they had to choose only one protocol. Other companies will offer their customers a choice between the two protocols *via* optional plug-in boards.

As mentioned earlier under modulation techniques, trellis encoding enhances modem efficiency with error-correction capabilities that identify and correct transmission errors with a technique called "forward-error correction." This entails the generation of redundant bits at the transmitting end, which are used at the receiving end to detect, locate, and correct most transmission errors before they can be passed on to the data-terminal equipment. This avoids having to retransmit information sent incorrectly.

While leased-line modems seem to be gravitating around these few link-level protocols, there are many more available for dial-up modems that are used for asynchronous data transfers over the public network. XMODEM, Kermit, ZMODEM, and Blast are a few of the most popular, each differing in level of complexity.

XMODEM was invented in 1977 by Ward Christiansen (Chicago, Ill.), originally for information transfers between microcomputers based on the CP/M operating systems, the forerunner of today's DOS and OS/2. Under this simple public-domain protocol, data is sent in 128-byte blocks. Error checking is accomplished with an additive check sum for each block. This scheme does not provide error checking for the transmission-control characters, which are sent individually. As the distance of the communication link increases, the probability of errors being introduced through line disturbances also increases. Because XMODEM is so simple, it does not lend itself to long-distance transfers.

An enhanced version of the product, XMODEM-CRC, incorporates cyclic-redundancy checking which catches data-block errors more effectively. Another variant, XMODEM-1K, transfers information in blocks of 1,024 bytes, but over longer distances the error-checking algorithm does not hold up well, providing no assurance that all the errors in the longer data blocks will be caught. WX-MODEM (Window XMODEM) permits faster file transfers through full-duplex communication. It does this by allowing the transmitting terminal to send data without having to wait for the receiving terminal to acknowledge data blocks and without compromising the integrity of the block checking. XMODEM and its derivatives, then, are simple protocols of reasonable reliability for use within the local-loop distance of fifteen miles. You risk data loss when using them for file transfers over longer distances.

Kermit was developed at Columbia University and is distributed through the school's Center for Computing Activities. It allows the exchange of text and binary files between any system over any network, including packet networks accessed *via* dial-up lines. Because Kermit can issue commands in protocol, the commands setting up and controlling the transmission can also be error free. This makes Kermit better suited for long-distance file transfers than the XMODEM-type protocols, particularly on noisy connections. Kermit can also be used to establish microcomputer links with IBM mainframes. Some of the more than 250 implementations of Kermit will even keep you informed of transfer status. Although reliable, the big disadvantage to Kermit is that it is slow.

ZMODEM uses a more powerful CRC (cyclic redundancy check) than XMODEM, which allows the transmission of larger blocks of data, but without risking uncaught characters. It has the same capability as Kermit to issue commands within the protocol and, instead of requiring acknowledgement of each data block, ZMODEM requires only that data blocks having errors be negatively acknowledged (NAK), otherwise the sending device assumes that everything went okay. This makes ZMODEM better than XMODEM for long-distance communication, especially over satellite links, where the error rate is generally low, but the end-to-end delay can be several seconds. If the last packet is dropped, however, you will not get an NAK. To protect against data loss in this instance, you would need a positive acknowledgement (ACK) upon closing the link or ending the transmission.

The proprietary product "Blast" (Blocked Asynchronous Transmission) from Communications Research Corporation (Baton Rouge, La.) offers a full-duplex, bit-oriented protocol that is much more resistant to line noise and can withstand satellite propagation delay. In addition to providing two-way simultaneous transmission, Blast includes some useful features. If a telephone line disconnects during transmission, Blast allows the file transfer to continue from the point at which the interruption occurred when the connection is finally reestablished. Blast also permits dial-up links to be established with IBM, DEC, or Data General mainframes, and during transfers, Blast keeps the user informed of its status.

3.1.5 Why Modem Design Is Important

At 19.2 kb/s, the tolerances of the voice network are being put to the test, which means that even slight differences in modem design can profoundly affect performance. When considering the purchase of a 19.2-kb/s diagnostic modem, you must delve into the basic design elements that have a direct bearing on performance. These design elements include the internal multiplexing technique used, the quality of the filters, the coding-decoding scheme, and the level of circuit integration.

Some 19.2-kb/s diagnostic modems use frequency-division multiplexing (FDM) to derive two separate bands (Figure 3.3). The lower frequencies of one band carry the diagnostic information, while the higher frequencies of the other band carry the production data.

Each side is protected from the excessive or unwanted variations of the other by its own filter, which sets up a buffer between the two channels. Without sharp filters, however, adjacent channel interference can result, causing errors on both channels. Although sharper filters can minimize this problem, the two-filter approach still squeezes production data between the voice-frequency (VF) channel filters and the modem-diagnostic channel. This approach also forces production data to be carried over the higher frequencies. While FDM is acceptable for transmission speeds of 14.4 kb/s and lower, it presents serious problems for diagnostic modems operating at 19.2 kb/s, even with D-type line-conditioning services.

When the transmission rate is increased to 19.2 kb/s, the buffer space between the two channels is materially reduced because more bandwidth is required to transmit at that speed. Not only is there more adjacent channel interference, but the primary channel also becomes more susceptible to the harmful effects of 2nd and 3rd harmonic frequencies caused by the fundamental frequencies of the diagnostic channel. These harmonic frequencies corrupt the production data, forcing the modem to downspeed to 14.4 kb/s or lower to protect the integrity of the data. This problem is exacerbated by line noise, which is not easily overcome at 19.2 kb/s even with D1 conditioning.

Recognizing this problem, AT&T has announced plans to introduce D6 line conditioning, ostensibly to help high-speed modems sustain their optimal transmission speed. This is not a totally altruistic gesture on AT&T's part, however. The move to D6 line conditioning may reflect concern over the performance of its

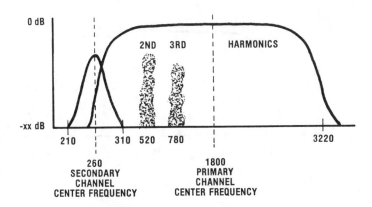

Figure 3.3 Bandwidth utilization with FDM (two filters).

Note: When a frequency is generated in a nonlinear medium, residual tones will be created, which are harmonically related. It is the 2nd and 3rd harmonic frequencies that are the most harmful, because they are the strongest. Assuming a secondary channel center frequency of 260 Hz (fundamental frequency), this will generate the following 2nd and 3rd harmonic residual tones:

$$2\text{nd harmonic is } 2 \times 260 = 520 \text{ Hz}$$
$$3\text{rd harmonic is } 3 \times 260 = 780 \text{ Hz}$$

The residual harmonic tones interfere with frequencies in the primary channel to produce errors.

own 19.2-kb/s modem, which was the result of a joint development effort with Fujitsu, whose data pumps are used by other modem manufacturers as well. Although no pricing has been established for D6 conditioning, it is reasonable to assume that it will cost more than other levels of conditioning, if only because there are fewer wire pairs available which exhibit immunity from generic noise and nonlinear distortion in high-density locations. It is quite possible that D6 conditioning will not become readily available in some metropolitan areas because of the scarcity of relatively "clean" pairs, and because conditioning levels of the local telephone companies do not always match up with those of AT&T, it is even conceivable that users may get D6 conditioning on the inter-LATA (Local Access and Transport Area) portion of the circuit, but not on the intra-LATA portion of the circuit.

You can avoid such problems by choosing a modem that incorporates time-division multiplexing (TDM) into its design. Codex and General DataComm, for example, combine diagnostic and production data onto a single wider band using TDM, and, through the use of a higher quality filter, that band is better protected from outside interference (Figure 3.4). In not having to use two filters to protect separate bands, TDM permits optimal bandwidth utilization. This means that the entire middle range of frequencies, from 315 to 3085 Hz, may be used to carry

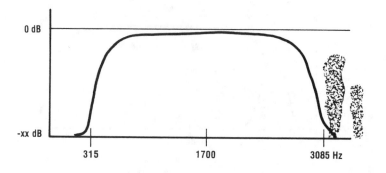

WITH NO SECONDARY SIDE CHANNEL, THERE IS NO POTENTIAL FOR ADJACENT CHANNEL INTERFERENCE. THE 2ND AND 3RD HARMONICS ASSOCIATED WITH THIS CHANNEL'S CENTER FREQUENCY AFFECT THE FREQUENCIES OF THE UNUSED BAND WIDTH OF THE CHANNEL, CAUSING NO ERRORS. THIS TRANSLATES INTO SUSTAINED HIGH THROUGHPUT.

Figure 3.4 Bandwidth utilization with TDM (one filter).

production data and diagnostic information. This does not degrade throughput, however, since the unit actually operates at 19.4 kb/s, which allows 200 b/s for diagnostics. This band is also less susceptible to the effects of harmonic distortion and signaling noise, which are the two enemies of high-speed modem transmission.

Another design element that affects the performance of today's 19.2-kb/s modems is the use of custom VLSI, which offers many hidden benefits. Not only does VLSI greatly reduce the parts count, it provides economies and efficiencies that extend well beyond those associated with high-speed transmission. For example, increasing the circuit density with VLSI results in much less power consumption. Some modems consume about as much power as a refrigerator light bulb (15 watts). The low power requirement also means that the circuitry throws off less heat, which greatly extends the useful life of the product, and less heat dissipation means that equipment-room utilities do not have to be upgraded.

VLSI also translates into streamlined packaging to permit more units to fit into a standard rack-mounted shelf. This saves valuable floor space in densely equipped installations.

When it comes to 19.2-kb/s diagnostic modems, the right choice will result in accelerated payback on your investment. The wrong choice, however, can result in throughput problems and higher operating costs that negate whatever economies and efficiencies you had hoped to achieve with the new investment.

3.1.6 Buying Tips

As with any other type of purchase, you have to take stock of the possible applications you have and try to find an appropriate product match. If you are in control of both ends of the transmission, this is easier to do, because you can specify the products for both ends, but you do not usually have anything to say about the types of products your bank, customers, or suppliers use. Even when communicating with them through a packet network, you must have the same type of modem that the service provider has on its dial-in ports. Tymnet, for example, uses V.32 modems that accommodate either MNP or X.PC, but without data compression. If you are about to buy modems that do not comply with V.32 and do not use MNP or X.PC link protocols, you may be unnecessarily limiting your connectivity potential. Ultimately, Tymnet and Telenet may be in a better position to dictate standards than either vendors or standards committees, if only because they have hundreds of thousands of subscribers.

If you plan to use the modems for international communication, find out whether the vendors have had their products certified by the PTTs to meet CCITT specifications. Ask them how they have tested their products for standards conformance. Any vendor that does business internationally will have this kind of information readily available. If you have reason to doubt the vendor's assurances, you should probe references or even consider getting an independent evaluation from a reputable service that specializes in testing modems for performance and conformance to standards.

Modem selection may also hinge on what type of DTE you use. V.32 modems provide communication at 9.6 kb/s in both directions simultaneously. Sophisticated echo-cancellation technology allows the modems at each location to ignore their own transmissions so that they can concentrate on receiving each other's signal. This makes V.32 modems ideally suited for bulk-file transfers and high-demand access to X.25 packet networks. With prices still hovering around the $2,500 range, however, V.32 modems may be overkill for most microcomputers, which typically do not run applications requiring simultaneous two-way communication. If interaction is desired, it may be provided at a lower cost with 9.6-kb/s modems that implement a fast turnaround, thus simulating full-duplex operation. Some modems, like the Hayes V-Series Smartmodem 9600, adhere to V.32 to the extent that they use trellis-encoded QAM, but since they leave out the echo-cancellation circuitry, they operate in only one direction at a time and, therefore, cost a lot less.

Do not confuse half-duplex operation with "asymmetrical" operation, which is really a variation of full-duplex. Instead of two-way simultaneous communication at 9.6 kb/s, the modem uses most of the available bandwidth for high-speed file transfer, but assigns 300 b/s to a separate channel for interactive keyboard operation and line turnaround signaling. This scheme is used by the US Robotics Courier HST, which, like the Hayes Smartmodem, uses V.32 trellis-encoded QAM, but without the expensive echo-cancellation circuitry.

Unfortunately, we could fill a separate book with virtually endless permutations of such schemes. Not only do vendors implement standards in different ways, but they also can choose to leave out some of the specifications included with the standards. Others claim to have found ways to improve on the standard. All of this points to the need for careful evaluation prior to a bulk purchase. A modem buy certainly does not lend itself to snap decision-making, and do not count on the vendor to be so sympathetic as to take the modems back and refund your money. In the real world of business, things never seem to work out that way.

You should also take a closer look at the speed factor. Some modems claim 19.2 kb/s over dial-up lines and will fall back to lower speeds in response to degrading line quality, but how they fall back is important. While some modems fall back gracefully to 16.8, 14.4, 12, and 9.6 kb/s, others fall immediately to 9.6 kb/s or lower. Others, like Telebit's TrailBlazer Plus, are capable of falling back in increments as small as 100 b/s. Since the slightest line impairment can force high-speed modems into the fall-back mode, you may find yourself operating at half speed most of the time. This is unacceptable, considering the high cost of these products.

If you are planning to use modems for remote access to corporate mainframes, you should evaluate modems for the security features they provide to guard against unauthorized access. For the most part, passwords alone are inadequate. It is not always easy to control who has them, even if you change them daily. Callback protection adds another level of security by requiring that the call be placed from a specific phone number. To access the remote system, the user enters the password. At that point, the remote modem disconnects. It then compares the password with a stored directory of phone numbers. Upon locating the phone number associated with the password, the remote modem calls back the modem. A higher level of protection may be achieved through data encryption. Some modem vendors like Codex, General DataComm, and Racal-Milgo offer data encryption through add-in boards.

Autodialing has been an integral feature of asynchronous modems since the early 1980s, eliminating the requirement for separate autodialers. In the last few years, this feature has been incorporated into synchronous modems as well. Both types of modem implement autodialing in a similar manner. They accept dialing commands *via* the same interface through which they pass data. Modem activity is separated into command mode and data mode. During the command mode, the modem accepts the dialing command, after which it switches to the data mode to implement the transmission. Until recently, however, if problems occurred with the transmission, the only option available was to use an escape sequence. This consisted merely of returning to the command mode to disconnect the call. Some new modems come with a voice-data toggling feature that lets the user discuss the problem with the remote user without having to hang up and redial.

If you plan to use modems in a multipoint, multidrop configuration, look into the network management features that can help you configure and diagnose them from a central location. Today's top-of-the-line networking modems offer some very sophisticated management capabilities. Some products constantly measure the actual line characteristics, thereby increasing the accuracy of VF impairment measurements. Other modems, like General DataComm's 19202 NMS, monitor the amplitude and delay characteristics of the line for real-time display at a network-management terminal, and, upon command from the network manager, loopback delay is measured between the master modem and its corresponding remote unit *via* the 200 b/s diagnostic channel.

This measurement is reported in milliseconds and may be used in a variety of ways. It may be used to configure tail circuits to insure that delay thresholds are not exceeded, which can degrade overall network performance. It may be used to isolate points of congestion on the network so that traffic may be rerouted. It may be used for planning alternate configurations, network expansion, and for justifying higher speed communications equipment and services. It can also be compared with the carrier's line specifications for data propagation, and, if throughput is slower than expected for some unexplained reason, it may be that the carrier has rerouted lines without your knowledge.

3.1.7 Conclusion

Innovations in data-encoding and compression techniques show no signs of slowing down. No sooner does one vendor proclaim itself the industry leader in a particular facet of modem technology than another vendor pulls the rug out from under him a week later with something smaller, cheaper, faster, or more reliable. The pace of innovation routinely overtakes standards-setting committees like CCITT. In fact, modem technology is progressing so rapidly that some industry observers are speculating about the possibility of modems eventually evolving into another species of equipment altogether as more and more functionality is built into them. There are already modems with integral-multiplexing, automatic dial-backup, and bandwidth-management capabilities. The first firmware-based modem was introduced by Omnitel, Inc. (Fremont, Calif.), early in 1988. This allows a microcomputer to be turned into a multiuser asynchronous-communications server. Up to four modem cards, having four modems each, can be loaded into the server to provide a total of sixteen ports.

Although exciting, unbridled innovation has its dark side. Faced with a decision to buy, users are confronted with multiple modulation techniques, error-correction schemes, and data-compression algorithms that are the source of compatibility problems among the products of different manufacturers. It would be nice if all modems could communicate with each other on the same network, but companies must compete in today's rough-and-tumble business world and cannot always afford to wait for standards, which seem to evolve at a glacial pace. Some manufacturers recognize that standards can stabilize modem prices and make them more universally applicable, but they also recognize that the technology is available to give customers what they want now and, if they do not do it, their competitors will. All of these factors make the job of selecting the right modem a very difficult task. Because of this, you can minimize the risk of a wrong decision by thoroughly testing modems in the planned-use environment before even considering a bulk purchase.

3.2 MULTIPLEXERS

The economic benefits of multiplexing are readily appreciated by most network managers. In optimizing the network through multiplexing, the number of voice-grade leased lines can be reduced, while high-capacity digital facilities may be utilized to full advantage. The resulting efficiencies can greatly reduce operating costs. After all, the monthly recurring costs for transmission facilities continue to consume the lion's share of most corporate communications budgets.

Even among smaller networks, the benefits of multiplexing can be quite substantial. For example, the line cost break-even point for 56-kb/s DDS (Dataphone Digital Service) is typically only three voice-grade private lines. If you presently use those lines for 9.6-kb/s data transmission, you can move to DDS and consolidate that traffic with a multiplexer. The result is equivalent to getting two free "lines," provided that the line-cost savings result in a reasonable payback period on the multiplexer.

The economies associated with T1 may be even more dramatic. Sometimes, just having five to eight voice-grade private lines is enough to justify the jump to T1. The actual break-even point, of course, is dependent upon which local-exchange carrier and interexchange carrier you use, as well as the distance of the T1 line. Additional savings may accrue from changes in tariffs, which are expected to continue their historical downward trend (Figure 3.5).

There are two basic categories of multiplexing technology, Frequency-Division Multiplexing (FDM) and Time-Division Multiplexing (TDM). FDM is the older of the two, having been used by telephone companies since the 1930s to consolidate many voice-grade lines onto fewer interoffice trunks. Utilizing this technique, the available transmission bandwidth of a given circuit may be divided by frequency into multiple narrower bands, each of which is used as a separate voice or data channel.

Although a simple method for deriving multiple channels, FDM poses a number of problems that restrict its usefulness in today's networks. For one, FDM

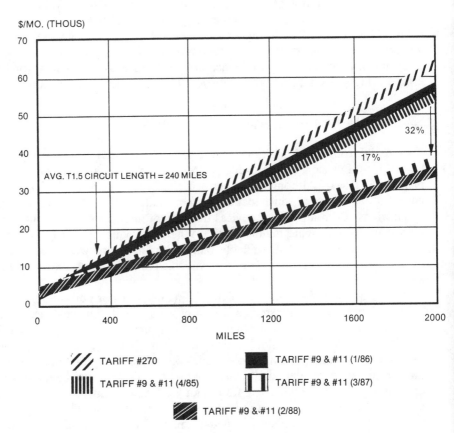

ACCUNET T1.5

TARIFF #270 vs. TARIFF #9 & #11

Figure 3.5 Tariff trends: an historical perspective.

is capable of supporting only a few input devices over a relatively low-speed link, and, compared to other multiplexing techniques, FDM uses up too much bandwidth per channel, which is not only inefficient, but costly as well.

Nevertheless, FDM is still in widespread use on point-to-point, voice-grade, leased lines. It is also used in some modems to derive separate channels for diagnostic data and production data. Although TDM is also used in such modems, some vendors claim that FDM provides faster "training," which is the time that equalizers in the modem spend in analyzing and adapting to the communications line. Coherent Communications Systems Corporation (Hauppauge, N.Y.) uses FDM in a line of private label modems, giving users the capability of combining one voice channel and one data signal for transmission over a dial-up or voice-grade leased line. This enables them to get the most out of their existing analog facilities, instead of having to lease separate lines for voice and data.

Any way you look at it, however, the big drawback to FDM is speed, which limits its use to voice and low-speed data on analog facilities. Even small networks are migrating to TDM for its higher speed and efficiency, and because it can handle voice as well as data in both synchronous and asynchronous formats. Except for some narrow applications, FDM has largely been abandoned in favor of more efficient, high-speed TDMs and Statistical Time Division Multiplexers (STDMs).

3.2.1 Time-Division *versus* Statistical Multiplexers

Aside from FDM, multiplexers are generally divided into two major product types: Time-Division Multiplexers (TDMs) and Statistical Time-Division Multiplexers (STDMs). Although they perform the same function, they go about it in entirely different ways, which makes one more suitable than the other for particular applications. Both are digital devices that accept multiple digital inputs and combine them over a single digital output. Whereas frequency-division multiplexing entails the partitioning of available bandwidth into narrower frequency bands, time-division multiplexing divides the available bandwidth into time slots to derive each channel.

In sequential fashion, the TDM continually scans or polls all input channels for any data awaiting transmission. In a process called "interleaving," the TDM accepts a bit or byte of data at a time, depending on the product type. Each channel is assigned its own "time slot" into which it can place data. If there are no data, that portion of the bandwidth goes unused, and scanning continues to the next input channel. A "frame slot" is inserted by the TDM at the beginning and end of each group of time slots. The frame slot is a bit in bit-interleaving multiplexers, and a byte in byte-interleaving multiplexers. The data collected by one complete scan of all the input channels is called a "frame." Each channel, then, occupies a unique time slot within a particular frame, with each frame separated by a frame slot.

Empty time slots are stuffed with bits or bytes as necessary to insure the uniformity of frames, which in turn helps to keep the transmission synchronized. The frame slots act as the primary synchronizing mechanism, enabling the remote multiplexer to pass received data to its proper destination at the proper time. Since framing formats differ from one manufacturer to another, the multiplexer of one may be incompatible with the multiplexer of another. Even multiplexers from the same manufacturer may employ different framing techniques, which underscores the need to ask the vendor whether its new product is backward compatible with those you may already have on your network, just to be sure that your previous investments are protected.

The framed data are transmitted over the digital facility to another multiplexer at the receiving end. There, they are "demultiplexed." This refers to the process of identifying the time slots containing particular bits or bytes of information from each input line so that they may be output to the appropriate device at the correct transmission speed.

Like TDMs, STDMs scan the input channels for data awaiting transmission, and each frame of data is loaded into a time slot. If no data are present on a particular channel, however, the available bandwidth is assigned to a channel that is ready to transmit data.

In most asynchronous, interactive data applications, it is unusual for all channels to be operating at full capacity simultaneously. Consequently, the sum of the active-channel data rates will typically not exceed the composite link rate. STDMs can handle peak traffic conditions better than TDMs. With a TDM, the aggregate input speed cannot exceed the composite link speed. With an STDM, on the other hand, the aggregate input speed can be exceeded because it makes use of a buffer that holds data during peak traffic conditions until bandwidth becomes available. To prevent data loss when the buffer fills up, the STDM invokes a procedure to regulate the output of the transmitting device. This is accomplished through the use of various flow-control protocols, such as XON/XOFF, DTR/DSR, CTS/RTS, and ENQ/ACK. With XON/XOFF, for example, the XOFF code stops data flow, while the XON code permits data flow. This buffering capability also enables high-speed terminals to communicate with low-speed terminals, since the data is buffered from the high-speed terminal for low-speed transmission to the remote terminal. To take advantage of this feature, however, you must insure that the multiplexer's buffer is large enough. Some vendors tout the 19.2-kb/s input-channel data rate of their products, but provide only 20 kilobytes of storage for the input buffer. This means that the buffer will fill rather quickly during heavy traffic periods, causing frequent flow-control signals to be issued to the terminals, resulting in substantially degraded throughput.

The ability of the STDM to accommodate variable length frames also contributes to bandwidth efficiency. This means that more than one byte of data can be accepted from a given terminal before the STDM moves on to the next active terminal. Other advantages of STDM include the capability to prioritize channel scanning and perform retransmissions upon error detection.

The principal advantage of STDM over TDM is efficiency of bandwidth. Despite this advantage, however, STDMs have some drawbacks that might make them unsuitable for certain applications. Using STDMs for voice would result in unacceptably low quality, as well as intolerable time delays, because the digitized voice signals do not arrive at their destinations in regular, predictable time slots as in TDM. Mixing voice with data is what TDMs do best.

Even in data-only applications, STDM may not be appropriate. In electronic brokerage transactions, for example, the priority scanning capability of STDMs would inhibit the timely flow of information. Imagine the outrage of users upon finding out that their customers' buy-sell instructions were delayed because a high-priority user representing a large institutional investor hogged all the bandwidth during a volatile trading session. Not only would individual users miss out on commissions, but they also would have to answer to irate customers who lost money because of the delay. The error-checking feature of STDMs can hold up the network with retransmissions, which could also impose intolerable delays in this type of environment.

TDMs, on the other hand, are totally transparent to the data that each channel is supplying; there is no analysis of the data to determine which protocol is being used. There is no priority scanning and no channel-based error checking. This means that all users have equal access to the available bandwidth. In the brokerage example above, the information bottleneck would not be the communications network. The volume of trading would be dependent on the transaction-processing capacity of the host.

Typically, TDMs are deployed as network "backbones" to insure the free flow of data between major nodes. STDMs are typically used at locations where many users require only sporadic access to the network, as when clustered terminals query a remote host in point-of-sale applications. In this case, the STDM makes the most efficient use of available bandwidth to "feed" the T1 backbone, or provide X.25 connection to a packet-switching network.

When confronted with the need to multiplex, you need not assume that you must choose between TDM and STDM. Actually, mixing and matching devices opens up all kinds of configuration possibilities that can enhance the efficiency of your network. Not only can TDMs network with other TDMs, they can interface with other types of hardware like STDMs, high-capacity T1 multiplexers, and data PBXs. An STDM, for example, can be used to concentrate data from a cluster of low-speed asynchronous terminals into a single-input channel of a TDM. The TDM combines that signal with other input signals for transmission over a 56-kb/s link. At a network node, a T1 multiplexer transmits the input of that 56-kb/s channel along with 23 other channels to another T1 multiplexer, where the 24 channels may be routed to their separate destinations. Some of those channels may be routed to a TDM, where one or more may go to an appropriately equipped data PBX, which acts as a demultiplexer.

The price difference between medium-speed (56 kb/s or less) TDMs and STDMs of equivalent channel capacity is virtually nil, with STDMs typically costing only $200 more than a TDM, but there is much more than cost that should enter into your decision to choose one type of multiplexer over the other. You need to compare such things as architecture, data and voice capabilities, synchronization characteristics, network-management features, and compatibility with standards.

From the start, you should anticipate growth. That means selecting products that are modular so you can start small with a minimum up-front cost, but add channel capacity incrementally as your needs require. With a stat mux, for example, you can start with a low-end product, which typically supports two to eight input channels in single-link point-to-point and provides transmission speeds of 9.6 kb/s to 19.2 kb/s. Later, you can graduate to the midrange, which handles from eight to thirty-two input channels. High-end stat muxes handle even more input channels and provide several high-speed links. They also incorporate sophisticated switching and control capabilities.

Modularity also applies to capabilities. Look for products that allow you to add specific capabilities to the multiplexer through plug-in boards or cartridges. These include boards that provide X.25 bridges, LAN gateways, V.35 interfaces for wideband modems, and T1 interfaces. Some vendors, like Micom Systems Inc. (Simi Valley, Calif.) and Case Communications (Columbia, Md.), offer cartridges that upgrade their products from simple point-to-point concentrators to switching statistical multiplexers. Other cartridges provide support for a variety of CCITT-recommended packet interfaces, including X.25, X.3, X.29, and X.121. Micom offers an Enhanced Statistical Multiplexer "Featurepak" cartridge, which supports built-in modems with automatic dial backup and per channel password protection. The same cartridge allows users to set different channel speeds and protocols at each end of the circuit.

Aside from cost savings, integrating modems with multiplexers simplifies the selection of equipment that is intended to work together, streamlines maintenance, and eliminates potential cabling and interfacing problems. Data concentrators may be integrated into stat muxes to strip out redundant bits, which increases the density of the data stream, thereby permitting greater throughput over cheaper, lower-speed lines. Compression devices may be integrated into TDMs *via* plug-in cards to increase the number of voice channels that can be carried over given circuits.

Some products allow users to pick the types of input channel they want: asynchronous channels to support terminals, printers, and modems; synchronous channels to interface with multiplexers and other high-speed devices; and isochronous channels for applications requiring some form of encryption to safeguard sensitive information during transmission.

STDMs offer some very sophisticated capabilities, such as channel prioritization, downline loading, alternate routing, and bandsplitting. With channel prioritization, you can designate one or more channels that will always be allowed to pass data before channels of lesser priority. Downline loading allows the host site to set the channel options at the remote site, eliminating the need for user intervention. Alternate routing is the capability of some high-end STDMs to invoke stored network configurations to bypass failed facilities or reconfigure the network by time of day. Alternate routing may be a simple redirection of a few channels, or a complete reconfiguration that involves every node in the network. Bandsplitting permits a composite synchronous data link to be divided into two paths, with both sharing the available bandwidth. This permits synchronous output from one multiplexer to be input into another multiplexer by sharing a composite link instead of having to be remultiplexed with data from other channels.

High-capacity TDM multiplexers, also known as T1 multiplexers, should be DS0 compatible to give you access to DACS, a switching facility which allows you to take advantage of a variety of carrier services with circuits that you can put up and take down as your needs dictate. These multiplexers should also support Extended Superframe Format (ESF), an AT&T-provided capability that permits users to perform nonintrusive (meaning that the circuit does not have to be taken out of service) circuit testing and diagnostics, as well as data-error correction. Many TDMs do not support ESF, nor can they be economically upgraded to support it. Others provide ESF support as an option, which is good, since there is no sense paying for it until you need it.

While subrate multiplexers are available from several vendors as stand-alone products for small users, some T1 multiplexers offer this capability as an add-in option. This allows users to feed lower-speed DDS circuits (2.4, 4.8, and 9.6 kb/s) to a central-office multiplexer *via* a single 56-kb/s or T1 link. You can use the subrate multiplexer to consolidate five 9.6-kb/s data channels from five different locations over a single high-speed link terminating at an AT&T point of presence (POP), for example. There, AT&T's multiplexer fans out the data onto several subrate DDS lines to their appropriate locations. Since the telephone company central-office or interexchange-carrier POP provides its own multiplexer, or appropriately equipped DACS, the user only requires a single-ended subrate multiplexing capability.

When it comes to selecting a T1 multiplexer, you should analyze the nodal intelligence of the system. Vertical Systems Group (Dedham, Mass.) offers this checklist to assist in the analysis [2]:

- How much manual intervention is required to define a circuit? Do you have to specify only circuit endpoints or all intermediary nodes as well?
- Does the router automatically adapt to changes in the network topology?
- Is there vulnerability because router intelligence is concentrated in only one node?
- How long does it take to reconnect all of the circuits on a failed internodal link? (NOTE: Use specific examples when comparing networking multiplexers. High-end products reroute traffic at speeds one to two magnitudes lower than less powerful products.)
- How is contention to internodal link resources handled? To channel resources?

3.2.2 Bit *versus* Byte Technology

As alluded to before, TDMs may package data in either bit or byte formats. A vendor whose product is built around a byte architecture will try to convince you that "bytes are best," while a vendor whose product is built around a bit architecture will try to convince you that "bits are best." When the TDM is a T1 multiplexer, the arguments on both sides are quite compelling, leaving buyers understandably confused.

Let us begin with some general guidelines: if you are primarily interested in moving voice traffic through the public network, a byte-oriented T1 multiplexer may be your best choice; if you are interested in moving data through private facilities with minimal delay, then a bit-oriented T1 multiplexer may be what you need, but these rules of thumb are really too simplistic to base a buying decision on them. More detailed analysis is required.

Byte-oriented multiplexers give you the level of connectivity you need to take advantage of current and future carrier services that require access to the public network through such services as DACS. The byte structure is also required if you want to take advantage of Extended Superframe Formatting, which is offered by AT&T on its Accunet T1.5 circuits. While it may be desirable to choose a byte-oriented multiplexer to take advantage of carrier service offerings, or even position your network to do so at some time in the future, byte-oriented multiplexers are not without some drawbacks. For instance, byte-oriented multiplexers waste bandwidth on overhead. Such multiplexers can consume as much as five percent of the T1 bandwidth for overhead functions. This may not appear to be very significant until you translate that loss into dollars. For example, a 712-mile Accunet T1.5 circuit between Chicago and New York costs approximately $16,000 a month, including local-loop charges at both ends. If overhead deprives you of five percent of the bandwidth, you are paying $800 for bandwidth you cannot use. That translates into a $9,600 loss per year on that circuit. Put another way, you are paying for 24 channels, but you can only use 23 channels. These losses are compounded if you have a multinode T1 network, and compounded further by the increases in distance. An Accunet T1.5 circuit between New York and San Francisco, for example, costs about $38,700 a month, including local loop charges at both ends. With five percent of the bandwidth consumed in overhead, you are paying $1,935 a month more for that circuit than you should, which equates to a $23,220 loss per year. These losses also prolong the payback period on the multiplexer. Beyond that, they give accountants attitude problems.

Let us say that you have several 56-kb/s circuits on your network and that you want to manage that bandwidth through a byte-oriented multiplexer. Assuming that the multiplexer abides by AT&T's specifications for subchannel derivation, you can get five 9.6-kb/s data channels out of the 56-kb/s circuit. That sounds nice, until you notice that this subrate scheme immediately deprives you of 14% of the bandwidth, or 8 kb/s. Add to that the 5% of lost bandwidth that the multiplexer uses, and you can see that you cannot really get the five 9.6-kb/s subchannels that you originally planned to get. That is 19% of unusable bandwidth on a 56-kb/s circuit! With multiple, long-haul, 56-kb/s links going through multiple, byte-oriented multiplexers, the dollar losses over a year's time are nothing less than astounding.

A bit-oriented multiplexer, on the other hand, does not present such problems because it uses all of the available bandwidth for production data, and when you use voice-compression techniques like Adaptive Differential Pulse Code Modulation (ADPCM) to double the amount of voice channels on the circuit, you can derive a full 48 channels, instead of only 44 under the byte orientation. Interestingly, some vendors of ADPCM equipment accommodate both bit and byte interleaving multiplexers, preferring to let users configure the device for compatibility with either format.

It is easy to trivialize the bit *versus* byte controversy by claiming that the decision to use one over the other boils down to which scheme best supports the applications running on your network. Indeed, if T1 multiplexers did not cost so much ($30,000 and up, depending on port capacity, features, and level of redundancy), you might feel comfortable letting the chips fall where they may, but networks do not remain static for very long. They must respond to tariff changes, to new carrier services, and, more importantly, to organizational needs. Network managers, then, are under the gun: knowing full well that they are accountable to top management for performance results, they must not only install systems on schedule to minimize costs, but also position their networks to reap additional cost savings from an optimal blend of public and private facilities.

First, however, let us elaborate on the nature and scope of the problem. Private networks are quite deficient architecturally. Unique vendor solutions offer little or no compatibility across the broad spectrum of products available in the marketplace. Although various data channels and voice channels can be consolidated onto DS1 facilities, there is still no way to route them individually through the central-office DACS or connect channels to public network elements, while maintaining autonomous user network management and control. In such cases, private networks still have to lease multiple voice-grade and DDS private lines to handle public-network voice and data separately, which is not only inefficient, but costly, as well. There is also no efficient way to integrate data channels with channels on the public network to enhance overall network management.

Until relatively recently, another limitation of the private network has been the conspicuous absence of a multiplexer to provide compatibility with public-network standards at the DS1 and DS0 levels for both the channel and aggregate (or internodal-trunk) connections. Typical private networks are conceived and installed from a perspective that DS0s are DTE (or channel) inputs that are not part of the transmission facility between nodes and, therefore, fall outside the domain of network management between nodes.

At first glance, this may not appear to be a serious problem, since many vendors claim that their multiplexers are "DS0 compatible." Channel DS0 compatibility, however, without consideration for internodal DS0 connectivity that allows supervisory communication within DS0s — or below DS1 — negates network management conveyance through the public DACS network. Vendor solutions with channel-side DS0 connectivity still require proprietary DS1 between nodes to establish network management and isolate DS0 connections from centralized, network-management schemes.

Until recently, no vendor has achieved the technological breakthrough that allows network management with either DS1 or DS0 internodal trunk connections. As a result, private networks really have not had the capability to transport data through the public DACS network efficiently.

Finally, there has been no mechanism available with which private networks could exploit the best of byte (DS0 frame structure) and bit (DS1 proprietary, or non-DS0 frame structure) technologies for optimal advantage. The byte framing structure permits cross connection of channels within the DACS network for maximum flexibility in implementing public-network interconnectivity, restoral, and alternate routing at the network-transport level. Bit structures within one or more DS0s can carry many channels, including the supervisory channel, at various rates instead of being limited to one channel per DS0 or synchronous-only subrate data within DS0s. In addition to permitting network management to be passed on all links — including inter-DACS links — the benefits of using bit technology include minimal nodal delays, high aggregate efficiency, and a broad selection of channel data types and rates.

The public network has three key inherent advantages that can be used to optimal advantage by private networks under a hybrid arrangement: uniformity, full interconnectivity, and interoperability. These advantages are demonstrated in the fact that all analog and digital central-office switches work together across the entire network spectrum, as do all multiplex elements. Also, all vendor products are compatible with each other when used over the public network, which makes for a high degree of interoperability, and the public network does an excellent job of supporting voice and synchronous data at the rates most in demand. It does not do a very good job, however, of supporting mixed voice or data services, asynchronous data, and data between 56 kb/s and the DS1 rate of 1.544 Mb/s, which are also in demand.

Now let us take a quick look at DACS. (More detailed information about DACS follows in Section 3.5.) As used by AT&T, the digital access and cross-connect system is a "front-end processor" to a central-office switch. Its function is to accept aggregates of channels *via* DS1 facilities and groom them for individual routing at maximum fill. With Customer Controlled Reconfiguration (CCR), users can improve network interconnectivity by implementing initial and changed circuit configurations from an authorized terminal, thereby attaining a level of control over their network without incurring additional costs for altering internodal backbone facilities. Other advantages of CCR include control of network load scheduling and disaster recovery.

Since the DACS actually performs the routing function on a channel-by-channel basis, the central-office switch is free to do other things, like process more calls or implement memory-intensive features. By off-loading the switch in this way, the cross-connect system allows service providers to establish overlay networks efficiently for dedicated services such as voice-grade private lines and DDS circuits, all of which are essentially permanent virtual circuits that would unnecessarily burden switching systems that are geared for continuous circuit set-up, supervision, and tear-down.

Central-office multiplexing services like M24 and M44 provide users with the means of adding remote locations to their backbone networks to extend DS1 economies of scale to ever greater portions of their networks. M24 adheres to the D4 frame format with 24 channel (DS0) compatibility. As such, it provides interconnectivity to virtually all network elements, as shown in Figure 3.6. M24-equivalent services are offered by all service providers under a variety of names. Since all service providers use D4-DS0 network elements, D4-DS0 has a universal application within the public network.

M44 is a tariffed service of AT&T which compresses PCM voice *via* ADPCM, allowing 44 voice channels to be sent over a single T1 circuit, but M44 is seriously deficient in that it is capable of accepting traffic only at the DS1 rate of 1.544 Mb/s, or in contiguous bundles of six DS0s at 384 kb/s. This limits its potential as a means of using the public and private networks to optimal advantage, since routing individual DS0 channels requires six DS0 "bundles" as a minimum requirement for routing.

M44, however, begs for DS1 public connections with tariffs structured for partial DS1 use in six or twelve DS0 bundles (768 kb/s) that could be defined as "Fractional T1" (FT1), similar to the fractional-service offerings of Nippon Telephone Telegraph (NTT) in Japan.

From a cost standpoint, FT1 offers very attractive economies. For example, the interexchange-carrier portion of a partially used, 1,350-mile, leased T1 line from New York City to Dallas, Texas, may cost a low-volume user as much as $21,000 per month. The same line tariffed and used as one-quarter FT1 could be offered to the customer at only 30% of that amount, or $6,300 per month. For large networks, FT1 would make a more economical backup to T1 in case of common equipment or link failures, permitting enough channel availability for the transmission of critical information.

A few domestic carriers have already started offering Fractional T1 services over their own DACS networks. Cable & Wireless Communications Inc. (Vienna, Vir.) sells DS0s, or multiple DS0s bundled to provide 128, 256, and 384 kb/s as part of its Intelli-Flex service. Williams Telecommunications Corporation (Tulsa, Okla.) also provides Fractional T1 services, and, at this writing, AT&T has announced its intention to tariff 64-kb/s service.

As network managers seek such hybrid network solutions, they should start evaluating the offerings of mux vendors according to their support of the following:

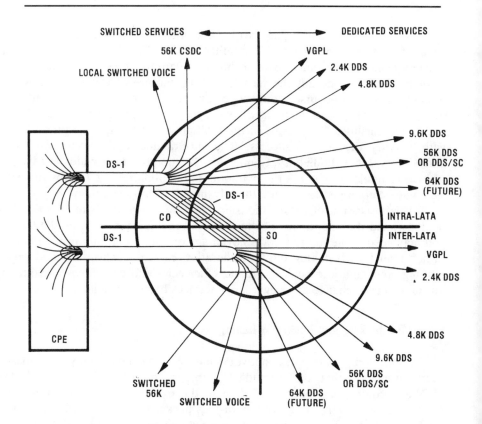

Figure 3.6 Public and private network interconnectivity.

- DS0 compatibility on the channel side (DTE side);
- DS0 compatibility and supervisory conveyance through DACS on the aggregate side (internodal trunk) — and without requiring proprietary DS1 between nodes;
- Optimal use of present and future service-provider offerings;
- Network-management compatibility with other vendor products (i.e., NetView, UNMA, DACS, and modems);
- Extended control of customer-owned (or leased) transmission facilities, as well as compatibility with public-transmission facilities;
- Capability to intermix public network-service elements;
- Capability to use DS1, FT1, or 56 kb/s as internodal trunks;
- Use of the product at CPE locations, as well as within the public network at service-provider offices.

Bit and byte technologies involve tradeoffs that impinge on the cost and flexibility of today's networks. While bit technology provides maximum bandwidth efficiency and cost savings, byte technology permits access to public networks through DACS to permit the most advantageous use of carrier offerings. The ultimate solution, of course, would be to have one multiplexer that takes advantage of both bit and byte architectures, allowing you to reap the benefits of both private and public facilities. Two rivals in the multiplexer arena have accomplished this with varying degrees of success. The bit-oriented multiplexers of General DataComm, Inc. (Middlebury, Conn.), may be equipped to provide the advantages of byte technology, whereas the byte-oriented multiplexers of Tellabs Inc. (Lisle, Ill.) may be equipped to provide the advantages of bit technology.

This discussion of the relationship between bit *versus* byte technology has demonstrated the importance of choosing a T1 multiplexer based on the flexibility you may need in the future to take advantage of both public and private facilities, rather than merely taking into consideration current applications and possibly limiting your ability to achieve greater efficiencies and economies as your organization's needs change and new carrier services become available.

3.2.3 Multiplexing and Packet Switching

Public and private packet networks have grown exponentially in the past few years, so much so that X.25 is rapidly becoming a "must-have" feature of STDMs. (TDMs accommodate packetized data to the extent that they allow it to pass through transparently on its way to various packet nodes.) STDMs can not only work in concert with packet switches, they can also be converted into access devices to feed packet-switching networks. In this way, stat muxes may be inexpensively configured for use as gateway devices to packet networks. Micom, for example, uses the same hardware building blocks for its stat mux and its X.25 PADs (Packet Assembler-Disassembler). It is the software feature pack that determines how the device will be used.

Some mux vendors are anticipating a convergence of stat-mux technology with packet-switching techniques, citing the fact that there is very little difference between the capabilities of a high-end stat mux and an X.25 PAD. After all, PADs are used merely to concentrate data for transmission over a packet network that conforms to the X.25 format established by the CCITT, and both types of device are capable of handling a mix of synchronous and asynchronous protocols. The convergence of the two technologies will become more pronounced as ISDN becomes more widely accepted.

ISDN services are already becoming available in select cities. Although widespread availability of such services is still a few years away, it is important to realize that ISDN relies on a packet-based technology for signaling and control. Therefore, if STDMs are to continue to function in an ISDN world, they must evolve to encompass packetlike characteristics. From a buyer's perspective, then, any new purchases of STDMs should be guided by the vendor's growth path for the product. Such "futures" information may be viewed by the vendor as proprietary, but your objective is to determine at least if a product development plan even exists, if only as a hedge against the early obsolescence of your planned investments.

3.2.4 Multiplexing in a Fiber World

With the increasing availability of fiber, network managers might question the need to multiplex at all. Current fiber networks already provide more bandwidth than there is a demand for, which is already leading to price wars among the major interexchange carriers.

As currently configured, however, fiber networks are essentially backbones that link major cities. Getting to the nearest backbone economically requires that multiplexing techniques be used, and, despite all the media attention being focused on bringing fiber to the customer premises, many corporate locations remain outside of planned fiber networks, and will remain so for the foreseeable future, if only because consumer demand in low-density areas does not yet justify the cost of installation.

Although major corporations are beginning to use fiber to link dispersed locations within a campuslike environment, or the vertical environment of tall office buildings, statistical multiplexers may still be used to tie a cluster of terminals into the fiber link. Furthermore, the use of statistical multiplexers within the fiber environment preserves intrabuilding configuration flexibility with respect to the allocation and placement of terminal equipment. The high transmission capacity of fiber, then, does not eliminate the need to multiplex, it creates the need to multiplex.

For longer distances, between buildings, for example, a fiber-optic multiplexer may be used. The fiber-optic multiplexers of Canoga Perkins (Chatsworth, Calif.) and Optelcom (Gaithersburg, Md.), for example, use TDM techniques to assign input data to time slots for transmission to an identical unit at the remote location. The primary difference between a fiber-optic mux and a conventional TDM is that the former supports many more input channels than the latter. The Canoga Perkins unit supports 144 input channels of 9.6 kb/s each, while the Optelcom unit supports 160 input channels of 4.8 kb/s each.

Fiber-optic multiplexers use a laser to emit the pulses of light which code the information for transmission over a fiber link. At the other end, a photodiode detects the light pulses for decoding. The differences in various fiber-optic multiplexers boil down to the quality of these elements at each end, which has a lot to do with how far the link may span before special repeaters are required to propagate the signal. Other factors influence transmission distance, such as the quality of the fiber and the signal loss caused by connectors. Notwithstanding these considerations, the higher the radiance of the laser LED (light-emitting diode), which emits the light pulses, the greater the distance the signal can travel before repeaters are required. Correspondingly, the photo diode at the receiving end must be more sensitive to detect the light pulse over the longer distance.

Aside from their great input-channel support capabilities, fiber-optic muxes do not stack up very well against conventional TDMs in one very important area: whereas TDMs may be remotely configured, most fiber-optic muxes use DIP switches for this purpose.

On the other hand, the diagnostic capabilities of the fiber-optic units do not have to be very sophisticated. Since fiber links are immune from most of the impairments that afflict leased or dial-up lines, there is no need for elaborate diagnostics. Simple local and remote loopbacking is all that is necessary to determine if the local and remote units are operating properly.

Because fiber-optic multiplexers are relatively new, they have not had a chance to evolve as many sophisticated capabilities as stat muxes. Therefore, the offerings of various manufacturers differ markedly. Some fiber muxes are limited to point-to-point applications only, whereas others may be used in multidrop or star configurations. Some support echoplexing (an error-checking scheme whereby characters received at a central host are echoed back to the terminal that originated them for output on a printer), while others do not, and in some fiber multiplexers there is no capability for even adding a redundant power supply.

3.2.5 Network Management

Most multiplexers are equipped standardly with a supervisory port through which users can monitor and control the network *via* a terminal. Although the number of management features that can be implemented through the supervisory port differs from product to product, they usually include alternate routing, port assignments, message broadcasting, and diagnostic testing. The tests typically include local and remote loopback tests, digital and analog tests, self-tests, and other tests designed to check for the proper operation of hardware and firmware. Some products allow users to retrieve network statistics through the supervisory port and set up event logs which record all significant network activities and alarms.

Features, functions, and capabilities are all important aspects in multiplexing, as are flexibility, expandability, and controllability. Network management encompasses all of these elements. Control of data-communication resources continues to be of critical importance to network managers. Without a central point of control, the pieces of the network are just that — pieces.

Network-management capabilities constitute the glue that holds the network building blocks together. This is why it is important to look for equipment that can be controlled remotely. With remote control, problems can be identified quickly and with precision — from the high-capacity backbone, all the way down to the individual modem circuits. This "neural" management capability is what turns mere communication resources into powerful competitive tools.

In evaluating vendors, look for one who offers both nodal and neural management capabilities; that is, a vendor who has strategically designed data sets and multiplexers to work together end to end. This is an important consideration because many vendors have merely tried to piece various network components together from other sources. As a result, their network offerings cannot be brought together under the same management umbrella.

Evaluate the product for its ability to accommodate expansion at the node, as well as the ability of the product to evolve to a higher level of networking, like ISDN. ISDN compatibility will allow T1 multiplexers to interact with central-office equipment to control the configuration of the channels available through the T1 pipe. For example, primary rate interface (23B + D) users will be able to route the 23 "B" channels to different destinations from the central-office switch. These channels can also be reconfigured to meet changing voice and data requirements. Although some mux makers have anticipated ISDN by building similar capabilities into their products, many have not, and the ones who have may be able to transmit information over the ISDN network, but not be able to take advantage of certain ISDN features, like channel reconfiguration from the central office.

Try to determine whether the vendor has the resources, the expertise, and the willingness to work with you to achieve your organization's long-term objectives, and who has demonstrated the ability to work with multiple vendors who may be supplying the various communications paths between LATAs.

Finally, look for a vendor who is independent of a computer manufacturer or telecommunications carrier. Such vendors are in the best position to provide unbiased service and advice. In dealing with independent vendors, you will not feel boxed in by the hidden agendas of parent firms. As a result, you will be more likely to get what you need, rather than what the vendor wants to sell.

With these considerations in mind, you can build a network that can be managed for greater efficiency, productivity, and peace of mind.

3.2.6 Special-Purpose Multiplexers

Channel banks have been in use for more than twenty years. They are similar to multiplexers in that they digitize analog signals into twenty-four 64-kb/s channels for consolidation onto a single DS1 circuit.

Channel banks employ a byte-interleaved framing technique to package analog information into DS0s. Because channel banks handle information a byte at a time, they are quite limited in what they can do. Unlike some multiplexers, which may use bit-interleaved technology to support a range of transmission speeds, channel banks are limited to supporting channel speeds of 64 kb/s, making them inefficient for sending lower speed data *via* modem.

Not only do channel banks lack switching capabilities, they also do not permit the user to allocate channels on an as-needed basis. Channel banks also lack extensive network-management capabilities. Since they are typically hard-wired devices, their functions cannot be remotely controlled. Although channel banks offer the user little flexibility in designing and maintaining networks, at less than $10,000 they are cheaper than multiplexers and require no technical staff to maintain.

Used in point-to-point applications, one channel bank may reside at the central office and another at the customer premises. Channel banks are also used for the economical multiplexing of analog lines at remote locations for long-haul transport to a high-capacity nodal multiplexer or cross-connect system.

Until recently, channel banks were considered a relatively old and dying technology, having been replaced by sophisticated multiplexers using newer electronic componentry and advanced technology for more efficiently transporting voice and data. Advances in technology, however, are being applied to channel banks as well, providing them with drop-and-insert capabilities, for example. Even with a digital PBX having a direct DS1 interface, a drop-and-insert type channel bank may be used to insert data channels in front of the switch. In addition, a number of hybrid products have appeared in recent years, which can be used as channel banks for point-to-point applications, or as drop-and-insert multiplexers for multidrop applications, depending on which hardware options you choose. You can start with a plain-vanilla channel bank, for example, and add plug-in boards to the common-equipment module that will give it drop-and-insert capabilities. The common equipment typically provides system power, T1-line interface circuitry, and control circuitry for the various channel units, as well as the timing source, framing patterns, and alarm-detection facilities.

For relatively simple networks, channel banks are emerging to become a viable alternative to multiplexers, serving the remote locations of large networks, as well as providing rudimentary multiplexing capabilities.

3.3 CHANNEL SERVICE–DATA SERVICE UNITS

Channel Service Units (CSUs) and Data Service Units (DSUs) provide the interface between on-premise data-terminal equipment (DTE) and digital transmission facilities. Intra-LATA digital services are provided by local telephone companies, whereas inter-LATA digital services are provided by such carriers as Cable & Wireless (Vienna, Vir.), ITT's United States Transmission Systems, Inc.

(Secaucus, N.J.), Western Union's Worldcom International Private Line Services (New York, N.Y.), and AT&T.

Of all the carriers offering inter-LATA digital services, AT&T has the most extensive digital network and holds the largest market share. Its offerings include Accunet T1.5 Service and Dataphone Digital Service (DDS). Accunet provides 1.544-Mb/s service, while DDS is available at 56 kb/s and subrates of 9.6, 4.8, and 2.4 kb/s. It is anticipated that AT&T will offer 19.2-kb/s service by early 1989.

In the predivestiture era, AT&T insisted that CSU-DSUs be used as an interface between customer-owned equipment and its digital transmission facilities. The FCC supported that requirement, which is still in force today. All of the former Bell Operating Companies (BOCs) and most of AT&T's inter-LATA competitors still require the use of CSU-DSUs by their customers. Over the years, CSU-DSUs have been labeled as commodity items because of their limited functionality. Conventional wisdom says that because one vendor's unit does not differ very much from another vendor's unit, price should be the determining factor in the purchase decision, but vendors like Tellabs Inc. (Lisle, Ill.), Verilink Corp. (San Jose, Calif.), Penril DataComm (Gaithersburg, Md.), and Paradyne Corp. (Largo, Fla.) have introduced new features, innovative testing capabilities, and higher levels of functionality into their products. Today, buying CSU-DSUs on price alone may not only impede your ability to improve the performance of your network, but also lock you out of emerging carrier services.

3.3.1 The Origins of CSU-DSUs

As originally designed by AT&T, the CSU provided isolation between the carrier's high-capacity digital facilities (1.544 Mb/s) and the user's data terminal equipment (DTE), which includes such devices as channel banks, T1 multiplexers, digital cross-connect systems, and digital PBXs. The purpose of the CSU, then, is to insure that the user's DTE does not interfere with the proper operation of the carrier's network-switching or billing facilities. The CSU also provides protection for technicians and the network from the potentially harmful effects of line voltages and lightning surges which may propagate beyond the DTE.

The FCC requires compliance with carrier technical references (AT&T as well as Bellcore), which define the signal bandwidth, power levels, and longitudinal voltages of CSUs to be connected to the public network. A minimum ones density and maximum number of consecutive zeros are also mandated by the FCC regulations. (It is worth noting at this point, that although T1 circuits may be leased for dedicated use, they are still considered a part of the public network.)

In 1988, the FCC withdrew the requirement that CSUs be powered from the line, since it was widely recognized that the CSU will not go into oscillation upon the loss of power. For the most part, users have the choice of paying the carrier a fee to cover its cost to provide power, or optioning the CSU to use the local 48-volt power source that may already be in use for CPE. All CSUs offered by vendors today allow for either line or local power, but if you are shopping the used equipment market, be aware that some older CSUs may not accommodate an external power source.

The FCC also requires that the user notify the carrier whenever disconnecting the CSU from the T1 facility so that central-office alarms will not be raised. In sum, FCC rules specify the connectivity requirements for T1 facilities. These specifications took several years to finalize and are still evolving as more carriers get involved in establishing end-to-end connections and as new communications technologies become available from an increasing number of vendors.

Whereas the FCC is primarily concerned with network protection, AT&T is concerned with network performance. In many cases, the commission uses AT&T or Bellcore technical references as part of its equipment registration requirements. For example, all CSUs must have a T1-repeater function to recover from the network low-level signals due to signal attenuation in the T1 line or the customer's in-house cabling. Also, a "keep-alive" signal must be sent toward the network upon loss of the signal from DTE to insure the proper operation of the span-line repeaters. Currently, three such keep-alive signals are in common use: all-ones with framing, all-ones without framing, and line loopback.

3.3.2 Basic Functionality

Since digital-transmission services are capable of carrying the signals from DTE nearer to their original form, there is no requirement for complex modulation-demodulation techniques to compress the serial data, as is the case with analog modems that are used to carry data over voice-grade leased lines. Instead, customer-provided CSUs or DSUs encode serial data from terminals or computers and perform wave-shaping of the transmit signal before it is sent over the digital facility. This insures an acceptable level of network performance. Since lower-speed (2.4 to 56 kb/s) data-terminal equipment requires standard RS-232C or V.35 interfaces, both the CSU and DSU functions are required for services like DDS. Only the CSU is required for high-capacity 1.544-Mb/s service, since the customer's T1 equipment (e.g., multiplexers, PBXs, *et cetera*) provides the DSU function. While CSUs and DSUs work in close proximity to assure proper network performance, each has a separate and distinct function.

The DSU's job is to convert the data stream generated by on-premise customer equipment into the bipolar format required for transmission over the digital network. It also performs such functions as data regeneration, control signaling, synchronous sampling, and timing.

The CSU functional element provides a physical termination point for the carrier's circuit. In addition to performing signal regeneration, 1.544-Mb/s CSUs monitor the incoming signal to detect bipolar violations, insure that the data stream does not contain more than 15 consecutive zeros, and that the ones-density rule that governs transmission over high-capacity 1.544-Mb/s services (at least three ones must be present in every 24 bits transmitted) are not violated. Because the CSU separates user equipment from carrier facilities, it provides a window on the network, allowing both the carrier and the customer to perform testing up to the same point. The position of the CSU on the network has prompted many vendors to equip it with increasingly sophisticated diagnostic and network management features, which go a long way toward enhancing user control and insuring high circuit quality.

Both the CSU and DSU fall into the category of customer-premises equipment (CPE). Because of the complementary functional relationship of the CSU and DSU for use over DDS facilities, many vendors have integrated CSU-DSU devices into a single unit. Such units are sometimes referred to as "data sets."

The CSU-DSU diagnostics are fairly similar. Just about all vendor offerings support loopback tests to check the integrity and performance of the line between the customer's equipment and the carrier's serving office, as well as the link between near-end and far-end CSU-DSUs, and most CSU-DSUs allow the user to perform a DTE loopback test to check the integrity of the link between the CSU-DSU and the originating terminal equipment. Since CSU-DSUs must adhere rigidly to network specifications and FCC regulations, you might think that one vendor's product is compatible with the products of another vendor, and that they can be mixed on the same network, and you would be right — up to a point. Some vendors have added proprietary-testing techniques to their units that cannot be used with the products of other vendors.

Recognizing that users would be implementing CSUs and DSUs for digital transmission in a manner analogous to modems over analog facilities, vendors started to add maintenance features and fault-isolation capabilities to their products as a means of differentiating their offerings from those of competitors. After all, these devices are strategically positioned on the network for implementing all kinds of tests: one side is the responsibility of the user and the other side, the responsibility of the carrier. With so many vendors involved in providing end-to-end connectivity, each providing its own brand of equipment and type of cabling, troubleshooting the link has become a complex and arduous task. CSU-DSUs are a natural and convenient point for such activities. With their built-in capabilities for self-diagnostics, as well as DTE loopback and network-loopback tests, problems can be isolated along any point of the network. These may be initiated either locally from the front panel or remotely through a network-management system, and since CSU-DSUs are separate from the DTE, their maintenance features may be invoked even when the DTE becomes inoperable. Building such capabilities into CSU-DSUs eliminates the requirement for expensive test equipment, which may, or may not, be available when you need it. For more extensive testing, some

CSU-DSUs provide jack access for external test equipment.

CSU-DSUs are equipped with various types of monitors. A digital readout provides information about T1-link performance over a 24-hour period by indicating error-free seconds. Front-panel LEDs, on the other hand, provide information about bipolar violations received from the network, loss of DTE, loss of network signal, and detection of an all-ones condition. Things work differently with a new enhancement to T1 service known as Extended Superframe Format (ESF). All ESF CSUs must compile performance data contained within ESF every 15 minutes and report it back to the network. Many carriers provide customer access to this data. This information, usually representing 24 hours of testing, may be accessed from a central-site microcomputer *via* an RS-232C port.

3.3.3 Testing Services

Aside from the self-diagnostics and loopback-test capabilities standardly available through CSU-DSUs, users can avail themselves of AT&T's Automatic Bit Access Test System (ABATS) and Customer Test Service (CTS). With the ABATS service, the user places a call to a test center which operates in conjunction with local telephone companies. There, technicians check the customer's entire network right down to the DTE interfaces of the CSU-DSUs, making this option particularly suited to smaller organizations lacking appropriate in-house expertise.

CTS is available over interstate DDS circuits only. It is accessed through an asynchronous terminal over a dial-up connection. The user draws upon the service provider's computer resources to identify a problem on the network and isolate its cause by initiating loopback tests in both directions, as well as self-diagnostics on the CSU-DSUs.

Tests initiated through ABATS and CTS, however, are disruptive, since they are conducted in-band. To lessen the impact, users may schedule circuit testing for off-peak hours, but to take advantage of these services, you must have CSU-DSUs that are compatible with them.

If your network is heavily reliant on DDS, however, you may not like the idea of having to rely further on the service provider to resolve problems on your network. It may take several calls just to reach the test center, and it may take half a day or more to get the test performed, during which time your network is operating with degraded service, or no service at all.

In areas where Dataphone Digital Service with Secondary Channel (DDS-SC) is available, services like CTS may not be necessary. Over the secondary channel, network surveillance can be performed on a continuous and nondisruptive basis. Depending on your location, though, you may have to be content with CTS until Secondary Channel becomes available. Even when this service becomes available in your area, you may have to replace your installed base of older CSU-DSUs, if they cannot be upgraded to achieve compatibility.

If CTS is inadequate and DDS-SC is not available, there are other options you may consider. Certain CSU-DSUs, like Datatel Inc.'s (Cherry Hill, N.J.) DCP 3800 and Teleprocessing Products Inc.'s (Simi Valley, Calif.) Multidrop Network Management (MNM) system, provide a virtual secondary channel on conventional DDS circuits by multiplexing test data along with user data. The multiplexing technique used by Datatel makes the tests nondisruptive at only a slight sacrifice in throughput, since bit-robbing is used to carry out the tests. The MNM system from Teleprocessing Products uses statistical multiplexing to derive the virtual secondary channel. During peak-traffic periods, though, this technique can interrupt testing when data are held in the buffer of the stat mux until enough bandwidth becomes available. Even though testing might not be continuous as with DDS-SC, the MNM system may be the closest thing to it.

Some vendors' 1.544-Mb/s CSU products support both Extended Superframe Format (ESF) and Binary 8-Zero Suppression (B8ZS), which are available on Accunet. ESF extends the normal 12-frame superframe structure of the T1 with one that is 24 frames long; hence the name "extended superframe format," or ESF. This creates two additional signaling bits in frames 18 and 24, which take a combined 6 kb/s from the T1 composite bandwidth. Of this, 2 kb/s are used for cyclic redundancy checking (CRC) and 4 kb/s for out-of-band signaling. With CRC, the circuit is continually monitored for errors, which are counted and output at a terminal or display unit. This information can be used to support complaints to the carrier about poor-quality lines. Some users balk at the notion of giving up an additional 6 kb/s to the carrier, over and above what they already give up to the carrier and lose to byte-interleaving multiplexers, but the loss of an additional 6 kb/s of bandwidth for ESF may be a small price to pay for a substantial gain in overall service quality.

B8ZS is a line-coding format used on T1 lines that uses bipolar violations to encode strings of eight consecutive zeros. This coding format insures that minimum pulse-density requirements are met without "pulse stuffing," even when there are long strings of zeros. As such, B8ZS is a method for getting around the 15-consecutive-zeros restriction. (No more than 15 consecutive zeros are permitted over T1, or timing problems will occur on the network.) The B8ZS facility monitors the data stream for zero-density violations and, when detected, substitutes the data with two consecutive 8-bit byte bipolar violations. Receiving equipment compatible with B8ZS recognizes those violations and converts the data back to its original transmitted format.

Before implementing B8ZS with an appropriately featured CSU, however, you must insure that the high-capacity transmission equipment in your area supports it. For example, M13 multiplexers are still in use at many AT&T serving offices, which may not yet have been converted to pass B8ZS.

On digital lines that are used for data, a CSU with a built-in capability to remove jitter prevents data errors that may be introduced by that type of line impairment. The permitted bit-error rate during transmission cannot exceed one error in one million bits, hence the need to remove jitter from the line. Voice is more tolerant of jitter, if only because the human "receivers" at each end can compensate for its effects to insure intelligible speech.

3.3.4 Buying Tips

Dynamic operating environments demand products with a comprehensive array of features and broad functionality. For high-density DDS applications, you may want to consider the benefits of combination CSU-DSU units. They can be used as a direct-connect DSU, in which case the internal CSU insures complete compliance with DDS transmission characteristics by providing all required equalization and line conditioning, as well as compatibility with carrier-initiated loopback tests. Alternatively, they may be used as DSUs for connection to DDS facilities *via* an external CSU, in which case the internal-CSU circuitry would be disabled. These units should also be able to handle asynchronous and synchronous applications.

The single unit CSU-DSU should accommodate any subscriber rate: 2.4, 4.8, 9.6, 19.2, or 56 kb/s. This will eliminate the need to upgrade equipment, should you ever decide to add new DDS services to your network. If Secondary Channel is not yet available in your area, or if you do not plan to take advantage of this service in the near future, you may want to base your purchase decision on the unit's upgradability to accommodate Secondary Channel *via* the addition of an option card. Be aware, though, that the Secondary Channel does not have to be used to carry diagnostic information or to download stored configurations to remote units. It may be used to provide an additional data stream to a host site, for example.

Many vendors already had CSU-DSU units that provided direct connection to 56-kb/s DDS service. When subrate services were tariffed in late 1986, new CSU-DSU units were introduced. You can save a lot of money by simply adding subrate CSU-DSU units to your network, rather than buying the latest products that accommodate all DDS rates. If you have to buy the subrate units from another manufacturer, insist on thorough compatibility testing before committing to a volume purchase. This applies even if the vendors of both products insist that they are "WECO compatible." (WECO refers to Western Electric Company, the pre-divestiture name for AT&T Technologies, which is AT&T's manufacturing arm.)

More and more functionality is being packed into the combination units. Datatel's DCP3056, for example, supports all DDS rates. Although its principal function is to interface with AT&T's Accunet Switched 56 Service, it may be used as an interface to basic DDS with either switched-56 or Megacom backup, or as a dialing unit for Megacom access. In addition, the unit may be upgraded to support 64-kb/s service on international circuits. The price is less than $2,000 per unit.

Whether you need single or combination units, you should also check for an automatic line equalizer to eliminate the need for manual adjustments, thereby simplifying installation and maximizing signal capture under all line conditions. The units should have clock circuitry that enables them to be connected to modem-extension tail circuits or limited-distance modems.

Look for DSUs that can invoke test procedures in a variety of ways: from a central location through a network-management system, locally by the front-panel switches on the units themselves, or from a data terminal *via* appropriate business-equipment interfaces, such as EIA RS-232C and RS-422A, as well as CCITT V.35. These interfaces may also be used to support various data channels. On Teltrend's (St. Charles, Ill.) DSU9000, for example, the RS-232 interface is capable of supporting a subrate DDS channel operating at speeds ranging from 1.2 to 19.2 kb/s, while the product's V.35 interface is capable of supporting a primary DDS channel operating at speeds ranging from 1.2 to 56 kb/s. Cylink Corporation's (Sunnyvale, Calif.) 4200 Advanced CSU even supports interface translation. It can convert the V.35 interface used on an IBM 3725 front-end processor's T1 port to the DS1 format supported by Accunet T1.5 Service. The Cylink unit can also perform speed conversions. The 768-kb/s output of videoconferencing equipment, for example, can be made compatible with DS1. It also converts unchannelized, high-speed data from local area networks to the DS1 format.

If the CSU-DSU cannot be centrally controlled by the vendor's network-management system, inquire as to how the unit may be upgraded for remote control, perhaps with a card-mounted receiver, and if you already use a management system on your PBX or multiplexer network, check into the feasibility of tying the CSU-DSUs into that system, rather than adding a master controller and terminal dedicated to the CSU-DSUs. While you are at it, check into the feasibility of using any V.32 modems you may already have as a backup to a CSU-DSU link.

Make sure the CSU can provide code translation, just in case the facility and the CPE are based on different line-coding techniques. That way, the CSU will continue to generate an error-free transmission that meets the ones-density restriction.

A switch-selectable choice of keep-alive signals is a useful feature of CSUs in multicarrier environments, or simply when the local carrier fails to specify which type of keep-alive signal to use. In the latter case, you can send the all-ones without framing signal, which will be interpreted by downstream equipment as the Alarm Indication Signal (AIS) representing an out-of-service condition with the trouble sectionalized to a remote part of the circuit. This will get the carrier's attention, and save you unnecessary downtime.

Look for other little extras that make network management a lot easier. General DataComm's remote digital loopback feature, for example, allows data at an unattended remote site to be looped back to the originating host location.

This feature provides fast fault isolation in point-to-point applications that do not use a network diagnostic system.

When it comes to installation, combination CSU-DSU units are obviously easier to install than two separate units, but other installation factors merit evaluation as well. If you have a high-density DDS location, you will want compact units that can be rack-mounted or cabinetized with minimum space requirements.

There are units that allow you to consolidate up to 96 units in about four square feet of floor space. In rack-mounted or cabinet installations, power requirements per unit should not exceed 15 watts. This will minimize heat dissipation and prolong the life of the equipment without the need for environmental controls.

Like modems, CSUs are available as optional plug-in cards for T1 multiplexers. This modular architecture eliminates the need to piece together network components from multiple vendors, which can reduce total system costs. It also permits incremental growth, which provides an economic way to keep pace with the growing demands being placed on your network.

For added installation flexibility, there are units that terminate lines up to 5,000 feet from the last repeater on the facility, which greatly exceeds the 3,000-foot industry standard.

CSU-DSUs should no longer be thought of as mere commodity items. Some products have been endowed with new features, innovative diagnostic capabilities, and higher levels of functionality which permit you to take advantage of new carrier services like ESF and DDS-SC. This means that buying on price alone may prove to be a costly mistake in the long run.

3.4 CIRCUIT-MULTIPLICATION DEVICES

Communications managers continue to wrestle with the problem of balancing cost with transmission quality in maintaining and expanding their corporate networks. One of the few reliable solutions for keeping pace with increases in voice and data traffic involves adding more line-termination equipment, like D4 channel banks, and leasing additional T1 lines for the backbone — all at substantial expense, but even when the funds are made available to add facilities, communications managers must grapple with the long lead times inherent in ordering, installing, and putting new facilities into service.

A variety of circuit-multiplication devices have emerged in recent years. They employ sophisticated voice-encoding schemes that double and quadruple the number of channels on the available bandwidth, allowing businesses to achieve substantial network economies quickly and easily with only a modest cost for hardware, and, depending on the application, they perform with little or no sacrifice in transmission quality. More complex encoding techniques provide as much as a tenfold increase in voice channels over a single T1 line, bringing the total number of voice channels available over T1 to 240, as with Republic Telecom Systems' (Boulder, Colo.) RLX product line which uses a proprietary algorithm for coding voice at a very low bit rate.

The circuit-multiplication capabilities are offered as optional plug-in boards for multiplexers by such vendors as General DataComm, Tellabs, and Timeplex. They are also offered as plug-ins to channel banks. The cost of such boards averages $600 each. Stand-alone circuit-multiplication systems are available for between $16,000 and $30,000, depending on the level of multiplication desired, the intended application, and the configuration.

Aside from the economic benefits that can accrue through the more efficient utilization of available bandwidth, circuit-multiplication devices provide communications managers with a fast, low-cost method of catering to their organizations' seemingly insatiable appetite for more bandwidth, and voice-compression technology is not just for large T1 networks; it may also prove advantageous for small networks utilizing only 56-kb/s DDS lines.

3.4.1 Voice Compression Basics

Despite all the innovation that is occurring in the area of voice compression, there is still only one technique that has been accepted by the CCITT as a worldwide standard (G.721) — Adaptive Differential Pulse Code Modulation (ADPCM). Although other compression techniques are vying for CCITT acceptance as worldwide standards, the reasons for ADPCM's acceptance will become apparent after a brief discussion of its relationship to Pulse Code Modulation (PCM).

A voice signal has a frequency range of 200 to 4000 cycles per second, which is expressed as Hertz (Hz). A voice signal takes the shape of a wave, with the top and the bottom of the wave composing the signal's frequency level, or amplitude. For a voice signal to be transmitted over a digital facility, like T1, it must first be converted into digital form by an encoding technique called PCM. Under PCM, voice signals are sampled at the minimum rate of two times the highest voice-frequency level, which is 4000 Hz. This translates into a rate of 8,000 samples per second. The amplitudes of the samples are encoded into binary form using enough bits per sample to keep the quantizing noise low, while still maintaining a high signal-to-noise ratio.

To accomplish this, the use of 8 binary bits per sample is required, which allows up to 256 discrete amplitude values (255 amplitude values are used in North America, with the 256th value used for polarity), or frequency levels. For quality reproduction, the required digital-transmission speed for 4-kHz voice signals works out to: 8,000 samples per second × 8 bits per sample = 64,000 b/s (64 kb/s).

The conversion of a voice signal to digital pulses is performed by the coder-decoder, or "codec," which is a key component of the D4 channel bank. The codec then multiplexes 24 channels together to form a 1.544-Mb/s signal (including 8 kb/s for control) suitable for transmission over the T1 line. PCM exhibits high quality, is robust enough for switching through the public network without suffering noticeable degradation, and is relatively simple to implement, but only allows for 24 voice channels within a T1.

ADPCM does not eliminate the 8,000 samples-per-second requirement for encoding 4-kHz voice signals. The ADPCM device, also called a "transcoder," accepts this sampling rate and uses a special algorithm to reduce the 8-bit samples to 4-bit words using 16 quantizing levels. These 4-bit words, however, no longer represent sample amplitudes, but only the difference between successive samples. This is all the information that is necessary to reconstruct the amplitudes at the distant end.

Integral to the transcoder is circuitry called the "adaptive predictor," which predicts the value of the next signal based only on the level of the previously sampled signal. Since the human voice does not change significantly from one sampling interval to the next, prediction accuracy is very high. A feedback loop used by the predictor insures that voice variations are followed with minimal deviation. Consequently, the high accuracy of the prediction means that the difference between the predicted and actual signal is very small and can be encoded with only four bits, rather than the eight bits used in PCM. In the event that successive samples vary widely, the algorithm adapts by increasing the range represented by the four bits. This "adaptation," however, will increase the noise for normal signals.

At the other end of the T1 line is another transcoder, where an identical predictor performs the process in reverse to reinsert the predicted signal and restore the original 8-bit code. Since the receiver's decoder must be able to track the transmitter's encoded signal, both must utilize identical algorithms (preferably conforming to CCITT recommendation G.721). It is worth noting at this point, that even though it is possible to use transcoders from different manufacturers at each end of the T1 link, you should avoid such arrangements if possible to insure the uniformity of remote maintenance capabilities. For large networks under centralized management, this is an important consideration, since it may lock you into one brand of transcoder, regardless of the CCITT's acceptance of the transcoder's compression algorithm.

Because voice and data patterns differ greatly, a speech-data detector is used in the transcoder to differentiate between voice and data. That way, data may pass through special data channels allocated for that purpose or switched to a noncompressed facility, while voice undergoes compression.

By halving the number of bits to encode a voice signal accurately, T1-transmission capacity is doubled from the original 24 channels to 48 channels, providing the user with a 2-for-1 cost savings on monthly charges for leased T1 lines.

3.4.2 Other Compression Schemes

There are other circuit-multiplication schemes that are commonly used over T1 facilities, such as Continuously Variable Slope Delta (CVSD) modulation and Time Assigned Speech Interpolation (TASI). Such encoding schemes have demonstrated their effectiveness in specific applications and configurations.

The higher the sampling rate, the smaller the average difference between amplitudes, and at a high enough sampling rate — 32,000 times a second in the case of 32-kb/s voice — the average difference is small enough to be represented by only one bit. This is the concept behind CVSD, where the one bit represents the change in the slope of the analog curve. Successive ones or zeros indicate that the slope should get steeper and steeper. This technique can result in very good voice quality if the sampling rate is fast enough.

CVSD, although an older technology, is still widely used. Like ADPCM, CVSD will yield 48 voice channels at 32 kb/s, but CVSD is more flexible than ADPCM in that it can provide 72 voice channels at 24 kb/s, or 96 voice channels at 16 kb/s. This is because the single-bit words are sampled at the signaling rate. Thus, to achieve 72 voice channels, the sampling rate is 24,000 times a second, while 96 voice channels takes only 16,000 samples per second. In reducing the sampling rate, however, the average difference between amplitudes becomes greater, and, since the greater difference between amplitudes is still represented by only one bit, there is a noticeable drop in voice quality. CVSD techniques can provide 192 voice channels at 8 kb/s, but the quality of voice is so poor that this level of compression is rarely even attempted. Although CVSD and ADPCM offer comparable voice quality at 32 kb/s, the compression technique of CVSD makes it more susceptible to noise. Therefore, ADPCM delivers consistently higher voice quality than CVSD.

Although ADPCM is slightly more complex to implement than PCM, it does have the robustness required to withstand limited compression-decompression as the signal goes through various switching centers of the public network, so much so that some telephone companies are offering compressed T1 to their customers.

Another advantage of ADPCM is that it maintains all signaling information when the channels are switched in the bundle mode through DACS (digital access and cross-connect system). This means that a single 1.544-Mb/s T1 pipe may yield 44 voice channels, with the remaining four channels reserved for supervisory signaling. For point-to-point applications, the robbed-bit mode may be invoked to yield the maximum throughput of 48 voice channels.

Just because two products use the same compression algorithm, however, does not mean they are compatible. There may be differences in what the vendors do with the signals after they are compressed. The multiplexers of Network Equipment Technologies (Redwood City, Calif.), for example, are not compatible with

AT&T's M44 service, which uses ADPCM to provide 44 voice channels over a single T1 circuit. NET, on the other hand, provides 47 channels.

An apparent limitation of ADPCM is that it passes modem traffic at only 4.8 kb/s or lower, but since ADPCM is used primarily for halving the cost of facilities used for voice communication, using modems on the same facility is not a wise use of bandwidth. This is because using a 32-kb/s channel for 4.8-kb/s modem traffic wastes 27.2 kb/s of bandwidth, which negates the efficient utilization of available bandwidth. To eliminate this concern, look for a feature that will detect data calls and switch them automatically to a multiplexer or noncompressed facility. Absent this feature, make sure the product at least offers the means to disable ADPCM locally at the front control panel or remotely *via* an administrative terminal connected to an RS-232C-V.24 supervisory port to allow noncompressible signals to be passed at their normal rate.

Unlike many ADPCM devices which offer the option of dedicating signaling channels for switched applications, most CVSD devices do not. Not only is CVSD unsuitable in circuit-switched environments, but, because CVSD implementations are vendor-specific, they also do not work with such AT&T services as DACS, even at 32 kb/s. Generally, CVSD is used in point-to-point applications where low cost is the overriding consideration.

Other circuit-multiplication techniques are statistical in nature. Since users are not normally able to talk and listen simultaneously, network efficiency at best is only 50%, and, since all human speech contains pauses which constitute wasted time, network efficiency is further reduced by as much as 10%, putting maximum network efficiency at only 40%.

Time Assignment Speech Interpolation (TASI) and its derivative, Digital Speech Interpolation (DSI), are statistical voice-compression techniques that take advantage of this quiet time by interleaving various other conversation segments together over the same channel. TASI-based systems actually seek out and detect the active speech on any line and assign only active talkers to the T1 facility. Thus, TASI makes more efficient utilization of "time" to double T1 capacity. At the distant end, the TASI system sorts out and reassembles the interwoven conversations on the line to which they were originally intended.

The drawback to statistical-compression methods is that they have difficulty maintaining consistent quality. This is because such techniques require a high number of channels, at least 100, from which a good statistical probability of usable quiet periods may be gleaned. With as few as 72 channels, however, a channel-gain ratio of 1.5-to-1 may be achieved. If the number of input channels is too few, as in small tie-line configurations, for example, a condition known as "clipping" may occur, in which speech signals are deformed by the cutting off of initial or final syllables. A closely related malady which afflicts this compression technique is "freeze out," which sometimes happens when all input channels are in use at the same time. In such cases, a sudden burst in speech energy can completely overwhelm the total available bandwidth, resulting in the loss of entire strings of syllables. Statistical compression techniques, then, work better in large configurations than in small ones.

Another drawback to statistical-compression techniques, even for large T1 users, is that they may easily be confounded by transmissions having too few quiet periods, such as FAX and music-on-hold, nor is the technique effective for use with voice-band, full-duplex data modems. These kinds of transmission have the cumulative effect of using twice the number of quiet periods as normal voice calls. Although statistical-compression techniques promise a 2-to-1 compaction ratio, it may be safer to hold this ratio to 1.5-to-1 to prevent noticeable degradation. In evaluating this kind of equipment for use on your network, you may want to limit your choices to the new generation of products that incorporate switching circuitry that recognizes virtually continuous transmissions and routes them off the DSI-TASI circuits.

Packetized voice is a statistical-compression technique that has received wide acclaim in the last year or two. It combines the advantages of packet switching and ADPCM. Briefly, packet-voice systems halve the amount of information to be transmitted by removing repetitive pitch patterns from the voice signal. Then voice is digitized with ADPCM and arranged into packets. Instead of dedicating an entire transmission channel to each user for the duration of the call or computer session, packet mixes all the traffic together, minus all the dead time. With each packet containing 192 bits (or 24 bytes), the required average capacity per voice channel is only 15.36 kb/s. Assuming that there is 1.536 Mb/s of usable capacity on the internodal T1 facility, each T1 line would be capable of carrying 100 voice channels utilizing packet technology (15.36 kb/s × 100 = 1.536 Mb/s).

Despite the inherent routing and networking advantages of packet technology, when it is applied to voice it has a serious drawback in the form of delay. Because the individual packets are dynamically routed to their destinations, they may arrive at slightly staggered intervals of 20 to 25 milliseconds or more, causing the listener at the distant end considerable annoyance from echo. In this case, costly echo cancellers must be deployed on the network. Like DSI-TASI, packetized voice is still vendor-specific; that is, their implementations are proprietary.

There have been many attempts to achieve much higher compression rates — not with much success, however. Although linear predictive coding (LPC) has been used by the military to produce intelligible 2.4-kb/s voice, speech quality is often quite poor, sounding more robotic than human.

The benefit of building a network around products that adhere to present and emerging standards is to protect your investment against early obsolescence. When applied to compression techniques, adherence to standards allows you to position your network to take optimal advantage of present and future carrier-based services, while you achieve bandwidth reduction goals. A network based on proprietary structures is, by default, tied to the migration path of the vendor.

3.4.3 Bottom-Line Benefits

Considering the high monthly cost of T1 lines, circuit-multiplication devices can provide a substantial savings to businesses operating private networks. For example, the current charge for a single 712-mile Accunet circuit from New York to Chicago runs about $16,000 per month ($192,000 a year), including local-loop charges at both ends. A single 2,564-mile Accunet circuit between New York and San Francisco costs about $38,700 per month ($464,400 a year), including local-loop charges at both ends. When the need arises to add circuits, compare these costs with the one-time expenditure of about $16,500 for a complete ADPCM system which can double the T1 channel capacity between these cities. In the case of the Accunet circuit between New York and Chicago, the payback period is only one month, *versus* the cost of an additional circuit from AT&T. With the New York to San Francisco circuit, the payback on a transcoder is about two weeks!

With ADPCM transcoders directly interfacing T1 facilities at the DS1 level, they are compatible with a broad range of digital-transmission media, including cable, fiber optics, microwave, and satellite. Additional savings can be achieved through the use of higher order digital multiplexers between ADPCM devices. For example, the output of two ADPCM devices can be multiplexed together over a single T1C carrier (3.152 Mb/s), increasing channel capacity fourfold. With a T2 microwave link (6.312 Mb/s), channel capacity may be increased eightfold to 192 channels and, over a T3 satellite link (44.736 Mb/s), it may be increased 56 times to provide 1,344 channels. ADPCM delivers exponential channel capacity to the largest private networks, resulting in even more savings from dramatically reduced transmission-facility and satellite-link investments.

All this is fine for large private networks, but what about smaller users who cannot justify T1 to begin with, let alone the cost for circuit-multiplication hardware? One possibility might be to use 56-kb/s service for voice transmission. A new class of products, like Republic Telecom Systems' RLX series, uses less than a full DS1 facility to provide multiple voice channels. Using Digital Speech Interpolation to eliminate silent periods, plus its own proprietary compression technique, eight conversations may be carried over a single 56-kb/s circuit. Eight voice-grade circuits from New York to San Francisco, for example, cost about $12,300 a month. A 56-kb/s circuit between these cities costs less than $6,600 a month, which roughly translates into a 50% savings in line costs. The cost of two RLX units to implement compression is approximately $30,000. The payback period on hardware is about five months, demonstrating that users can take advantage of voice compression, even if their bandwidth requirements fall far short of T1.

The cost comparison between eight voice-grade lines and a 56-kb/s link from New York to Chicago is quite revealing. The 56-kb/s circuit costs less than $3,700 per month, whereas eight voice-grade circuits between the two cities cost about $7,800. With a 50% savings in line costs, the payback period on the necessary hardware is about 7.5 months, demonstrating that voice compression is worthwhile for smaller networks, even over shorter distances.

3.4.4 Other Considerations

With the addition of circuit-multiplication devices to the private network, one might suspect that the testing of T1 facilities becomes more complicated, since the source of a fault may be with the line itself or with the added equipment. After all, a lot of man-hours may be consumed before the source of the fault is finally determined and appropriate restoral action taken. To insure that this does not happen, you should compare the various products for their integral diagnostic and self-test capabilities.

Some ADPCM devices, for example, feature "loopback" points that allow operational verification of individual T1 channels or the entire T1 link. Using a specialized test mode, the user may verify that the entire near-end ADPCM system and individual channels are operational. In another test mode, the user may verify near-end system operation as well as the operation of the entire transmission link. Still another test mode may allow verification of both near and distant transcoders *and* the entire link. End-to-end voice quality may be verified through an analog-digital-analog (A-D-A) voice port.

Look for self-test diagnostics that may be performed at any time with no disruption of voice, and, if self-testing disrupts data throughput, find out to what extent. A two-to-four millisecond delay imposed by self-test implementation is considered acceptable.

Another feature to look for in a circuit-multiplication device is an automatic internal bypass which insures that the original 24 channels of a T1 line remain available for use in the event of catastrophic system failure. This feature is especially important when one considers the revenues and productive working hours that may be lost during downtime and the confusion that often results when the transmission of critical information is disrupted.

In evaluating such products, look for the ability to configure the system's operating parameters, monitor the status of those parameters and alarm conditions, and access the system's loopback modes and self-test diagnostics. All of these should be implemented from the unit's front panel, as well as from a remote terminal *via* an RS-232C/V.24 port. The data-transfer rates for that port should be user-selectable from 300 b/s to 9.6 kb/s. Make sure that port can also be used to daisy-chain the devices so that they can be controlled from a single terminal. This will also simplify cabling when a number of compression devices are required and ease installation of additional units.

If your configuration includes long-haul terrestrial or satellite voice circuits, find out if the vendor offers a per-channel, digital, echo-cancellation option that can be used at both ends to provide for two-way cancellation of the talkers' echo.

Cost savings aside, most circuit-multiplication devices are easy to order and install, as well as to operate and maintain. Depending on location, some T1 circuits may take as long as 3 months to order and put into service, compared with 30-day off-the-shelf delivery for most compression devices.

This points to another benefit of compression devices: they permit greater latitude in expanding T1 facilities, since channels may be added without the need to order new facilities. Because most of these products are modular, T1 capacity can be increased as needed simply by adding system modules. This eliminates the long lead times required to order T1 circuits to meet forecasted requirements, and if future transmission requirements fail to materialize as forecasted, opting for the circuit-multiplication technique will reduce the effect of the error, since a penalty charge may be levied if ordered T1 lines are canceled prior to the start of service. Canceling an Accunet T1.5 circuit after test and acceptance, but before the due date, for example, will result in a penalty charge of $2,065, consisting of $1,335 for the interoffice channel and $365 for each of the two central-office connections. Over the years, AT&T has been quite persistent in seeking FCC approval for proposed tariff changes that would increase such charges.

At the same time, voice compression products can be relocated quickly to ease unanticipated network congestion. Installation of compression devices may be as simple as mounting an additional shelf onto the line-termination equipment rack, inserting the required system modules, and wiring these modules to the channel bank assembly — an easy task for an experienced technician. For an extra charge, some vendors will prewire an equipment rack when customers order both D4 line-termination equipment and the transcoders, and then pretest the entire system before shipment.

What about quality and reliability? High-tech concoctions have been touted in the past, with vendors promising huge cost savings and very attractive payback periods that were eventually negated through poor product performance. With statistical methods of circuit multiplication, quality depends on many factors, including the total number of channels and the nature of the communication on those channels. As a result, quality may be high, just not consistently high. With CVSD at 16, 24, or 32 kb/s, as well as 10-to-1 compression techniques, quality is really a matter of perception. Since perception will determine the level of acceptance of the product among your network users, you should work with the vendor to install a demonstration system. This will give you the opportunity to check the vendor's performance claims, while allowing you to gauge the acceptance level of network users. After a reasonable prove-in period of about 30 days, you can either buy the product or tell the vendor to pull the plug and get out. At the same time, consider the advantages of compression for routine, intracompany voice communication, while allowing communication of a critical nature to pass uncompressed.

Of the many compression techniques currently in use, only ADPCM has been accepted as a standard. Thousands of subjective tests conducted by Bell Labs and independent research firms have confirmed that the compressed (4-bit) signals used in ADPCM are virtually indistinguishable in quality from the original 8-bit signals under PCM, and ADPCM holds up well in the multinode environment, where it may undergo compression and decompression several times before arriving at its final destination. The same cannot be said of other compression techniques, which may still prove worthwhile for point-to-point applications.

Although users may experience performance problems with transcoders from time to time, the source of the problem is not the ADPCM concept itself, but, ultimately, poor product design. This points to the need to evaluate the vendor carefully before committing to a purchase. Also keep in mind that, as in any other type of product, transcoders differ from vendor to vendor in such things as features and ease of installation, as well as price and warranty period. Just because ADPCM is a standard does not mean any vendor's transcoder will do.

Circuit-multiplication solutions may be considered viable if they meet the following criteria:

- Result in accelerated payback;
- Meet end-user needs efficiently;
- Provide an acceptable level of voice quality for the specific application;
- Are immediately available.

Some questions that may arise during the purchase process are, what will happen to this investment when ISDN becomes widely available and will the bandwidth glut created by the emerging fiber-optic networks eliminate the need for circuit-multiplication techniques altogether?

ISDN promises intelligent, end-to-end, digital connectivity on a network that will support simultaneous voice and data. Bell company field trials have been going on since 1986, and basic service (2B + D) lines have been in operation in several regions on a limited basis since early 1988.

Under AT&T's concept of "nodal architecture," ISDN services will become available through switching nodes accessible to end users *via* T1 links. The 24 channels available over T1 correspond to the 23B + D primary-rate service of ISDN. Under ISDN, each of the 24 channels can be defined as a special type of circuit to suit the constantly changing needs of individual users, not only giving customers total control over bandwidth allocation, but service definition as well — all with minimal reliance on the telephone company. Thus, some channels may be defined one minute as AT&T Megacom circuits and be changed the next minute to Megacom 800 circuits. AT&T calls this ISDN service, Call-by-Call Service Selection. Another ISDN offering from AT&T is Info-2, which lets users identify incoming calls by telephone number. Both services were announced in April 1988, with initial availability in 18 cities. Eventually more services will be added to ISDN.

The thing to remember about ISDN is that it will be implemented over switched facilities. T1 already allows private-network users to allocate bandwidth any way they choose — and ADPCM enhances the flexibility of network planning very economically. While ISDN will become a powerful tool for managing corporate voice and data communications, it should be viewed, at least initially, as just another option with which to design, manage, and maintain the high-volume digital backbone network. Although service providers could conceivably make ISDN so attractive as to make reliance on T1-based private networks economically unjustifiable, the emergence of fiber-optic networks offering virtually unlimited bandwidth should act as a deterrent.

Even though ISDN promises to evolve into an economical and viable tool with which end users can gain more control over their networks, another factor to consider about ISDN is that it will be a service provided largely by telephone companies. In using ISDN, private-network users may have to give up some of the operational freedom that they have grown accustomed to and be at the mercy of telephone companies and interexchange carriers for timely and reliable installation, repair, and maintenance services, as well as become dependent for features and network options. Despite the fact that some large companies have already committed themselves to ISDN, for the vast majority of companies it will be viewed as a pilot technology until the mid-1990s. Even then, ISDN will involve high up-front costs, mostly for premises equipment, including digital interfaces and adapters for nondigital equipment.

Users who allow telephone companies and interexchange carriers to lure them away from private networks to switched services may find themselves locked into a single-vendor solution and lose the bargaining power they now have with private networks. As with T1 circuits, corporations would be at the mercy of telephone companies and interexchange carriers for priorities and schedules for ISDN-circuit installation and cutover.

Becoming overly dependent on any one service provider is a prospect few private-network users would welcome. It would not be the wisest decision, therefore, suddenly to abandon T1 in favor of ISDN, even if that were to become possible. The reality of ISDN is that such services will evolve gradually, taking into consideration the available capital investments of the service providers. Consequently, T1 and ISDN will coexist far into the future, making voice-compression products all the more important.

As fiber-optic networks continue to evolve and become operational, the number of T1 circuits sending voice, data, and video around the country will skyrocket. This portends lower rates for long-haul inter-LATA T1 circuits, which may prolong the payback period of circuit-multiplication systems. The fiber-optic networks, however, as currently layed out, are only "backbone" networks that will link major cities. Such networks may not coincide with the backbone facilities of private networks; not even fiber is economical for reaching into every nook and cranny of the country. Circuit-multiplication systems will continue to be cost-justifiable for far-flung private networks and small users for the foreseeable future. In fact, compression products may play an expanded role in providing economic short-haul access to fiber facilities.

For businesses burdened with the escalating costs of operating their own networks, implementation of voice-compression technology represents a reliable, economical alternative to leasing additional facilities. On the flip side, compression technology provides a high degree of flexibility to network planning — with only minimal reliance on telephone companies.

3.5 DIGITAL CROSS-CONNECT SYSTEMS

The proliferation of T1 links over the years has created the need for devices to simplify their administration and control, facilitate the ability to access and test individual circuits, switch channels from one link to another, and reconfigure circuits to achieve optimal efficiency and cost savings. The Digital Access and Cross-Connect System (DACS) does all this, and more.

Developed by AT&T in 1981 to streamline its internal engineering and operations activities, the DACS was designed to automate the entire process of circuit provisioning. Instead of having a technician manually patch the access line to the long-haul transport facility, for example, the DACS allows customer-ordered circuits to be set up between two points from a remote location *via* keyboard command, and, since connections are software-defined, customer reconfigurations can be implemented in a matter of minutes instead of days or weeks.

3.5.1 AT&T's DACS-CCR Service

The scheme worked so well that in 1985 AT&T started offering these capabilities to its Accunet-T1.5 subscribers through a tariffed service called Customer Controlled Reconfiguration (CCR). Instead of waiting for service orders to be processed before they could rearrange their networks, users now have the capability to do it themselves *via* a terminal connected to AT&T's central CCR-control system.

The control system is the heart of the DACS network. It communicates instructions from the user to the various DACS switches that will be involved in implementing the routing changes so that the necessary cross-connections may be performed. This capability is particularly useful for disaster recovery, bypassing critical systems so that scheduled preventive maintenance may be performed, and for handling peak-traffic demand.

The reconfiguration activity is typically done at 1.2 kb/s *via* a separate leased-line or dial-up connection. The changes now take effect within five minutes of being entered by the customer, rather than thirty minutes later, as when CCR was first introduced. The central control's memory holds network routing maps which allow users to invoke predefined alternate configurations with only a few keystrokes at the CCR terminal. Customer routing maps are stored in Freehold, New Jersey, in a central computer which is capable of supporting every DACS with CCR in AT&T's network.

Another benefit of DACS is that it may be used in conjunction with other AT&T services like M24, a tariffed, central-office multiplexing service. The user may consolidate traffic from multiple locations and transport it to an M24 site *via* a T1 link. From there, an Accunet T1.5 access line brings the traffic to the DACS, where the individual channels may be switched to as many as 24 separate locations. This assumes, of course, that the user has a multiplexer at his premise that is compatible with the M24 service and that traffic patterns justify this kind of arrangement.

DACS can also be used to improve transmission quality by removing "jitter" from channels, an impairment that affects the integrity of the bit stream as it moves through repeaters and switching equipment. Jitter is the difference between where a pulse is and where it should be. It has a variety of causes, including transmission-system noise and transmission delay. Although the DACS rids the line of jitter by resynchronizing the transmission to improve the quality of the bit stream, this capability should not be the major factor in your decision to subscribe to AT&T's DACS or a telco-provided cross-connect service because other special-purpose products are available to solve impairment problems at far less cost.

Unlike the central-office switch, which sets up, supervises, and tears down communication paths every time a call is placed, DACS allows communication paths to remain in place for continuous use over a period of months, or even years, but for networks with constantly changing needs, circuits may be added, deleted, or rearranged as demand warrants. This capability can result in tremendous cost savings, because it eliminates the tendency to over-order facilities to meet any contingency. AT&T charges $387.50 a month for each circuit termination using CCR.

Since the introduction of DACS, it has been enhanced with myriad sophisticated features and capabilities that have made it an indispensible management tool for large private networks. The most sophisticated cross-connects provide three levels of switching. At the DS1 level, the entire composite of 24 channels may be switched from one T1 facility to another. At the DS0 level, individual 64-kb/s channels may be switched from one DS1 stream to another, and others inserted in their place. In other words, one or more DS0s may be dropped off at an intermediate location, while the others continue on to the distant location. While some DS0s are dropped at the intermediate location, others may be inserted into the bit stream at that time for transmission to the distant location.

Some cross-connect systems even perform switching at the sub-DS0 level, allowing individual subchannels operating at 2.4, 4.8, or 9.6 kb/s to be bundled into one 64-kb/s channel and then separated at another cross-connect for individual routing to their respective destinations. Together, these capabilities provide network managers with the same level of control over their private facilities that, until a few years ago, resided solely with AT&T.

It is quite common for corporations to use a mix of switched and nonswitched services on the same network. Cross-connect systems may be used both to consolidate different types of traffic over a single facility and to separate the mix of traffic for routing over specialized outgoing links. For example, a T1 link coming into the cross-connect system carrying a mix of seventeen switched channels and five nonswitched channels (e.g., Dataphone Digital Service) can be "groomed" so that the switched channels are routed through the outgoing facility dedicated to that type of service, while the nonswitched channels are combined with those of the same type from another source onto a facility dedicated to such services. Separating channels in this way increases the "fill" of specialized facilities for direct routing to a particular destination. This makes for a more efficient utilization of bandwidth, which results in greater cost savings.

By 1986, more than a dozen other manufacturers were offering cross-connect systems, not only to telephone companies for service resale, but to private networks as well. In fact, the sophisticated capabilities of today's new-generation digital cross-connect systems has made them a valuable adjunct to nodal T1 multiplexers, and the modularity of these products permits port capacity and features to be added incrementally as needs require. Scaled-down cross-connect systems make them affordable to even small networks.

3.5.2 Subscribing to Carrier-Provided Service

DACS-CCR or some variation of AT&T's service is now offered by interexchange carriers and telephone companies, marketed under different names, of course. U.S. West's subsidiary telephone companies, for instance, offer a cross-connect service called "Command-A-Link." Not only does the cross-connect make circuit provisioning more reliable and less costly than manual methods, it also takes some of the processing load off the central-office switch, enabling the telephone company to offer a variety of new services to business customers without diminishing the quality of service to residential customers. In this sense, the cross-connect may be viewed as a front end to the central-office switch.

Since the cross-connect actually performs the routing function, the central-office switch is free to do other things, like implement more features or process more calls. With a cross-connect system, telephone companies can postpone expensive upgrades to digital switches, which might normally be required to add call-handling capacity, lines, or custom-calling features. For telephone companies with analog or first-generation digital switches, the addition of a cross-connect system can put off huge capital outlays for new switches indefinitely, while adding diagnostic precision and control to the network *via* remote-test access, and since the cross-connect system is separate from the central-office switch, telephone companies are eager to provide customers with direct access through CCR, letting them assume responsibility for their own leased facilities.

Further, a single cross-connect system may be shared among many customers, allowing the telephone company to offer "virtual private networks" even to small companies, many of which appreciate the benefits of private networks, but still cannot afford to implement multinode configurations of their own.

Hubbing off the central-office cross-connect system might make a viable alternative to setting up a separate network node of your own, because the arrangement can provide all of the control and flexibility of a private network, but without the huge capital investment in equipment and the risk of obsolescence.

The carrier-provided service offers a high degree of configuration flexibility. Instead of connecting all sales offices in a region *via* dedicated facilities, for example, a single T1 span from each site to the local central office would allow company-wide voice, data, and video communication through the telephone company's cross-connect system, thus facilitating the flow of vital information. *Via* a multipoint bridge, the headquarters office can be connected to all sales offices at once for E-Mail broadcasting of current price and product-availability information, or for voice conferencing with branch managers. Corporate reports requiring immediate action may be batch FAXed to all locations simultaneously, and instead of gathering the regional sales force at a central location, video transmissions can relay product demonstrations and installation procedures at a fraction of the cost. Because today's business environment requires that information be shared among multiple locations, companies that can do this in a timely and efficient manner have the competitive advantage over those that cannot.

Telecom personnel at the company's main office would be able to alter the circuit configurations as needed to control the cost of telecommunications. During peak hours, for example, the cross-connect may be configured to make more bandwidth available to accommodate the increase in traffic. If traffic reports indicate trends in usage by hour-of-day or day-of-week, this information can be stored in a microcomputer. Primary, secondary, and tertiary configurations may then be uploaded to the cross-connect system with only a few keystrokes.

In addition to obtaining a suitable hubbing location for your private network at tremendous cost savings, you have access to standby digital facilities through the cross-connect in the event of catastrophic failure. This "bandwidth on demand" eliminates the need (and the cost) for dial-backup systems, which are typically configured to give you access to the public switched network in case of deteriorating conditions or line outages. In such cases, the cross-connect enables the telephone company to bill you for bandwidth actually used, instead of having to settle for the uncertain line quality inherent in the public network, and, if you do not have the diagnostic know-how to get the facility back into service, the telephone company may access and test your downed line through the DACS to isolate the cause of the failure. For a charge, the telco will perform tests and verification procedures to insure the proper operation of circuits before turning them back to you.

The cross-connect systems and associated services offered by carriers are typically shared among several customers. For this reason, they offer password protection to prevent one user from inadvertently accessing the network of another user. Passwords also protect each customer's network against unauthorized intrusions from within. In being able to partition the cross-connect system in this way, carriers can provide businesses, large and small, with the means to control their own facilities without the capital outlay for nodal systems — and without having to invest too heavily in a whole team of experts to maintain them.

In subscribing to a telephone company's cross-connect system, companies with expensive digital PBXs would be able to achieve even greater economies. The PBX could be used merely to pipe all voice and data traffic directly to the central office *via* T1, where the cross-connect routes the individual channels to their appropriate destinations. This would greatly reduce the switching burden of the PBX, especially during peak hours, while permitting its more efficient utilization for other memory-intensive, call-handling features, as well as sophisticated options like voice mail.

In determining whether to subscribe to a carrier's cross-connect service, you should inquire about the degree of modularity inherent in the system. This kind of modularity is important to you, because it will affect the carrier's ability to meet your changing needs and your ability to take advantage of new types of service as they become available. Of course, when buying a cross-connect of your own, it is advisable to select a product with a high degree of modularity for the same reason.

The modularity of today's sophisticated cross-connects allows carriers to buy only those features and ports that they can immediately sell to customers. If you have the requirement for low to high speed DDS, for example, the carrier can purchase subrate multiplexer modules that multiplex and demultiplex channels running as slow as 2.4 kb/s. The Tellabs 5332 Subrate Multiplexer Module, for example, may be software-configured to provide three types of subrate services *via* the Tellabs 532 TCS T-Carrier Cross-Connect System:

- With *Type 1* service, the module multiplexes and demultiplexes twenty 2.4-kb/s DS0A channels onto a single 64-kb/s DS0B channel.
- With *Type 2* service, the module multiplexes and demultiplexes ten 4.8-kb/s DS0A channels onto a single DS0B channel. In this mode, the module functions as two independent 4.8-kb/s subrate multiplexers.
- With *Type 3* service, the module multiplexes and demultiplexes five 9.6-kb/s DS0As onto a single DS0B, functioning as four independent 9.6-kb/s subrate multiplexers.

The Tellabs 5332 Subrate Multiplexer also provides the cross-connect system with the ability to perform subrate cross-connections. For example, after the 5332 Module has demultiplexed a DS0B channel into several DS0A channels, the DS0A channels are switched by the cross-connect matrix and then multiplexed into the appropriate DS0B channel by another 5332 circuit.

In another example of cross-connect modularity, the Tellabs T1C Interface Module allows the cross-connect system to interface directly with T1C (3.152 Mb/s) lines, eliminating the requirement for an external M1C multiplexer. Not only is this interface module compatible with M1C multiplexers and D4 channel banks operating in the asynchronous T1C format, a software-selectable option allows the module to operate in synchronous T1C format as well.

Also available for the Tellabs 532 T-Carrier Cross-Connect System is an Asynchronous DS3 Interface which multiplexes and demultiplexes up to 28 T1 lines over a single DS3 link. In addition to integrating DS1 and DS3 facilities, this interface offers single-point network control for all DS0, DS1, and DS3 circuits.

Although DACS-CCR is becoming a popular service offering among telephone companies, some of them are still locked into predivestiture modes of thinking that may inhibit their responsiveness to customer needs. Some telephone companies may even try to convince you that some DACS implementations are not possible, citing "conflicts with long-standing operational philosophy." Therefore, you may have to do some gentle arm-twisting to get what you really want. In discussing your specific needs with the telephone company, you may open their eyes to a service offering that can be marketed to the rest of their corporate customers. After all, telephone companies are continually looking for ways to enhance their reputations for offering advanced technology and innovative services, both to attract new business and to compete with each other for the attention of potential investors. You can leverage this situation to your best advantage.

The experience of Hewlett-Packard is illustrative of the new-found leverage that users have with telephone companies in the postdivestiture era. In upgrading its Northwest area network between 1986 and 1987, Hewlett-Packard found itself without a suitable hub location from which it could economically route traffic among its numerous locations in Washington and Oregon and consolidate traffic from among those locations for long-haul transport to its headquarters in Palo Alto, California. The company considered the possibility of hubbing off a Pacific Northwest Bell (PNB) centrex exchange in Portland, which included a cross-connect capability that would allow corporate customers the means to change the routing paths of individual T1 subchannels, channels, or groups of channels, according to immediate or long-term needs.

PNB's cross-connect system, however, was partitioned among several corporate customers, and it did not permit the separation of maintenance and diagnostic information by individual subscribers. This meant that PNB could not pass this information to Hewlett-Packard without violating the confidentiality of the other cross-connect users. Without real-time access to the cross-connect's alarms and status reports, however, Hewlett-Packard would not have had the means to identify deteriorating line performance and to avoid catastrophic failures.

In acquainting PNB with its problem, Hewlett-Packard proposed leasing floor space at the central office for its own cross-connect system that it had planned to purchase from Tellabs. PNB, however, was not sure about the legal ramifications of colocating CPE. Furthermore, it had not yet devised a plan for implementing Open Network Architecture (ONA), which was a relatively new concept.

Appreciating the need to offer creative solutions to grow and to protect its base of corporate customers, PNB offered Hewlett-Packard a deal it could not refuse: PNB would buy the Tellabs cross-connect for exclusive use by Hewlett-Packard under a lease agreement. The arrangement has worked so well that PNB is eager to serve other companies in the same way. The Hewlett-Packard experience demonstrates that the peripheral issues of security, confidentiality, and lines of responsibility can be resolved for the mutual benefit of the telephone company and the customer, even without ONA.

To succeed in an operating environment that promises only to become more competitive, telephone companies are becoming more market-oriented, not necessarily with the idea of increasing market share, but with the idea of protecting their revenue bases against further erosion. In subscribing to a telco-provided cross-connect service, however, there are several factors to consider.

Be aware that you are entering into a marriage with the telephone company (or interexchange carrier). Consequently, you cannot suddenly exit the arrangement if things do not go as anticipated without seriously disrupting your organization's communications system. Furthermore, the long lead times required to plan, order, and install your own cross-connect system, as well as find a suitable location for it, can make dependence on the telephone company's service a long-term affair. To minimize your risks, you must try to anticipate future problems by thoroughly discussing issues of installation, maintenance, operation, and expansion with the telephone company. At the same time, you must know the cost of withdrawing from the arrangement if it is in your best interest to do so. You cannot afford to take anything for granted.

You must also define and agree on the scope and level of responsibility of the telephone company. For example, if you plan to rely on the telephone company's digital facilities as a backup link for mainframe access in the event one of your facilities fails, you must not only inquire about the availability of such links, but also about what recourse you have if a backup facility is not available when you need it. Despite assurances, there is really no way a telephone company or carrier can guarantee that it will have the capacity available to switch all your T1 circuits in the event of disaster. Even if it has disaster-recovery swapping arrangements with other carriers over their fiber networks, there is really no guarantee that they will have all the bandwidth you need when you need it, and, since it is possible that the telephone company's cross-connect may suffer a catastrophic failure, however brief, you should also find out what measures have been taken to guard against that eventuality and what penalties can be invoked in case you lose critical facilities as a result of a cross-connect failure.

Some leading-edge cross-connects, like AT&T's DACS III, can automatically reconfigure around a failed link in as little as 60 milliseconds and switch back to the restored circuit in 250 microseconds, but before opting for this feature you have to decide whether you really want to give up this level of control on your dedicated circuits. The risk is that the carrier may reroute to a poor-quality circuit or that the circuitous route may introduce intolerable delays which can disrupt both voice and data traffic, all of which may just compound your original problem. Beyond that, when the carrier switches back to the restored circuit, the delay, however slight, may be enough to disrupt transmissions in progress.

You also have to be cautious about making direct connections to any cross-connect system operated by a carrier. This is because you may be forced to take network timing from that carrier. If you use multiple carriers for dedicated circuits, be aware that each carrier may have a slightly different timing network, in which case jitter can become a real problem.

In subscribing to a telco or carrier-provided DACS service, be prepared to do without real-time management information in the form of status reports. For the most part, service providers do not want you to know what is going on with equipment located on their premises. Furthermore, the M13 multiplexer that resides on the front end of most cross-connect systems is very limited in the kind of information it provides. Until telcos and carriers changeout the M13s in favor of more sophisticated multiplexers that can be partitioned among many users to yield detailed status information, you will be giving up a good deal of control over your leased facilities. Fortunately, the M13 changeouts are well under way.

Thoroughly evaluate the potential for lost opportunities before committing to hubbing off the central-office cross-connect system. This means comparing the cross-connect services of AT&T, as well as the other interexchange carriers, to find out what additional features and capabilities they will offer within the next year or two. You should also inquire as to the locations of future DACS nodes, and the timetables for their cutover. Look into the features and capabilities of vendor products for the purpose of weighing the advantages of implementing your own cross-connect system. There is nothing so frustrating as making a snap decision based on a telephone company's sales presentation, only to find out a few weeks after implementation that another carrier is about to offer more cross-connect functionality and reach more locations — and at a better price than you are currently paying.

Make sure that the telephone company will provide the necessary training on the use of the CCR terminal to order new facilities and reconfigure existing facilities. Although most cross-connect systems feature CCR systems that are menu-driven, it may take a while for operators to get comfortable using such systems. After all, taking responsibility for rearranging an entire network, or even a small segment, with only a few keystrokes can overwhelm anyone, until they get accustomed to the idea. Determine to what extent follow-up training is available from telco service personnel. A customer support hotline should also be available so that you can call for assistance at any time of the day or night.

3.5.3 Private-Network Cross-Connect Systems

AT&T DACS and telco-provided cross-connect services are not yet available in many areas. Even when they are available, they may not offer real-time access to diagnostic information and status reports. In such cases, there are a variety of scaled-down cross-connect systems available that may fit your needs.

Check into the cross-connect's modularity, as applied to features as well as port capacity, to insure that it meets your immediate and long-term needs. You can purchase cross-connect systems with only a single digroup, which equals twenty-four 64-kb/s DS0 channels. It can be expanded at the rate of one digroup at a time, making this kind of modularity equally advantageous for both high-density locations where bandwidth demand may proceed in unpredictable start-stop fashion and low-density locations characterized by relatively limited demand.

Establish a node that is strategically positioned to decrease the distance of interstate T1 links. At the same time, recognize that it is not necessarily the biggest site that counts most. The most important factor in locating your cross-connect site may not be distance at all, but the volume of traffic and the patterns of that traffic.

Some cross-connect systems, like Coastcom's (Concord, Calif.) DXC-8, have enough memory to store up to ten network configurations so that circuit rearrangements may be invoked automatically by time of day using an internal clock as the controlling mechanism, similar to the way in which data switches activate mainframe port reconfigurations by time of day. Telecommunication Technology, Inc. (Sunnyvale, Calif.), even has three levels of memory backup to protect the cross-connect maps, which typically take hours of meticulous planning to prepare. Even if external power is disrupted, the map restores itself automatically from nonvolatile memory when power is restored.

Find out if the vendor's cross-connect has redundancy in all critical circuits, including the switching-network modules, system clock, and synchronization circuits, as well as the power converters. Check for self-monitoring software that continually verifies system performance and facility integrity.

In evaluating various cross-connect systems, you should be looking for an administration terminal capable of performing a range of network-management tasks. These tasks fall into several basic categories:

- *Facility monitoring,* which lets you request slip- and framing-loss counts and reset slip- and framing-loss counters for individual digroups, or for all digroups terminated on the cross-connect. The system should monitor and automatically report the occurrence of a facility failure.
- *Diagnostics,* which includes the ability to initiate tests while circuits are in service or out of service.
- *Connection service,* which includes the ability to set up or remove connections between individual DS0 channels or sequential groups of channels of various cross-connect digroups.
- *Test access,* which defines the test ports on the cross-connect system, invokes the test-port configuration, and makes or removes test connections.
- *Removal and restoral of service,* which removes circuits from service and restores them to service when removing or inserting various cross-connect system modules. During restoral of service diagnostics should be run automatically.
- *Utilities,* which request status reports on facility performance and system status.
- *System configuration,* which allows you to add or remove T1 interfaces and switching-network modules in preparation for expanding or contracting the network or network segment.
- *Miscellaneous,* which includes the ability to perform such tasks as managing data bases, retrieving audit reports, and recovering the cross-connect system from error conditions.

Network Equipment Technologies (Redwood City, Calif.) makes adding cross-connect systems to the network a very simple undertaking. When a node is added, it automatically configures itself by interrogating each of the attached links for their routing tables. With this information, the node builds its own routing table. This application of distributed control requires no human intervention, which greatly reduces operating costs.

Finally, do not overlook the benefits of using the smaller cross-connects on your network in conjunction with AT&T's DACS or telco-provided cross-connect systems to build a two-level hierarchical network. Under this arrangement, several low-end cross-connects would multiplex and switch bit streams onto a backbone network controlled by only a few high-end cross-connects, but even if this is not an immediate concern, you should position your network for that possibility by choosing hardware that is DS0-compatible, has a future growth path, and has a solid performance record in both public and private networks.

3.5.4 Innovative Management Capabilities

Cross-connect systems will continue to grow in capacity, while integrating more features and functionality in response to the demands of a rapidly growing user community. Already, cross-connect systems are available that switch channels between T1 and T3 facilities, but innovation continues at the low-end as well, as demonstrated by AT&T's offering of Bandwidth Management Service (BMS), a new Accunet T1.5 option that allows users to reconfigure their networks quickly, as well as monitor and test their T1 facilities.

BMS provides Accunet T1.5 customers with many of the same features as CCR, but only offers control at the DS0 level, instead of at the sub-DS0 level. Instead of the 1.2-kb/s link required with CCR, BMS permits communication between the customer's terminal and the system controller at the AT&T serving office at 9.6 kb/s. This cuts the average time to implement reconfigurations from five minutes under CCR to only one minute under BMS. This makes BMS more appealing even to large customers because of the number of time-critical applications they have, which cannot wait five minutes.

Customers may even prioritize channels within the DS1 bit stream. If a facility is degrading, for example, high-speed data channels can be reassigned to another facility, while lower-speed modem channels, which can better withstand impairments, continue to be transported over the same facility.

3.6 NETWORK-MANAGEMENT SYSTEMS

Since the breakup of AT&T, organizations have had to assume more responsibility for managing their own networks. That, however, has not been easy. Complicating the task of management is the plethora of equipment types that must be interconnected on the same network to achieve the organization's business objectives, and it is not only equipment types that complicate network management. The variety of transmission services that have become available to facilitate the economical growth of today's networks also complicates the task. Finding qualified technicians to oversee the equipment and lines in a dynamic operating environment only adds to the problem of keeping networks up and running, and once you find such people, you have to worry about keeping them. In less than a decade, organizations have moved from a state of comfortable dependence on Ma Bell to a state of "communications glasnost" that has left many organizations ill-prepared to deal with the sheer diversity and complexity of their sprawling networks.

An all-encompassing network-management system, however, is hard to come by. Users cannot understand why vendors cannot get together to work out common interfaces that will allow them to interconnect diverse network components so they can be managed from a single terminal. Vendors claim that users' problems stem from their apparent inability to plan their networks with more foresight, instead of growing them in a hodgepodge manner just to meet immediate needs. Furthermore, say vendors, if users would just buy products from them, they would have no problems. Users counter with the argument that it is not possible to have a single-vendor network. At this writing, no single vendor offers all the pieces of the network puzzle, nor can users implement all the pieces at once. Also, technology is moving fast; users have to buy now to serve customers and stay competitive. Finally, the continual cycle of the acquisitions, divestitures, and new ventures that are so much a part of doing business today insures a mixed-vendor environment. That means users must contend daily with compatibility issues, installation coordination, and finger-pointing among vendors when problems arise. In such situations, organizations are forced to rely on a variety of tools to track and solve problems, bypass failing facilities, and plan distribution.

In setting up a network-management system, users are faced with choosing the best of available products that will get the job done at a reasonable price, but it is not easy to sort out the claims and counterclaims of vendors. One source for the confusion is that network management means different things to different people. To the MIS manager, network management means having the ability to monitor the host for proper operation and spot problems with front-end processors and mainframe applications. To a LAN manager, network management might mean having the ability to configure ports on front-end processors or data switches to control access to various distributed data bases, bridges, gateways, and other shared resources like high-speed printers, disk storage, and modems. The data communications manager is preoccupied with monitoring the quality of transmission paths to insure data integrity and maximum network availability. This includes having the ability to take appropriate action to restore malfunctioning equipment or bypass failing facilities. The telecommunications manager, on the other hand, thinks of network management in terms of capacity planning and usage monitoring to insure that the PBX has enough trunks to handle peak-hour traffic without frustrating too many users, but not so many as to waste the company's money on underutilized facilities. All of these views are valid, of course, but by themselves they do not address the needs of the organization as a whole.

Interexchange carriers, too, have their own versions of network management, which includes the ability to monitor usage and implement reroutes to offload traffic from congested nodes, and vendors have their own definitions of network management. Modem manufacturers, for example, think of network management in terms of the ability to monitor, test, and control a network of modems from a central location. Multiplexer and cross-connect system manufacturers go one step further in the network hierarchy, enabling the user to manage T1 bandwidth. Computer vendors offer users the ability to monitor the performance of host-based systems through a single terminal, like IBM's NetView, which is essentially the result of consolidating other management schemes under a single umbrella. Not to be outdone, AT&T has announced its own network-management system, Unified Network Management Architecture (UNMA), which promises to integrate a variety of management products for modems, multiplexers, wide-area networks, advanced 800 service, and Dataphone Digital Service (DDS). Each of these approaches, however, reflects the traditional strengths of the vendors. IBM's NetView is focused on the protocol and diagnostic side of mainframe systems, while AT&T's UNMA will focus on the management of transmission facilities.

The point is that everybody — users and vendors — approaches network management from a relatively narrow perspective. For vendors, that perspective is limited by the pieces of the network they are capable of providing. For users, that perspective is limited by the scope of their responsibilities. Carriers, too, have a myopic perspective. For the most part, they could not care less about what happens outside of their own fiefdoms, and that is the crux of the problem — the absence of a truly global perspective to network management.

There is more to network management than meets the eye. Network management begins with such basic hardware components as modems, multiplexers, and dial-backup units. Each component must have the ability to monitor, self-test, and diagnose problems regarding its own operation. The next level of management is the ability to test components on the other end of a point-to-point configuration. On multipoint and multidrop configurations, the ability to test, diagnose, and resolve problems from a central location greatly facilitates network management, which helps insure maximum network availability. It also minimizes the need and the cost for skilled personnel.

3.6.1 Network-Management Functionality

Despite the absence of a global perspective to network management, a consensus is emerging about what such systems should do. The basic functions you can expect to find in a network-management system may be categorized into administration, control and diagnostics, configuration management, and performance measurement.

The "administration" element allows you to take stock of a network so that you know what hardware you have and where it is located. It also tells you what facilities are serving various locations, and what capabilities exist for alternative routing. With "control and diagnostic" capabilities, you can determine from various alarms what problems have occurred on the network, and pinpoint the sources of those problems so that alternate facilities can be manually or automatically put into service. This capability has two distinct aspects: problem determination and system restoral. "Configuration management" gives you the ability to add, remove, or rearrange lines, paths, terminals, and access levels on a daily basis as business circumstances may warrant. "Performance measurement" is the ability of the system to gauge the response time of the network, as well as the quality of information being conveyed throughout the network. Let us discuss some of the possible applications of these elements.

From an administration perspective, there must be specialized data bases that relate to each other, hence the term "relational data base." One of these data bases accumulates trouble-ticket information. A trouble ticket contains such information as the date and time the problem occurred, the specific devices and facilities involved, the vendor from which it has been purchased or leased, and the service contact. It will also contain the name of the operator who responded to the alarm, any short-term actions taken to resolve the problem, and space for recording follow-up information, such as visits from the vendor's service personnel, dates on which parts were returned for repair, serial numbers of spares installed, and the date of the problem's final resolution.

A trouble-ticket data base can be used for long-term planning. The network manager can call up reports on all outstanding trouble tickets, trouble tickets involving particular segments of the network, trouble tickets recorded or resolved within a given period, trouble tickets involving a specific type of device or vendor, and even trouble tickets over a given period not resolved within a specific time frame. Some network-management systems allow the user to customize such reports in other ways to meet unique needs.

With these reports, a network manager can have a better understanding of the reliability of a given operator, the performance record of various network components, the timeliness of on-site vendor maintenance and repair services, and the potential of certain segments of the network to fail, and, with information on both active and spare parts, network managers have objective backup for decisions on purchasing and expansion. In conjunction with the trouble-ticket function, some network-management systems even provide cost and depreciation information on the network's components.

Through the control and diagnostic capabilities of the network-management system, alarms are triggered from malfunctioning network components and failing facilities. When a local or remote monitor detects a problem, it sets off an alarm at the operator's console. Front-end processors, modems, switches, and multiplexers typically provide "negative" alarms, which consist of the disappearance of various lights on their front panels. The carrier-detect light, for example, indicates an active line, but when it goes off, it may go unnoticed by the operator. Network-management systems, on the other hand, provide a positive alarm signal, such as a message at the operator's terminal, which not only indicates that a failure has occurred, but also provides information on the nature of the failure and its location.

Most network-management systems offer a color display, which makes it easier for the console operator to spot a problem and instantly obtain a reading on its severity through the use of multilevel alarm indicators. Red, for example, usually indicates a failure. Green may indicate normal operation, while yellow may be used to indicate a deteriorating condition. Some network-management systems come only with monochrome displays, making them less effective for indicating problem severity. If the choice is yours, always opt for the color display; there is something about seeing a red light that gets a console operator's adrenalin flowing, especially on the late shift.

After determining that a problem exists, the next step is system restoral, which has two aspects: in the short term, the network manager must find a way around the problem; in the long term, the failed component must be repaired or replaced. Some network-management systems incorporate a fallback switching mechanism by which an operator can immediately replace a malfunctioning device with another device that is ready to go on-line at a moment's notice. Such devices are said to be on "hot" standby. In large networks certain modems, multiplexers, and front-end processors may be standing idle until called into action. For communication links, the backup facility is usually the public switched network, although redundant private facilities may also be used.

Restoral, however, does not end with fallback switching; stopping at that point constitutes only a band- aid approach to problem solving. You should consider a network-management system that provides assistance for the long-term repair or replacement of faulty components, such as a built-in capability for detailed testing further to isolate the problems that were responsible for generating the alarms. This capability saves time and money over traditional manual methods, whereby a technician was dispatched to a site with portable test equipment to determine the cause of a problem, and then returned later with the appropriate spare part or replacement unit. If the problem was not hardware-related, a service call was placed to the carrier, who would typically begin the testing process from "square one."

Another function of a network-management system is performance measurement, which has two aspects: response time and network availability. Many network-management systems measure response time at the local end, from the time the monitoring unit receives a "start of transmission" (STX) or "end of transmission" (EOT) signal from a given unit. Other systems can measure end-to-end response time at the remote unit. In either case, the network-management system displays and records response-time information, and generates user-specified response-time statistics for a particular terminal, line, network segment, or the network as a whole. This information may be reported in real-time or, when needed, for a specified time frame. Systems with color-graphics consoles can display elaborate, multi-colored network schematics with specific colors assigned to specific levels of response time, with green for normal operation, yellow for degrading response time, and red for abnormally high response time. Network managers can use this information to track down the cause of the delay. When an application overruns its allotted response time, the network manager can decide whether to reallocate terminals, place more restrictions on access, or install faster communications equipment to improve response time.

Availability is a measure of actual network uptime, either as a whole, or by segments. Such information may be compiled into statistical reports that summarize such measures as total hours available over time, average hours available within a specified time, and MTBF. Some network-management systems allow you to customize such reports to suit your own organizational requirements.

With long-term response-time and availability statistics, calculated and formatted by the network-management system, managers can establish current trends in network use, predict future trends, and plan the assignment of resources for specific present and future locations and applications. Response and availability information from a network-management system, then, provides objective tools to support such decisions.

Without a network-management system to isolate problems and facilitate restoral, users must depend on equipment vendors and carriers to find the problem. During the interim, employee productivity suffers and revenue losses mount, if only because the employees must still be payed for waiting idle at terminals until service is restored. A network-management system, then, enables an organization to control its own destiny. In Darwinian terms: any organization that can control its own environment better than its competitors will win out in the struggle for survival.

3.6.2 Network-Management System Components

A minimal network-management system consists of a central processing unit, a system controller, a storage device, an operator's console, and the vendor's own line of diagnostic modems. The central processor may consist of a minicomputer or microcomputer. High-end products usually feature a color cathode-ray tube (CRT) at the operator's console, whereas low-end products still use monochrome CRTs. The system controller, the heart of the network-management system, continuously monitors the network and generates status reports from data received from various network components. The system controller also isolates network faults and restores segments of the network that have failed, or which are in the process of degrading. In most cases, any printer may be used with the vendor's network-management system.

The nature of the monitoring devices in a network-management system depends on the type of vendor. Systems from modem vendors, for example, have diagnostic monitoring facilities built into their products. Some vendors offer "wrap" devices that tie nondiagnostic modems — or the modems of competitors — together under the same management umbrella. These devices are called "wraps," because they interconnect with the installed base of nondiagnostic modems to surround the digital and voice-frequency interfaces electrically. This permits continuous, noninterfering, secondary-channel network surveillance and testing under the supervision of the network-management system. The value of the wrap box is that it protects the organization's investment in modems, which may be quite substantial. The single-unit price of a wrap box is about $1,000, compared to buying a new 9.6-kb/s diagnostic modem costing about $3,000 or a 14.4-kb/s modem costing $4,495. (Although 19.2-kb/s modems are now available, at this writing there are no wrap boxes that are capable of interconnecting with them.) Codex Corporation

(Mansfield, Mass.) and General DataComm, Inc. (Middlebury, Conn.), are among the vendors who provide both alternatives: integral diagnostics in their own modems and wrap devices that consolidate different manufacturers' modems under their respective management schemes. The wrap product of General DataComm goes even further in functionality by offering two inbound and two outbound control signals that can be used to protect the security of remote sites. These control signals allow lights to be turned on and off, for example, and security systems to be enabled or disabled from the central site.

Some diagnostic modems, like Racal Milgo's (Sunrise, Calif.) Omnimode line, can yield a wealth of measurement data on VF parameters and send them to the network-management system for graphic display at the operator's console. Some products can be initialized at the time of installation to compensate for amplitude and delay distortions on the line. With these products, however, any subsequent adjustments that may be necessary must be done manually by an on-site technician. Other products feature automatic adaptive equalization, which continuously compensates for line impairments to maintain a consistent level of performance over varying line conditions. This, in turn, provides significantly more accurate data. At this writing, General DataComm appears to offer the most comprehensive compromise-equalization capability for networking modems, which is available on its line of NMS modems. This capability minimizes the need to reroute traffic and may even allow more drops on multipoint circuits. Also, in matching modem performance to line characteristics, carrier charges for conditioning consistently noisy lines may be eliminated.

Modems communicate with the network-management system by one of two techniques. With the "mainstream" technique, favored by IBM and other vendors of host-based network-management systems, the central processing unit polls the remote devices (modems or wrap units) in a dedicated time slot over the main data channel. This technique lends itself to end-to-end monitoring at the application level, since it is often closely coupled with the network's equipment, from host processor to modems and terminals, but host-based mainstream systems have several disadvantages. The polling process eats up the host's processing power. When the host or front-end processor is taken down for repair or preventive maintenance, network management is nonexistent, and when the host crashes, the whole network-management system goes down with it. In distributed-processing environments, however, the network can support a high level of activity even in the absence of a controlling host. With fault-tolerant hosts, the level of activity can remain unaffected, despite the degraded performance of critical subsystems. All of these considerations merit careful evaluation during the purchasing process.

With the "sidestream" technique, favored by most modem vendors, the remote devices transmit diagnostic information asynchronously over a special, low-speed data channel that is multiplexed onto the same facility as the main data channel. This data channel may operate as low as 75 b/s. This technique is also referred to as the diagnostic channel, or "secondary channel." In utilizing a separate, low-speed data channel, status information may pass intact over a degrading link, enabling the system controller or console operator to implement an alternate configuration. Sidestream systems usually report modem, terminal, and communication-link failures much faster than mainstream systems, since the modems do not have to wait for their turn in a polling arrangement to report trouble.

Although some vendors claim to use the sidestream technique, and hype the noninterfering aspects of a separate diagnostic channel, what they really do is combine diagnostic data and production data *via* a statistical time-division multiplexer (STDM), which constitutes the core of their network-management systems. During periods of low to medium usage, this scheme is acceptable, but during hours of peak usage, system tests and monitoring information will be delayed in the buffer until bandwidth becomes available. Furthermore, STDM-based products are protocol sensitive. The Multidrop Network Management System from Teleprocessing Products, Inc. (Simi Valley, Calif.), for example, is limited to supporting only IBM's SDLC and BSC. The network-management system is only compatible with its own Data Channel Units, and does not even provide restoral capabilities for DDS facilities.

3.6.3 Monitoring Links for Quality

The network-management system you choose must have the capability of monitoring the *quality* of information on the communication path, not just its presence. Under ideal conditions, you are assured of *both* the presence and the quality of information, but ideal conditions do not usually continue for long. A variety of line impairments can corrupt the data at any point along the transmission path.

One way to guard against data corruption is to implement an error-checking scheme, but this may adversely affect throughput, since many retransmissions may be necessary to get the data intact to their destination. (Although forward-error correction is available with some asynchronous modems, they are not typically used in conjunction with proprietary network-management systems.) Another way to guard against data corruption is to use transmission devices that will automatically downspeed to a point at which data will not be affected by line impairments, and then reestablish normal speed as line condition improves, but this suffers from the same limitation as error checking — reduced throughput.

As alluded to before in the discussion about automatic compromise equalization, the only way to guard against data corruption without sacrificing throughput is with the proactive approach of monitoring the performance of communications links through a network-management system. Specifically, what you need is a network-management system that integrates the tools for measuring the quality of information with the tools for measuring the presence of information so that you can make intelligent judgments on how to reorganize or reroute the system in the event data-error rates exceed predetermined thresholds.

The time to initiate restoral functions or to reconfigure circuits is not when there is an outage, but when the error rate has become high enough to start degrading throughput. (With DDS services, degradation occurs so quickly that the line may go dead before there is time to react appropriately to prevent the loss of data. Some newer DDS services, however, allow bit-error rate testing over the secondary channel, which provides the means to implement continuous nonintrusive testing to identify problems that portend a link failure. DDS with secondary channel, then, allows enough time to implement a reconfiguration according to the probability that a link failure will occur. Such continuous nonintrusive testing is available over Accunet T1.5 circuits that have been upgraded with the Extended Superframe Format.) Error-rate measurements, then, are meaningful only when monitored on an end-to-end basis at the premises — from the data set associated with the communications path. To make a judgment on how and when to reroute the system, you must take into consideration the *quality* of the lines, not just the presence of information being carried over them.

When the error rate forces you to reroute portions of your network, you are faced with the same factors that you must consider when traveling by car: Would you be willing to go out of your way to get to an interstate highway to save time, or would you travel the shortest physical distance, even if it meant having to go through many stoplights?

Similar considerations apply to networks. Upon detecting high error rates or isolating a fault on a particular link of the network, are you going to reroute through the nearest node, even though it may already be congested, or are you going to reroute along a path that has the shortest delay?

Delay is an important consideration when choosing network products like modems and multiplexers. The network delay is the sum of the propagation delay plus the multiplexing delays. Since a poll requires a response, it is the round-trip network delay that must be considered and added to the modem-turnaround delay to arrive at the total network delay. If an end-to-end error-correction scheme is employed, delays compound and could degrade network performance beyond acceptable limits.

Most multiplexers use a byte architecture, which is an inherent source of delay. Some products produce as much as 56 bits of delay per node, *versus* a bit-oriented multiplexer, which may produce only an 8-bit delay per node. At a channel rate of 2400 b/s, this translates into a delay of 23 milliseconds per node in one direction for the byte-oriented product, *versus* 3.3 milliseconds for the bit-oriented product. Assuming you had four multiplex nodes in your network, the multiplexing delays would be just over 90 milliseconds each way when using the byte-based product. That is 180 milliseconds round-trip. If the multiplex channel drives a 10-

drop multipoint line, the response time required to service any terminal is increased by 1.8 seconds. Since computer response-time objectives are in the order of just a few seconds, the added multiplexing delay may have eroded any response tolerance to the point where the response-time objectives simply cannot be met.

You must also consider what will happen if a data link goes down. The propagation plus multiplexing delays of the reroute path may be twice those of the normal path. In that case, there would be a difference in throughput of two to one. How do you correct this situation? There are two possible alternatives (Figure 3.7), neither of which are very attractive economically. You could buy higher speed modems (4.8 kb/s) to better compensate for the presence of delay, but that would mean incurring the expense of new modems at each drop location. Notwithstanding the additional cost, doubling the speed of the modems would reduce the capacity of the multiplexer by 50%.

Another possible solution might be to use the 2.4-kb/s modems, but reconfigure the drop locations so that they are more evenly distributed among two master controllers. This, too, can be quite expensive, assuming that you even have a spare port available on the multiplexer.

You could easily have avoided this costly mistake, however, by choosing a vendor who makes both modems and multiplexers with these considerations built into its products, and who has a network-management system capable of taking delay measurements.

Returning to the previous travel-by-car analogy, upon reaching a detour, you do not just randomly pick another route because it happens to be available. You must check all reasonable routes for sources of delay, and then pick the one that will get you to your destination on time. In accounting for delay, your network will not suffer performance degradation upon service restoral.

Recognizing that time delays are critical to network performance, some vendors use a bit architecture for their multiplexers. At the subrates — rates below 56 kb/s — the bit architecture provides greater system throughput for data. With the bit architecture, it is possible to keep nodal delay below one millisecond at 9.6 kb/s. As stated before, multiplexers using a byte architecture provide as much as 23 milliseconds of delay at 2.4 kb/s.

Telecommunications managers, too, must consider path or propagation delay, and the effect it can have on voice-transmission quality. The disturbing effect (or amount of annoyance) of echoes is determined by the combination of amplitude and delay. While telephone users can tolerate a relatively high degree of delay (up to 40 milliseconds round-trip delay if the echo is not too loud), this same amount of delay can cause supervisory problems for PBX tie lines, creating a condition known as "glare." This occurs when both ends of a trunk are seized at the same time by different users, thereby blocking the call. Wide disparities in delay among various user locations can make glare a frequent and particularly annoying problem.

SOME ALTERNATIVES FOR
CORRECTING THE EFFECTS OF DELAY

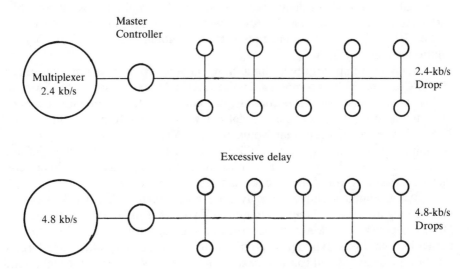

Replacing 2.4-kb/s modems with 4.8-kb/s modems results in a 50% reduction in throughput of the multiplexer.

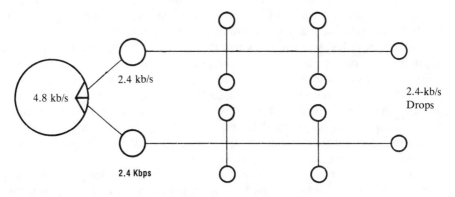

Solution: Assuming availability of spare multiplexer port, retain 2.4-kb/s modems, but at the cost of another master controller.

Figure 3.7 Some alternatives for correcting the effects of delay.

One-way propagation delay is approximately one millisecond per hundred miles. Along a 500-mile route between two multiplexers, for example, the path delay is about five milliseconds. With multiplexers in front of the PBXs, you would have five milliseconds of delay at each node of a four-node network, which works out to 20 milliseconds of delay. Add to that the five milliseconds attributable to path delay, and you have 25 milliseconds, which requires the use of an echo canceler at the PBX. With the additional equipment and cabling, the cost of this band-aid solution can go sky high.

Delay also imposes severe limitations on voice transmission in that only 64-kb/s (PCM) and 32-kb/s compressed voice (ADPCM) can be supported. Although 16-kb/s voice transmission is becoming quite common, the compression techniques cannot tolerate a 3.5-millisecond nodal delay (56 bits at 16 kb/s), much less the 17.5 milliseconds of delay that would be encountered over the public network, assuming that the transmission would traverse at least five nodes (with 3.5 milliseconds per node). From a design standpoint, network managers should anticipate that transmissions would traverse as many as nine nodes.

Byte-oriented multiplexers may pose no problems when used in point-to-point configurations, and they are required if you want to interface with the public DACS network. It is when these products are used for nationwide multi-node networks that users experience problems with throughput, which are largely a function of delay. If you can minimize nodal delays by choosing the right vendor for network products, you give yourself the ability to tolerate more internodal delays.

The capability to tolerate and monitor delay is important for another reason. In ordering transmission facilities from interexchange carriers, you obtain the ability to communicate between two or more points, but unless you have the ability to monitor for delay, you will not know if changes in throughput are being caused by the carrier, or you just have too many users on your system at once. With the ability to measure internodal delay through your network-management system, you could not only request a better route from the carrier, but also monitor the result of that change. For these reasons, you should look for a vendor who offers products capable of providing delay measurements, and with this capability integrated into the network-management system, you will have a powerful tool for determining primary and alternate routes.

When buying both T1 multiplexers and modems, look for products that provide minimal delay. Beyond that, look for a vendor with a global perspective to the network — one whose products reflect a strategic plan. This means looking for a vendor who is network-oriented, rather than product-oriented. This approach insures that you get products that complement each other in providing the measurements you need to determine the quality and the presence of information at every point along your network.

This level of integration cannot be obtained merely by a larger firm's acquisition of a smaller company, or even through OEM arrangements between vendors. The network-management system must be closely coupled with a vendor's modems and multiplexers, the key ingredient that is missing from such vendor relationships. A vendor's success with network-management products is the result of design, not afterthought. Instead of providing either multiplexers or modems (or computers and multiplexers), the two companies together only provide a hodgepodge of products that were not designed to work together. Consequently, the new entity cannot bring anything new to the table that was not there to begin with.

It may not be enough to acquire the comprehensive tools for network management. If the vendor can provide network management in the multivendor operating environment, apart from computer manufacturers and interexchange carriers, so much the better. Then you can be reasonably assured of the vendor's objectivity. This can result in maximum savings in the amount your organization spends annually on communications.

3.6.4 Buying Tips

As networks continue to grow and become more complex, the need to manage them from a central location, preferably from a single terminal, becomes more apparent. The communications industry is trying to move in that direction through Open Systems Interconnection (OSI). Eventually, OSI standards will make it possible for different vendors to exchange network-management information such as alarms, performance measurements, usage statistics, and diagnostic-test results in a standard format. The current OSI standards, however, are very unstable and subject to change.

Until there is agreement on an OSI standard, users must make do with the products already available. Rather than wait, vendors are competing with each other in a new arena: that of supporting both IBM's NetView and AT&T's UNMA from their own proprietary management systems. At this writing, AT&T is even considering the possibility of providing an interface to NetView.

There are now network-management systems for every segment of the network, but until there is agreement on standards, there is not likely to be a comprehensive management system capable of embracing the huge variety of components available for today's networks from a multitude of vendors. Some vendors insist that they have come quite close to doing that, but their solutions fall far short of OSI.

AT&T's UNMA is only a piece of its Universal Operations Systems (UOS) architecture, which is proprietary in nature and not compatible with OSI in its present form. All that is known at this writing is that AT&T plans to adapt a distributed processing approach, as opposed to the centralized approach of NetView. It is also known that AT&T has not received very much support in its efforts, which means that UNMA is likely to remain proprietary for the foreseeable future.

Although not yet complete, NetView has serious limitations. Aside from the drawbacks inherent in host-based management systems mentioned earlier, the IBM product is not aimed at finding transmission problems. IBM's product is useful for finding problems on the local computer network, but treats remote nodes of the network as peripherals. With its mid-1988 purchase of PacTel Spectrum Services (Walnut Creek, Calif.), IBM hopes to be able to support T1 networks and extend NetView's capabilities to voice management. It is still too soon to determine IBM's likelihood for success in that venture, however.

In choosing a network-management system, you should inquire about the vendor's plans to offer interfaces to NetView and UNMA with the idea of building future flexibility into your network. That way, you will be positioned to take advantage of the best of all worlds: the networking expertise of the data-communications vendor, plus the host-oriented management expertise of IBM and the public-network management expertise of AT&T.

This brings us to a probe of some of the nuts-and-bolts items to look for when evaluating network-management systems. There are several network-management systems available that support point-to-point and multipoint circuits, with or without drop locations. As alluded to earlier, they employ a built-in bit-error rate tester (BERT) to perform analog and digital loopback tests. The test data is looped back to the system controller, where it is compared against a data base of user-defined performance parameters. The network-management systems allow modems, with and without wrap boxes, to initiate self-diagnostics. They will isolate a defective modem and perform the switch to a hot spare when the primary unit dies. The real point of differentiation between these network-management systems, however, is in their level of component isolation and control.

Some network-management products have very extensive isolation and control capabilities. For example, in isolating a streaming terminal (a terminal that is putting signals on the line because it lacks flow control), the network-management system can remove it from the network, and then put its associated modem back into operation for use by another terminal. Other network-management systems do not have such extensive capabilities; instead, they take the modem out of operation along with the terminal, depriving other terminals of being able to share it. In removing the modem from the configuration, the network-management system will not have the means of detecting when the faulty terminal has been repaired and put back on line. If your network has a large installed base of shared modems, you will want a network-management system that is capable of isolating malfunctioning terminals, while putting the modems back on line.

Network-management systems also differ in the method by which modems may be disconnected from the network. Some permit both automatic and manual procedures, while others require manual intervention. The problem with the manual-only method is that until the console operator notices that a problem has occurred, service will be disrupted until he does. Meanwhile, users fume. In the absence of an automatic disconnection feature, insure that the system is at least capable of generating an audible alarm to get the operator's attention.

A comprehensive network-management system should be able to perform an extensive array of voice-frequency (VF) impairment measurements, and report alarms whenever selected parameters exceed user defined thresholds. Some systems are quite limited, offering only the capability of measuring VF-receive level, signal-to-noise ratio, and phase jitter. There are, however, a variety of other common VF line impairments that can disrupt data transmissions. You should look for a network-management system that is capable of taking these additional measurements on point-to-point circuits:

- Overall signal quality;
- VF receive-transmit levels;
- Loss of data-carrier detect;
- Dropouts;
- Nonlinear distortion;
- Gain, impulse, and phase hits;
- Frequency offset.

When VF impairments reach levels that degrade communication links to the point where primary data-transmission rates cannot be sustained, fallback rates must be invoked. Network-management systems differ on how fallback rates are implemented; some initiate fallback automatically, but only if their own modems are tied back to the system controller. Other network-management systems require that fallback be operator initiated, even on their own modems. The problem with the operator-initiated approach is that it takes time for the operator to notice that there is a problem, and it may take even more time to respond appropriately. By the time fallback is initiated, data loss can be quite extensive. For example, if the modem requiring fallback is attached to a multiplexer, the multiplexer's channels must be reconfigured to accommodate the change in modem speed. Some network-management systems allow all the channels to be reconfigured simultaneously with a single command from the operator's console. Other network-management systems require that each channel be reconfigured with separate commands. This can be very time-consuming. Until all the channels are reconfigured, the multiplexer will operate in a degraded mode.

Some network-management systems can generate alarms to report the loss of any of the following conditions:

- Modem-terminal power;
- Transmit-receive data;
- Transmit-receive clock;
- Data terminal ready;

- Request to send;
- Receive carrier.

Look for a network-management system that enhances your control over the network by detecting:

- Modem power recovery;
- Streaming terminals;
- Corrupt configurations;
- Front-panel tampering.

Dial backup is an option with most network-management systems. This feature may be added to modems and wrap units with a plug-in module that allows you to reroute your private-line communications through the public switched network temporarily. Dial backup may be implemented automatically by the network-management system in response to specified alarm conditions, but, more often than not, such products require that dial backup be initiated by the operator upon receipt of the alarm.

Through the use of various loopbacks and tests, faults can be isolated to a particular terminal, modem, or line. In addition to invoking modem self-tests from the operator console, a network-management system should have the capabilities to perform the following:

- End-to-end error-rate test;
- Polling test;
- Outbound error-rate test;
- Circuit-quality test;
- Loopback search.

If you are also concerned about saving space, most firms offer rack-mounted versions of their diagnostic modems. When considering the rack-mounted products, however, there may be some quirky problems to contend with, such as partially obscured front panels, or the requirement for a single-panel display for each rack. In some cases, shelf power must be turned off to pull a single modem card. Sometimes the cabling must be disconnected to remove a modem from the rack. In a rack-mounted configuration, check into the requirement for fans to dissipate heat.

To get the most out of a network-management system, the vendor should offer a complete range of modems, including analog, digital, and wire-line, as well as wrap and dial-backup units. Both synchronous and asynchronous data should be supported. The network-management system should also support a range of modem speeds — preferably 1200 b/s to 19.2 kb/s.

Analog modems should be supported by the network-management system. They should, however, be able to accept and store configuration instructions transmitted to them from the network controller, permitting completely unattended operation. In addition, they should have the ability to send configurations back to the network controller for verification. The only hard option on an analog modem should be a ground strap. Front-panel test switches should be for use only by technicians to perform on-site testing during installation; not for daily use in the management of the network. Even switches of that type, however, should be enabled and disabled through the management system.

Make sure the vendor also offers Data Service Units (DSUs) and Channel Service Units (CSUs). DSUs provide direct connection to AT&T's Dataphone Digital Service (DDS). They should accommodate subrates, as well as 56 kb/s, and be compatible with DDS with Secondary Channel (DDS-SC). DDS-SC provides a separate signaling path on digital lines, which can be used for communication between a remote site and the centrally located network-management system. This arrangement allows continuous, noninterfering monitoring of digital EIA leads, data-set status, DDS alarms, and data-set alarms. CSUs are used with DSUs when the terminal lacks a digital-line interface. Some vendors package the DSU and CSU together in a single package, but these all-in-one devices may lack the extensive diagnostic capabilities of stand-alone units.

With a network-management system capable of supporting the broadest range of speeds, protocols, facilities, and equipment types, your network will be "future proofed" to handle any emerging requirement.

Some network-management systems, however, limit your flexibility in selecting future modems and multiplexers for your network. AT&T and NEC, for example, presently support only synchronous operation on their network-management systems, as do pre-NetView management systems from IBM. Paradyne and Racal Milgo claim to provide asynchronous support, but as an extra-cost option. AT&T, NEC, and Paradyne offer no wrap capability. NEC, a relative newcomer in the network-management marketplace, provides no multiplexer support.

Do not commit to a network-management system thinking that your installed base of modems can be tied into it with the vendor's wrap box. You should have the vendor thoroughly test its wrap unit on each type of modem you use. The vendor will require at least two modems to test its wrap unit for point-to-point and three for multipoint configurations, including appropriate line connectors and cables. Such testing should only take two to three weeks at the vendor's test lab. Insist that the vendor perform such tests, even if it claims to have tested the same types of modem for other customers.

Check into the configuration-storage capabilities of the modems. Some require battery backup to hold stored instructions during a power outage and, in the case of AT&T, information may be stored in the unit without power for up to four days. No matter how glowingly the vendor touts the advantages of battery backup, this feature can impose an administrative burden if you have a large installed base of modems. You may be better off with modems that hold instructions in nonvolatile storage media, such as electronically erasable programmable read-only memory (EEPROM).

There is also a wide disparity in the power requirements of network modems, from 20 watts to 75 watts. This may be an important consideration if you are centralizing a large pool of rack-mounted modems, because you may have to make special provisions for utilities. The physical size of the modems is also important, because it determines how many units will fit into a rack and how many racks you can squeeze into a room without imposing maintenance access restrictions. Minimal packaging densities also give you room to grow, without tearing down equipment-room walls to free up more space. Despite advances in large-scale integration (LSI) and the widespread application of this technology to modems, the products of some vendors are still three to four times the size of their competitors' offerings.

It is important to determine the capabilities of the network controller before committing to volume purchases of the vendor's modems. Some vendors have been known to bait customers by discounting the network controller by as much as 50% in return for a volume purchase of modems. Although the network controller supports the latest modem version the customer wants to buy, in reality it is only a jury-rigged version of a mature product, offering capabilities that are quite limited by today's standards. For example, some of these controllers still use floppy disks as the storage medium and, consequently, cannot accommodate the data bases you might expect in even a bargain-basement network-management system, such as historical alarm files and trouble tickets. They usually have a maximum capacity of less than 20 diagnostic ports, and do not allow for modular increases in port capacity. Some controllers are capable of generating only a limited range of reports, or none at all. In addition, there may be no graphics capabilities, no provisions for multiple CRTs, and no support for a multiplexer or peripheral equipment such as a printer or external storage device.

When looking for a top-of-the-line network-management system, and satisfying yourself that you have found exactly what you are looking for, you can minimize risk by determining where the individual components come from. For example, if the network-management operating system is based on UNIX, while the vendor's network controller is based on minicomputer hardware from a company that does not deal direct with end users, chances are that you will get embroiled in third-party maintenance disputes down the road. You can avoid potential finger-pointing by purchasing a network-management system from a vendor capable of maintaining the product from top to bottom, regardless of where the components come from.

Whether you run the network-management system on your own computer or buy a dedicated system from the vendor, make sure the screen displays are not overly cryptic. If so, users will take longer to become proficient in running the system. Also determine if on-screen HELP is available with the package. You want to avoid unnecessary network downtime while the operator searches through a set of manuals for instruction on what to do next. Also make sure the system comes with enough network-controller storage capacity to format the management reports the way you want them.

When choosing a network-management system, there is just no substitute for seeing one in action at a customer site. Ask the vendor for at least five references. Talk to all, and select one for a visit. Make it a point to ask references how they think the system can be improved, and then ask the vendor to account for these shortcomings.

To recap, a comprehensive network-management system should provide the capabilities to:

- Generate alarms from devices at remote locations in response to user-defined thresholds, and test their validity;
- Display alarm information on a console for the operator's attention and response;
- Maintain data bases for statistical analysis, as well as for network configuration and inventory control;
- Generate management reports based on the information in those data bases;
- Monitor the quality of transmission paths so that channels may be down-speeded to avoid the harmful effects of line impairments;
- Restore service if a problem is detected by invoking reconfigurations to bypass failing links or malfunctioning equipment;
- Reroute traffic around points of congestion.

These capabilities are important because they insure maximum network availability, thus minimizing revenue loss. Being able to do all this from a central location is important because most organizations cannot afford the delay associated with sending a technician from site to site to monitor equipment-indicator panels or flip reset switches, nor can an organization afford to let its lines fail before implementing bypass. Network-management systems permit a proactive approach to service restoral by allowing measurements to be taken of a line's quality so that bypass may be invoked before the link dies, and being able to monitor all network components from a single terminal is important because it enhances the organization's ability to manage its communications resources. In fact, it can be argued that a communications network, no matter how expansive, is not really a strategic asset unless it can be used to fulfill organizational objectives. You will not know to what extent organizational objectives are being fulfilled, unless the network can be centrally managed.

3.7 DIAL-BACKUP SYSTEMS

When communications are interrupted, for whatever reason, corporate executives mentally tally the loss of thousands of dollars for every minute of downtime and agonize over the real or potential loss of customer goodwill. Network managers, mindful of the unpleasantness to come, typically scurry about, implementing appropriate restoral actions, which may, or may not, produce the desired results within a reasonable time.

One way to minimize private-network downtime is to have redundant facilities ready and waiting. When a link goes down, the standby facility may be quickly activated, while the failed link is diagnosed and restored. Although reliable, having duplicate facilities is an expensive insurance policy that most companies would prefer to do without.

For most companies, dial-backup systems may be a viable alternative to redundant facilities. Such systems allow data communications to be temporarily transferred to the public switched network, accommodating modems operating at speeds up to 19.2 kb/s and providing integral Data Access Arrangement (DAA) functions that are still required for direct connection to the public network. Leading-edge dial-backup systems are laden with features to provide maximum applications flexibility, simple call initiation, automatic tone or pulse dialing, status reporting, and easy installation.

3.7.1 Basic Functionality

In providing the option for one or two-call dial backup, all 2-wire and 4-wire private lines are restored over the public switched network. The 4-wire, one-call restoral method is particularly useful in international applications where switched-network circuits may be expensive or difficult to obtain. For this type of restoral, the modems must have the inherent capability for 2-wire, half-duplex operation. Of course, the standard two-call restoral method may be used to restore 4-wire private lines. This method entails making two calls from the central site, thus requiring two dial-up lines, one for transmitting and another for receiving.

There are two possible methods for initiating dial backup over the public switched network: manual originate and auto originate. In the manual-originate mode, the operator sets up a dial-backup circuit for a failed data line. This is accomplished by entering the backup mode, dialing a remote station from a standard telephone set, and switching to the backup circuit *via* a pushbutton on the dial-backup unit's front panel. In the auto-originate mode, the dial-backup unit recognizes a leased-line failure or signal-quality degradation and automatically establishes communications *via* the public switched network to the remote site.

For unattended locations, an auto-answer feature provides the means to answer a dial-backup call automatically. An auto-terminate feature disconnects the backup call when the originating site ceases to transmit due to the restoral of the primary link. An auto-abort feature disconnects the call if the handshake protocol is not properly completed. This prevents "wrong-number" calls from disturbing the network and minimizes its vulnerability to hackers.

Transmit and receive levels may be set in a variety of ways, providing dial-backup units with the means to handle any switched-network application. The originating unit offers two transmit-level modes: programmable and permissive.

In the programmable mode, the dial-backup system's modem-transmit level is limited to a signal level of -12 dBm measured at the central office. This is accomplished by the level-setting resister within the standard jack arrangement that is provided by the telephone company. The value of the telephone company–installed resister is selected based on loop-loss measurements performed by the telephone company. This programmable connection is recommended for the optimal performance of the dial-backup system. Receive levels may also be selected to amplify the signal by 0 or $+6$ dBm to meet a variety of switched-network applications.

In the permissive mode, the dial-backup system is set to transmit at a signal level of -9 dBm, which complies with telephone-company requirements for modems that access the public switched network from behind a customer-premise switch. On facilities using lower speed modems (2400 b/s or less), the telephone company's permissive installation is adequate.

Another useful feature of today's dial-backup systems is their selectable switched-network impedence levels of 600 or 900 Ω, making them compliant with the 600-Ω requirement used throughout the United States and Europe, as well as in Canada where 900 Ω is the national standard. Private-line impedance, however, is fixed at 600 Ω, which satisfies all markets.

3.7.2 Dial-Backup Implementation

Dial-backup systems have demonstrated their value in a variety of applications, particularly with 4-wire, full-duplex private lines. In this application, the dial-backup system is optioned for automatic or manual originate at the central site, while the remote unit is set for the auto-answer mode. When dial backup is required, the central-site operator will initiate two calls, *via* separate telephone sets, to the remote site. If optioned for auto-originate, the dial-backup unit will dial two phone numbers stored in nonvolatile memory. The unattended remote unit will complete the handshake sequence and then enter the backup mode, thereby establishing the communications link. Of note is that the operating protocol of the modem and communications system does not change upon implementation of dial backup. The only change involves the lines over which data are transmitted, and they are transparent to the user.

Upon restoral of the failed line, dial backup is terminated. The return to leased-line service can be accomplished in either of two ways. In the manual mode, the operator at the central site simply presses a button on the dial-backup unit, which disconnects both dial-up lines. Alternatively, the dial-backup unit disconnects the dial-up lines when it detects restoral of the leased line, and automatically switches back to the leased line.

In the auto-answer mode, the remote dial-backup unit is activated without the need for an attendant at the remote site. Upon the failure of the primary line, operation over the dial-backup network can be manually or automatically initiated.

At the remote site, both calls are answered automatically by the dial-backup unit. When the handshake sequence is completed and the dial-backup connection is established, the flow of data resumes. Upon restoral of the failed line, dial backup is terminated in one of two ways: by the central-site attendant, who manually releases the backup switch on the dial-backup unit; or, when in the automatic mode, the dial-backup unit reestablishes the leased line upon detection of acceptable signal quality. With this action, the remote site's dial-backup unit automatically disconnects, and all communication is transferred to the primary line.

Dial backup is implemented on private lines in much the same way, except that only one call is used, which sets into motion some interesting behind-the-scenes events. In this application, the dial-backup unit must convert a full-duplex private line to a half-duplex switched-network link upon implementation. This operation is only used when the front-end processors and terminals are using bisync protocol. Dial backup will not work when other protocols like HDLC (High-Level Data Link Control) and SDLC (Synchronous Data Link Control) are used, because they will not automatically transfer to a half-duplex link.

Although the conversion from full-duplex to half-duplex will cause a reduction in throughput, the benefits of dial backup still outweigh the liabilities, especially when you consider that private lines at international locations are at a premium because of cost or availability.

Because 4-wire private lines may be quite expensive between international locations, substantial savings may accrue through the use of 2-wire lines. In such cases, the half or full-duplex 2-wire link is simply transferred to another pair of wires over the public switched network.

When evaluating dial-backup systems for purchase, it is important to understand a few of their limitations:

- When using the switched network for dial backup, you can expect poorer overall line quality and greater fluctuations in receive levels from line to line.
- Noise levels on the switched network are such that it may take several phone calls to achieve a minimum acceptable-quality level, especially for high-speed applications of 9.6 kb/s and above.
- Given these factors, anticipate using fallback speeds when using dial backup to support high-speed applications.

Another thing to consider is that dial-backup systems may not be practical for large networks, where the possibility exists of a few hundred lines on a single segment going down at once. Such extensive use of dial backup can be both costly and hard to manage.

To support applications requiring 56 kb/s or 1.544 Mb/s, dial backup over the public switched network is not possible, of course. In such cases, you may want to look into AT&T's Switched 56-kb/s Service, or Accunet T1.5 Reserved Service for disaster recovery. With T1.5 Reserved Service, you would pay the monthly charges for local-loop connections, but pay for the standby interexchange

T1 link only when you use it. The only drawback to using this service for disaster recovery is that it entails calling an 800 number to request the connection, which may take as long as 45 minutes to implement. This may not be all that bad, however, when faced with loss of network availability for an indeterminate period. With Switched 56-kb/s Service, you establish the connection yourself *via* a 700 number.

3.7.3 Intelligent Calling Systems

While dial-backup systems are adequate for rerouting communications to the public switched network from a relatively small number of failed private lines, they can be quite cumbersome to implement when a large number of lines are knocked out of service at once. In such circumstances, the ability to reroute users rapidly from the main computer center to a secondary host site can save organizations from financial disaster. Large private networks require "intelligent calling systems" that can be tightly coupled with their network-management systems. Such calling systems represent the next generation in dial-backup arrangements. They are distinguished from their plain-vanilla predecessors by their ability to restore rapidly and efficiently AT&T's Dataphone Digital Service (DDS) at subrates of up to 19.2 kb/s, as well as analog private lines. Although DDS circuit interruptions are rare, they tend to be long in duration when they do occur, lasting as long as two or three days. Since many of these circuits carry data that are vital to the overall operation of the user's business, even a short interruption in DDS service may be intolerable. To keep vital information flowing, special remote dial-backup units are required to open up alternative pipelines.

Intelligent calling systems also have the ability to work in conjunction with multiple remote drops on multipoint as well as point-to-point lines. When disaster strikes, disaster recovery plans can be implemented quickly with a few keyboard commands at the network-management system console, rerouting the failed circuits to alternate facilities.

Upon notification of a failure, the intelligent calling system rapidly restores the faulty line or lines. The system initiates two calls (or multiple calls for as many drops) from a central or nodal site, with two dial-up lines, one for transmitting and another for receiving.

Stored telephone numbers are passed from the network-management system to the intelligent calling system, which performs the automatic dialing. The intelligent calling system notifies the network operator of the successful or unsuccessful establishment of the dial-backup link by monitoring call-progress tones. The system continues to monitor the link for status reporting to the operator.

Many networks today contain both analog lines and DDS circuits. DDS users should consider a vendor who provides intelligent calling systems for both analog and DDS lines under the same network-management umbrella. In this way, the widest possible range of requirements will be satisfied with maximum efficiency and economy.

Today's intelligent calling systems offer microprocessor control, common logic, and internal, electronic, cross-connect capabilities that permit the local end of a failed line or group of lines to be switched to any available dial-up facility. These systems are economical, simple to install, and easy to manage, in turn making disaster recovery plans easier to implement, manage, and control.

3.8 VSAT

Introduced in 1981, very small aperture terminals (VSATs) have since become a well-established, reliable transmission technology for today's private networks. In fact, advances in technology have combined to bring about smaller, more powerful and economical satellite dishes that offer considerable efficiency, convenience, and flexibility over terrestrial, private, leased lines, including fiber-optic transmission systems. VSAT networks are particularly suited for far-flung networks and for a variety of applications, including:

- Retail: point-of-sale (POS) inventory control; credit-card verification; real-time pricing control; payroll accounting; order fulfillment; customer profiling; video broadcasting for product introductions, sales promotions, and employee training.
- Financial: stock, bond, and commodity transactions for individual and institutional investors; on-line data-base access; automated-teller machine (ATM) transactions.
- Travel and lodging: hotel, motel, airline, and rental-car reservations and ticketing; travel-agent access to multiple data bases.
- Utility management: monitoring and control of electrical usage; multisite environmental-control monitoring; security services.
- Process control: supervisory control and data acquisition from remote natural-gas and oil-pumping stations, platforms, and pipelines.
- Private networks: economic alternative to terrestrial leased lines from telephone companies, offering high performance, fast setup time, network control, and streamlined network configuration; backup to terrestrial leased lines.

VSAT technology has made possible the concept of specialized business networks, such as the Hospital Satellite Network, which offers a 24-hour-a-day health-care programming and teleconferencing service to hospitals. Other specialized VSAT networks include the Continuing Legal Education Satellite Network and the Institutional Research Network dedicated to the investment community. College campuses across the country even offer continuing education programs to businesses willing to invest in the cost of a VSAT for their rooftops. That way, employees can continue their education during, or after, the workday without having to commute to the campus. With interactive communication, the give and take between students and instructors closely emulates the classroom environment.

VSATs are also helping companies overcome the burden of local-access charges, which are increasing in some metropolitan areas by as much as 15% to 25% a year. Such increases seriously hamper the ability of network managers to plan and, in some cases, to predict how much their companies will be spending from month to month. In moving from private leased lines to VSATs, companies can stabilize monthly transmission costs, instead of having to deal with increases in private-line charges. In the process, network expansion may be accomplished much easier than with a terrestrial network. Instead of contracting with the local telephone company or interexchange carrier for installation of a new line, for example, capacity may be added to the network in a matter of minutes by switching to a higher data speed.

3.8.1 Network Components

VSAT networks may be classified into three types: point-to-point, point-to-multipoint broadcast, and two-way interactive systems. The first type involves a centralized hub earth station that broadcasts video, voice, or packetized data, or any combination of the three, to all or to a selected group of remote receive-only VSATs. In point-to-point systems, two VSATs communicate voice, data, and images between each other without the need for a hub. The Gemini 56 systems from Hughes Network Systems, for example, provide single-channel digital voice and data on a point-to-point basis. In this case, the VSATs at each site communicate with the satellite directly.

VSAT circuits may be interfaced with any standard, synchronous, 56-kb/s data source such as a host computer, CAD-CAM equipment, codecs, or multiplexers. Two-way interactive or "intelligent" VSATs communicate to a hub in a star arrangement, or among themselves through the hub in a mesh arrangement.

The hub is a master earth station (MES), usually located at the company's headquarters. It can communicate with hundreds or thousands of geographically dispersed locations equipped with smaller, relatively inexpensive VSATs. The MES bounces signals off a satellite parked in a geostationary orbit 22,300 miles above the earth's equator. The bandwidth available with VSATs closely parallels what is available through leased lines. Although VSATs are more expensive than leased telephone lines in terms of up-front costs, ongoing costs are fixed and relatively low, so, in the long run, they can be more economical than leased lines. Building nationwide networks using leased lines to every location may not only be expensive, but also impractical from a logistical standpoint, if only because of the number of carriers and telephone companies that would be involved. Beyond that, VSATs can be used to provide communication to locations that are not reachable by

Table 3.1 Satellite Hub-Station Costs.*

Item	Cost Range
VSAT hub controller	$400,000–$1.2 million
Related equipment	$150,000–$600,000
Uninterruptible power supply	$ 80,000–$120,000
Generator-related equipment	$ 80,000–$100,000
Computer-related equipment	$100,000–$200,000
Engineering and installation	$180,000–$350,000
Building and real estate	$180,000–$600,000
Taxes (sales and property)	$ 80,000–$180,000
Licensing and legal fees	$ 30,000–$ 50,000
Total	$1.28 million–$3.40 million

* Cost of VSATs not included.

terrestrial leased lines, like oil platforms in the Gulf of Mexico or along the Pacific Coast. Satellite channels are leased from such carriers as AT&T, GTE Spacenet (McLean, Vir.), Contel ASC (Rockville, Md.), and about a dozen other vendors, including regional Teleports. In this way, data, voice, and video may be sent and received among all locations.

Since the MES constitutes the bulk of the cost for setting up the VSAT network (Table 3.1), from $1 million to $3 million or more, shared-hub arrangements may be implemented through such vendors as Contel ASC (Rockville, Md.), Cylix Communications Corporation (Memphis, Tenn.), GTE Spacenet, and Hughes Network Systems (Rockville, Md.), formerly M/A-Com Telecommunications, who offer nontariffed bundled services at fixed monthly rates that may be locked in for as long as five to seven years. That way, users who want to take advantage of VSAT networks may do so, even if they do not have a large number of remote locations to justify the expense of establishing a private hub. The shared-hub arrangement may be economical for users with only 40 to 50 locations, in which case the applications may justify using VSATs. Monthly costs vary from a few hundred dollars to a few thousand dollars per VSAT node, according to response-time objectives, traffic volume, and sizes of transactions. Aside from low start-up costs, sharing a hub has many practical benefits, including:

- Contemporary facilities: the service provider's hub is equipped with the most sophisticated transmission and tracking systems to maintain signal quality under the most adverse weather conditions.
- Redundancy: each hub-station subsystem is fully redundant, with backup equipment operating in the hot-standby mode. This eliminates the added expense of protecting your own hub.
- Network management: all VSAT transmissions are monitored and circuits

maintained 24 hours a day, 365 days a year. The service provider insures that a stringent schedule of testing is carried out by trained, experienced personnel, who understand the operation of the MES itself as well as the VSATs and associated equipment at each user site. A continuous connection is maintained with each VSAT to monitor such things as signal levels, signal frequency, transmission accuracy, and proper operation of the entire network. These activities are transparent to the user and do not inhibit the proper operation of user equipment. Escalation procedures insure that even minor problems get prompt attention. This arrangement eliminates the need to staff your own installation and incur further expenses in ongoing training and test equipment.

- Emergency power: the service provider supplies expensive emergency-power supplies to insure that interruptions in power do not affect network performance.

Whether owned or leased, another component of the hub station is the central-control facility. In a VSAT network that is essentially an X.25-packet network, the hub would have a packet switch with PADs (packet assembler-disassemblers) at each VSAT. If the network is essentially an IBM environment, each VSAT would emulate a 3725-cluster controller.

Similar to leased-line terrestrial networks, VSAT networks require a capital investment in customer-premises equipment (CPE) and entail recurring monthly charges based on satellite-channel capacity and shared-hub facilities.

The components associated with each VSAT include the parabolic reflector (dish), which is available in diameters of 1.2 to 2.4 meters. Each VSAT comes with a radio-frequency (RF) power unit that supplies one to five watts, which is all that is needed to support communication at up to 1.544 Mb/s over the C band or Ku band, which are the two bands most commonly used for VSAT transmission. Although one or the other is typically used for a VSAT network, it is possible to build hybrid networks that use both the C band and Ku band together.

The C band was used with the first Intelsat communications satellites in the 1960s. The uplink frequencies of the C band range from 5.9 to 6.4 GHz, while the downlink frequencies range from 3.7 to 4.2 GHz. The C band is the same as that used for terrestrial microwave systems and is still used for satellite applications involving TV distribution and international voice and data communications. Some of the benefits of using the C band are that the equipment is readily available, satellite transponder capacity is also available, and the C band is robust enough to withstand such climatic effects as solar interference, high winds, heavy rain, and ice. When used with VSAT networks, however, the C band's wavelength is such that a dish antenna of at least 4.5 meters is required for interactive operation, unless spread-spectrum modulation is employed, in which the average energy of the transmitted signal is spread over a bandwidth that is much wider than the information bandwidth. This technique eliminates most of the interference, permitting the use of a much smaller antenna. This is precisely the method used by Equatorial Communications Inc. (Mountain View, Calif.), which was bought by Contel ASC (Rockville, Md.) in 1987. The spread-spectrum technique limits interference from adjacent satellite transmissions. Without using this technique or using a large enough antenna, frequencies must be coordinated to avoid the possibility of interference.

Ku-band transmission was developed by Satellite Business Systems (now absorbed into MCI) in the early 1980s for use with its own satellites. Many other satellite companies have since adapted the Ku band for use with their own satellites. The Ku band's uplink frequencies range from 14 to 14.5 GHz, while the downlink frequencies range from 11.7 to 12.2 GHz. The Ku band was specifically developed for use with very small aperture terminals.

Both the C band and the Ku band offer ample bandwidth capacity, but both require that users obtain an FCC license for operation. With Ku-band systems, users may obtain a single, blanket license for network operation, provided that all the VSATs on the network are of the same type. Otherwise, each site is required to be licensed individually, a process that may take as long as six months or more. With C-band systems using spread-spectrum technology, each site must be licensed individually, but in recognition of the C band's lower interference potential, the FCC allows satellite transmission to begin immediately upon the submission of a short application form. License fees for C-band and Ku-band systems having a dish diameter of two meters or less consist of a one-time network fee of $3,000, plus $30 for each terminal on the network. Licensing is the vehicle by which the FCC controls the use of frequencies to prevent interference problems.

A digital interface unit (DIU) is also included with the VSAT system, which differs in port capacity according to manufacturer. The ports of most DIUs may be configured for up to 9.6-kb/s asynchronous or up to 64-kb/s synchronous data and provide RS-232C, RS-422, or V.35 interfaces. Voice ports may support 32-kb/s and 16-kb/s compressed voice. The VSAT interfaces directly with a PBX four-wire E&M trunk or the two-wire loop *via* RJ-11 connections.

Other VSATs support the full range of leased-line transmission speeds, including T1. For example, NEC America Inc.'s Radio and Transmission Systems Group (Herndon, Vir.) offers the Nextar CL VSAT which operates at the standard 1.544-Mb/s T1 transmission speed, while its enhanced version also allows the user to select from a menu of other data rates from 9.6 kb/s to 56 kb/s. Both products are available at the European T1-equivalent speed of 2.048 Mb/s. Some VSAT vendors even customize networks to provide nonstandard or "Fractional T1" rates, such as 384 kb/s and 512 kb/s, to support specific applications.

A bare-bones, two-meter, receive-only VSAT may be purchased for under $1,000, while the cost of sophisticated interactive systems ranges between $10,000 and $15,000. With high-end systems, network-management software is usually bundled into the hardware price. VSAT installation costs vary from only a few hundred dollars per location to a few thousand dollars per location. In most cases, a VSAT can be installed and operational within 30 to 45 days from the date of purchase, offering another advantage over terrestrial leased lines, which may take two to three months or longer to put into service. This brings up another advantage of VSAT networks — the monthly costs are distance-independent. A 56-kb/s point-to-point satellite link, for example, goes for around $1,000 a month, while a 56-kb/s DDS circuit from New York to San Francisco costs $3,700, including local-loop charges at each end.

Within the VSAT network itself, protocol handling is a key function. The scope and functionality of protocol handling, however, differ markedly among VSAT-network providers. There are two aspects of a VSAT-network protocol that merit your attention. The first is the backbone network protocol, which is responsible for flow control, retransmissions of packets delivered with errors, and running concurrent multiple sessions. The second is related to the user-application interface, which provides a compatible translation between the backbone protocol and the user protocol. Several data-communication protocols are used in VSAT networks, such as SNA-SDLC, BSC 3270, X.25, and Houston Automatic Spooling Program (HASP). Multiple protocols may be in use at the same VSAT location. Specialized protocols and less frequently used protocols, like Burrough's Poll Select, may be accommodated by some VSAT vendors. In fact, every port may be configured for its own protocol and transmission rate. Ports may be reconfigured locally, or remotely from the network hub. Some VSATs are designed only for instantaneous, low-speed, inquiry-response traffic and do not support protocols for channel setup. Consequently, they may not be used for any other application. Although most VSATs available today support terminals operating in the SNA-SDLC environment, some may not support the different versions of SNA-SDLC at the same hub. If you plan on using a specialized protocol, you may incur additional costs for optimization algorithms and external protocol converters.

Because some vendors have had difficulty implementing communications protocols, you have to address this issue in the performance-test criteria, which you should include in your purchase contract as well as your RFP. That way, the burden is on the vendor to get things running properly. One VSAT startup, Telecom General Corporation (San Jose, Calif.), has gone out of business because of network software difficulties, but do not think that the size and reputation of the vendor precludes software-related problems. One of the reasons Hughes purchased M/A-Com Telecommunications in 1987 was to shore up its apparent deficiencies in the ground-segment of VSAT technology, including such areas as digital-signal processing, switching, and protocol handling.

There are several access protocols that may be in use over the VSAT network. Time-division multiple access (TDMA), for example, is used to provide each VSAT with access to the hub *via* the satellite by the bursting of digital information onto its assigned radio-frequency carrier. Each VSAT bursts at its assigned time relative to the other VSATs on the network. Dividing access in this way — by time slots — allows capacity to be allocated in real-time according to demand.

A number of subaccess methods are available to suit differing applications. For example, slotted Aloha is best suited for short, bursty messages in applications that require fast response times, such as point-of-sale credit-card verification and ATM transactions. This contention technique allows VSATs to transmit at any time, and to continue transmitting if they receive acknowledgement that no other station is sending, but this method requires that channel utilization be limited to around 37% to avoid an unacceptable number of collisions, which increases the delay caused by retransmissions. With pure Aloha, channel utilization is limited to 17% before exhibiting this kind of instability.

With the transaction-reservation access mode, the host reserves slots for each terminal for the transmission of an assigned number of packets. In assigning bandwidth to match message duration, more efficient use is made of the space segment than the random-access method, thus improving throughput. A drawback to reservation access is that it requires more time for channel setup, which is a source of delay. This access method is best suited for long-duration transactions of variable message sizes.

The stream-access method involves dedicated time slots that are preassigned to one transmitter without contention from the others. This means that bandwidth is assigned to the terminal for the duration of the call. This access method is typically used in applications requiring long connect times, such as circuit-switched voice and high-speed data transfer.

Also available are adaptive techniques that change the method of access to coincide with changes in the characteristics of the traffic originating from the VSAT. Thus, the VSAT could be operating in slotted Aloha and then switch to reservation or stream access when traffic becomes heavy or message lengths expand beyond the allocated Aloha slot size. The hub station controller carries out traffic monitoring, switching, and coordination tasks that make such adaptive techniques work.

Determining the most appropriate access method for your application or applications brings into play trade-offs between response time and space-segment utilization. Because the choice of access method affects overall network performance, this aspect of VSAT operation will enter into your selection of vendor. It is interesting to note that some large users who have moved from multidrop leased-line configurations to VSATs have done so out of the need to reduce the amount of delay on their networks, as well as to obtain more control over communications costs. The Holiday Inns' (Memphis, Tenn.) Holidex reservation system, for example, routinely experienced polling delays of 10 to 25 seconds. With the VSAT network, which will include 1,200 sites by the end of 1988, delay has been reduced to five seconds or less, inclusive of host-processing time, satellite delay, and equipment-processing delay.

For security-conscious VSAT users, there are two types of encryption generally available: source-to-destination and link. In the former case, data is encrypted at the point of origin and decrypted at the point of final destination, providing security throughout most of the communications process. In the latter case, data are encrypted right before transmission over the link and decrypted as they arrive at the other end of the link. The degree of security is usually dictated by the application. Financial data such as fund transfers, credit-card numbers, and automatic-teller machine-access codes are examples of the kind of information that typically warrants maximum protection through source-to-destination encryption.

Aside from the cost of FCC licenses and building and zoning permits, when evaluating VSAT suppliers, do not overlook the possible "hidden" costs of the system. Since the terminal end requires cabling, find out if cabling is included in the purchase price, how much is required, and what is involved to add more later. Also, if security is important to you, data encryption is another hidden cost of the VSAT network, requiring an investment of several thousand dollars per site.

Many who are considering VSAT networks think that adding transmission capabilities only involves adding appropriate satellite channels. That is only half the story, however. Keep in mind that adding capabilities to your VSAT network will require appropriate investments in componentry. For example, if you originally designed your VSAT network for transaction-based data communications and want to add the capability to carry voice traffic, additional space capacity may be required so as not to reduce the response times for data. You will also need special voice cards at the DIU to perform the conversion to analog, unless your telephones are all-digital, and if you want to add the capability to transmit video, you will require special equipment like codecs and monitors, as well as additional satellite channels.

As new capabilities and VSAT locations are added to the network, you will require additional satellite channel capacity, but this may not be so easy to come by. Since space is usually sold on an as-available basis, you may not get the same transponder, in which case you will have to shell out more money for up and down converters.

You should find out how your space capacity is protected. If the transponder holding your channels goes out, will a spare transponder take over to continue service? If the whole satellite fails, will you have access to a transponder on another satellite? These are important questions, because there are premium charges for such protection, and you will have to determine whether they are worth the cost.

If you are having difficulty justifying the up-front cost of implementing a VSAT network, do not overlook the value-added benefits, which may outweigh economic considerations. Typically, a VSAT network will deliver a higher level of performance, availability, and reliability than terrestrial networks. Moreover, a VSAT network can more easily accommodate new applications and services than a terrestrial network, and do so within a much shorter time frame. Also consider the revenue-generating opportunities that a VSAT network may present. You can allow smaller companies in your area to connect to your VSAT to access special services, for example.

Once the VSAT is installed and operational, the cost to add other services is very small. Additional savings may accrue from eliminating the need to buy more modems, multiplexers, and data sets, which are necessary for achieving economies and efficiencies over terrestrial networks.

Determine the charges for such things as site surveys and electrical work. Make sure you know in advance what the costs are for special construction required to mount, protect, or hide the dish. Flush out any equipment shipping and handling costs, as well as any costs that may be required for schedule implementation. If the vendor views your purchase as a large order, you can negotiate away a lot of these miscellaneous charges.

If you are sharing a hub, find out the location of the vendor's facility. You will probably need a leased line from your corporate headquarters to the hub to perform data-base updates and access management reports. The cost of one or more leased lines, plus modems, are expenses usually overlooked in the overall scheme of things.

Also look into what skill levels are necessary to operate and maintain the particular product you are buying. Some systems are highly complex, requiring the most skilled operators. Others are very user-friendly, and can be manned with less technically oriented personnel. For this reason, pay particular attention to the network-management system the vendor provides. It is a good idea to have potential operators with you on a site visit so you can gauge their readiness to learn the management system. Their questions, interest level, and interactivity with operators at the customer site will provide you with some indication of their ability to learn that particular system.

If you like the idea of one-stop shopping, you will have to determine how many pieces of the VSAT network the vendor is really capable of providing. Many VSAT-network providers are not diversified enough to provide all of the services required to get the system installed and operational. Very few vendors own their own satellites, for example, and must coordinate with those who do. Some vendors lack a national service presence, and must subcontract for installation and maintenance. Others do not even manufacture their own hardware or develop their own software. Because there are potentially so many different players involved in putting together a VSAT network, it is important to establish who is responsible for what when troubles arise, and to nail down response times in writing.

3.8.2 Network Management

The performance of the VSAT network is continuously monitored at the host location by a network-control system, which receives the same signals as the remote sites. A failure anywhere on the network automatically alerts the network-control operator, who can reconfigure capacity among individual VSATs by reprogramming the microprocessor interfaces at the remote sites — *via* dial-up lines, if necessary.

Overall link performance is determined by the bit-error rate (BER), network availability, and response time. Because of the huge amount of information transmitted by the hub station, uplink performance requirements are more stringent. A combination of uplink and downlink availability, coupled with BER and response time, provides the network-control operator with overall network-performance information on a continuous basis.

Reconfiguring the VSAT may be done at each remote site *via* a special hardware-software package. As part of its Intelligent VSAT, Scientific-Atlanta Satellite Communications Inc. (Atlanta, Ga.), for instance, provides users with the means to access the network data base at the hub site *via* a dial-up or leased line. Upon linkup, remote users can reconfigure their VSATs without involving the hub-site operator. Through menu-driven software running on a microcomputer, users can do such things as add or delete device addresses at their own VSAT locations, configure a remote VSAT-to-hub link, and create permanent or switchable virtual circuits between VSATs. Users can also change the protocols supported at their VSAT location.

A network-management system should be able to indicate whether power failures are local or remote. It should also be able to find the source of communications problems, and whether the trouble is with the software or hardware. Such capabilities can eliminate the expense of dispatching technicians to remote locations and make sure that when technicians must be dispatched, they will have appropriate replacement parts, test gear, software patch, and documentation with them.

To get the most out of your VSAT network, the management system should offer a full range of accounting, maintenance, and data-flow statistics, including inbound *versus* outbound data flow, peak period statistics, and total traffic volume by node.

When evaluating the network-management systems of particular vendors, look for the capabilities that improve efficiency. When the controller polls the remote VSATs for status information, for example, make sure it includes a report-by-exception capability. That way, only changes from previous reports are sent to the printer or terminal display. This makes managing the network much easier, because redundant information is left out of the reports.

If you are plagued by high personnel turnover or personnel with limited technical skills, you may want to consider a network-management system that can handle key operations without too much operator intervention, such as identifying fault conditions, performing diagnostics, and initiating restoral procedures. The network-management system should have the capability of downloading software to remote locations for system upgrades and reconfiguring the network. Remote users should have limited access to the network data base to add, delete, or change terminal equipment addresses belonging to their locations.

3.8.3 Planning Considerations

If you are thinking about the possibility of a VSAT network for your organization, the first thing to understand is that it is one of the most planning-intensive tasks you may ever encounter. Even before you start checking into vendor offerings, it is a good idea to do some probing into possible sites for the VSAT dishes. If your organization owns its own buildings or real estate, locating a suitable location will probably not be a problem. In this case, all that is required is compliance with local zoning ordinances and that the VSAT site have an unobstructed "view" of the satellite. VSAT vendors and service providers will conduct extensive site surveys to find suitable places to install the VSAT. They will determine whether the dish is best mounted on the ground, roof, a pole, or a tower. Because the VSAT can be installed in a variety of places using numerous standard mounts, finding a line of sight is usually the least of your problems.

If you lease your building or offices in a metropolitan area, however, there are some potentially sticky problems to resolve before committing yourself to a VSAT network. The first potential problem is getting permission to install a dish on the roof of the building. Some landlords will not permit rooftop installations under any circumstances, believing that such equipment is a blight on their building which can diminish its value. Others are worried about possible lawsuits if the mount should ever fail and the detached dish cause damage to property or injury to people.

Still others recognize the profit potential of renting roof rights, but try to capitalize on it by charging exorbitant rates. In such cases, you might find yourself paying more for the square footage on the roof than for office space. Some VSAT network providers like GTE Spacenet can lend a helping hand, providing a complete instruction manual on how to explain to a landlord what a VSAT is and what it is used for. It even includes a sample "dear-landlord" letter, so you can get negotiations off to a good start. Despite the assistance of vendors in these matters, the onus is on you to prepare the way for VSAT installation. Even when vendors have provided you with assurances that they are in compliance with all rules and regulations, you must verify everything.

If the landlord is not the problem, you may still encounter resistance from historic-preservation groups, who believe the dishes go too far in changing the character of the building or neighborhood. Zoning boards may object to the dishes, claiming that they are unsightly and must be hidden from public view. Some zoning boards are so adamantly against dishes of any kind that they have placed restrictions on installation that make it too expensive for the user to consider any further.

Even architects may try to block dish installation on the grounds that it detracts from the building's original design.

Your best risk-avoidance strategy is to face up to these issues head-on. Find out what problems other VSAT users in your area have experienced and how they have handled them. Research local zoning ordinances; some cities along the Gulf Coast, for example, require that dishes withstand 120-mile-an-hour winds, which can make the cost of installation prohibitive. Even if VSATs have been installed in your area very recently, you must review the local ordinances and building codes for any changes. Review building plans for possible VSAT locations and alternatives. Check into possible insurance coverage and building-permit requirements. A good source of information about these matters are the VSAT vendors themselves, if they are currently conducting site surveys for dish installations in your area. Failure to do your homework in these vital areas can result in costly litigation for your company or additional expenses to bring the VSAT installation into full compliance with local ordinances and building codes. Only after these issues have been thoroughly explored, should you get serious about committing to a specific vendor.

In evaluating vendors, make sure that the company is well capitalized and financially stable, particularly if it is less than three years old or is the subsidiary of a larger firm. In this segment of the communications industry, sales cycles are typically very long, while product-development efforts are very expensive. You must be sure that the vendor has the resources to ride out cyclic demand and rough-and-tumble competition. In the case of subsidiary companies, you will have to determine the degree of commitment and financial support the parent is providing. In this regard, it is interesting to note K-Mart's selection of GTE Spacenet to provide its nationwide VSAT network. K-Mart began planning its 2,100-node network in 1986 for phased-in implementation through 1989. Among its reasons for going with GTE were:

- Commitment to developing new satellite technology and the $500 million it had already invested in five satellites.
- Leadership in supplying public and private networks, including the public-packet network operated by GTE Telenet.
- Eighth largest telephone company in the world; second largest in the nation.
- Assets of $14.4 billion.

VSATs should not necessarily be evaluated as stand-alone networks, but as key components of the overall network mix, particularly as it relates to wide-area networking. Although T1 lines, intelligent multiplexers, and cross-connect systems have made it fairly easy to build economical hybrid networks that blend the best of public and private networks for optimal advantage, VSATs offer the means to serve isolated nodes economically, providing a gateway to the backbone T1 network.

Despite the apparent advantages of VSAT networks and their fit with a variety of applications, it must be understood that they are not the answer to every networking need. The user must do a thorough evaluation of VSATs as they relate to current and foreseeable organizational needs. You should also take into consideration that the VSAT market fluctuates wildly. One moment there is a lot of demand, and the next moment there is virtually no growth, forcing vendors to backpeddle to stay alive, and if they do not backpeddle fast enough and market conditions do not improve, they can quite easily go out of business, leaving you without ongoing support.

Of course, as in any other market, VSAT vendors favor their own products over those of their competitors. So do not look for vendors to provide objective information concerning such matters as system design. You cannot expect vendors to understand your total network operation; for the most part, they are capable of understanding only a small piece of it — the part they provide. A better source of information is a systems integrator, who offers the equipment of several vendors and understands how the pieces of a network operate together. You can also turn to network design firms, which will research your needs, outline the design options for you, and make an appropriate recommendation.

If you are staffed to design the network in-house, you might want to consider having a consultant review your plans. A second opinion may reveal a design flaw or stimulate discussion of a better way of doing things.

While VSATs may prove to be an economical solution for wide-area networking, digital microwave is better suited for short-haul transmission involving distances of 30 to 40 miles. For greater distances, repeaters may be added, the cost of which can run as high as 80% of the price for two antennas. Digital microwave offers a range of transmission facilities, including T1 (1.544 Mb/s), T2 (6.312 Mb/s), and T3 (44.736 Mb/s). Like VSATs, microwave systems require a license from the FCC, which is not difficult to obtain, if you follow the procedures.

If you are considering adding microwave* to your network, much of the previous discussion about VSAT planning, installation, and vendor selection also applies, except that the availability of transponders is no longer an issue, because microwave does not involve transmission *via* satellite. What you will have to do is evaluate microwave equipment in terms of transmission capacity, reliability, expansion, costs, management, availability of field service, as well as vendor stability.

---

* For additional information about microwave, I recommend Steve Ditto's book, *Buying Short-Haul Microwave: The Official Guide to Choosing, Acquiring and Using a Short-Haul Microwave System in North America,* published in 1988 by The Telecom Library (New York, New York).

REFERENCES

[1]   *Datamation,* March 15, 1988, p. 86.

[2]   "T-1 Multiplexer Industry Analysis 1988," Vertical Systems Group (Dedham, Massachusetts), 1988.

Chapter 4
Office Automation

4.1 WHAT IS OFFICE AUTOMATION?

There was a time when the term "office automation" meant "word processing" with the added capabilities for simple graphics, internal messaging, and telecommunications *via* modem. This was accomplished by tying dumb terminals to a specialized minicomputer *via* coaxial cabling. (Today, a variety of cabling schemes may be used, including twisted pair and RS-232C.) Such a system was introduced by Wang Laboratories, Inc. (Lowell, Mass.), in the late 1970s. By today's standards, it was relatively simple, yet it conjured up images of greatly improved office productivity. The media even speculated about the imminence of the "paperless" office, a vision of the future that has not only proven naive, but laughable.

These primitive office systems used proprietary hardware and software, which meant that you could not mix and match components from different vendors. If you bought the computer from Wang, for example, you had to buy terminals and printers from Wang as well. You were also dependent on Wang for port expansion, upgrades, and options. The only things you did not have to purchase from Wang were blank eight-inch diskettes, print wheels and ribbons, and dust covers for the equipment.

In its formative years, office automation was a concept aimed at helping secretaries improve their administrative-support capabilities. Vendors even targeted secretaries for their sales pitches with the expectation that they would sell the concept to their bosses.

Today's concept of office automation is multidimensional; it goes beyond providing administrative support to include decision support as well. No longer limited to text editing and calendar management, office automation now brings under its umbrella just about every conceivable application, communications medium, and technology that can be brought to bear on organizational performance, and it is a concept that is no longer aimed at secretaries alone; it is also aimed at improving the performance of managers and technical people.

Office automation is still a very fluid concept, so much so that the term may have lost its meaning, if it ever had one. Pick up any trade magazine, for example, and you will be overwhelmed by articles and advertisements that run the gamut from office shredders, postage meters, FAX, electronic mail, binding and lettering machines, telephone systems, copiers, electronic typewriters, to microfiche systems. Office automation, then, appears to be quite an open-ended term that may include any conceivable office task that can be automated, including hole punching and pencil sharpening. In the absence of a consensus definition of the term, this chapter will confine itself to a discussion of the basic building blocks of office automation: computers, printers, storage media, desktop publishing, and electronic mail. Other facets of office automation such as LANs, data switches, PBXs, and modems are covered in separate chapters.

The approaches to achieving office automation are numerous, depending on the needs of the organization *and* its installed base of computers and networks. Here are some possibilities:

- Dumb terminals connected to a mainframe *via* a server, administered by Management Information Services (MIS);
- A combination of microcomputers and dumb terminals connected with each other through a local area network or a minicomputer *via* a server, administered by MIS;
- A combination of microcomputers and dumb terminals connected to a minicomputer *via* a server, administered by a department;
- A combination of microcomputers and dumb terminals connected to a minicomputer *via* a server, administered by a work-group;
- Stand-alone microcomputers that share peripheral devices *via* direct connections or inexpensive front-end switches, administered by a department or a work-group.

The configuration possibilities for office automation are virtually endless. This chapter focuses on the increasing use of microcomputers and related peripherals for office automation.

4.2 DETERMINING NEEDS

The essence of office automation is communication. You can determine the need for office technology by borrowing a page from journalism school, by asking "who, what, where, when, why, and how." Specifically, get the answers to the following basic questions to guide your efforts in selecting the most appropriate products and systems: [1]

- *Who* has to communicate, and with whom?

- *What* is the nature of their communications?
- *Where* do their communications go?
- *When,* and how frequently, do they communicate?
- *Why* do they communicate?
- *How* is the best way for them to communicate?

As applied to microcomputers, for example, the question of who has to communicate is very important in determining the type of equipment to buy. If the answer involves your organization's field salespeople, the appropriate equipment might be laptop computers that are lightweight and easily carried during extended periods of travel. A manager might need a networking capability so that information may be gathered from a variety of data bases for the purpose of putting together a business proposal or to stay in touch with other managers working together on a common project. For a financial person, a desktop computer with 40 Mbytes of hard-disk storage for spreadsheets might be appropriate. A secretary might only need a microcomputer with 20 Mbytes of hard-disk storage and a monochrome screen.

As applied to traveling salespeople, the question of what is the nature of their communications determines how their laptops will be equipped. Salespeople in the field require capabilities for batch and interactive communications. For this, a modem is required, preferably internal. The question of where their communications go has already been determined: to the corporate headquarters. If the data bases are distributed, however, each may have its own access number. This would indicate the need for a modem with an autodial capability. In establishing when and how frequently the salespeople communicate, you may discover that they have the need to communicate with various corporate data bases many times on a daily basis. This would call for an autodial modem that can handle macro programs for automatically entering the various log-on procedures upon establishing the connection. Why salespeople must communicate with the mainframe is not difficult to determine; the most likely reasons include posting sales call reports and entering orders. Order entry may entail interactive communication in that the salesperson enters the order as responses to prompts issued by the mainframe. Upon completion, the order is acknowledged. Alternatively, the salesperson fills out the order using a template stored in the laptop. At the most convenient time, accumulated sales orders can be batch loaded to the mainframe. Retrieving current price and inventory information is definitely a batch operation, which may require that the laptop have a hard-disk capacity of at least 20 Mbytes. Having analyzed the situation this far, the question of how best to communicate must be addressed. Some of the choices might include dial-up communication at 300 to 1200 b/s *via* the public switched network, electronic mail, or packet networks like Telenet Communications Corporation's (Reston, Vir.) Telenet or McDonnell Douglas Network Systems Company's (San Jose, Calif.) Tymnet. Of course, if the communications from the field are not considered urgent, the question of how best to communicate may include postal mail.

It is important to establish the needs of each category of user before recommending the purchase of hardware or software. Since top management holds the purse strings — at the least, line item veto power — a thorough needs assessment is oftentimes required to justify the purchase. Sometimes, a needs assessment can reveal other opportunities within the office environment to increase efficiency and improve performance.

4.3 MICROCOMPUTERS

Barriers to office-management efficiency and to productivity continue to be broken, largely through the introduction of faster processors, greater storage availability, and improved operating systems, which allow today's microcomputer vendors to boast of multitasking, multifunctional, and multiuser capabilities. Such developments are of particular significance to small and medium-sized companies which appreciate the competitive aspects of maintaining high levels of office efficiency, but which nevertheless can neither spend lavishly for dedicated staff, nor tolerate unnecessary redundancy in computer hardware and software. Although large companies still favor the centralized approach to office automation, largely because of their huge investments in data-processing systems, small and medium-sized firms seem to favor the decentralized approach for its immediately available power, sophistication, and low start-up cost. Further, microcomputer-based office systems are easily modified for multifunctional use and have the flexibility to run mainframe-type applications, as well as hundreds of commercially available programs.

There is, however, more than technology driving the trend toward decentralization. In the last decade, office workers have won more say in how their jobs are structured and have assumed more responsibility for their own performance. With the arrival of microcomputers in the workplace, it was inevitable that users became computer literate. With more autonomy on the job and the general availability of these productivity tools, users have become more self-sufficient and less dependent on MIS, which was viewed, rightly or wrongly, as a bottleneck to office productivity. Users got the support of their managers — mostly through *faits accomplis* — who gradually came to appreciate that they could, and should, be responsible for their department's own destiny, instead of being at the mercy of MIS, which had its own priorities and hidden agendas. At the risk of sounding Marxist, office managers and users teamed up to take control of the means of production, relying on MIS only for such support functions as planning, research, product evaluation, maintenance, and networking. In taking control of their own destinies, however, office managers and users had to accept some of the responsibilities that went with their newfound independence, such as learning to cope with differing file formats and noninteroperable software.

It is now apparent that microcomputers are rapidly displacing minicomputers and mainframes as the cornerstone of office automation, if they have not already done so. At the same time, the microcomputer has become truly a commodity item, with an installed base of over 20 million nationwide [2]. The increasing demand for microcomputers and the quickening pace of innovation have combined to produce literally hundreds of models to choose from. This, in turn, has complicated the purchasing process and increased the element of risk with respect to choosing the vendor.

Shopping for the right microcomputer is not an easy task. The buyer must become educated about microcomputers in general and about their many possible applications within the office environment. Beyond these considerations, the buyer must determine what software is available to address the specific applications desired, and understand that the choice of software should drive the choice of hardware. The temptation to make a hardware-driven decision increases the chance that the software will not meet your overall requirements. At the same time, you must protect your investment against early obsolescence, and position yourself to take advantage of the vast opportunities that new operating systems and applications programs will bring. These considerations point to the need for properly timing the purchase of new equipment.

More often than not, this education process takes up more time than inexperienced buyers can usually afford. Even experienced buyers can become quickly overwhelmed by the avalanche of information about new developments in software and hardware, as well as the claims and counterclaims of vendors. In such cases, the selection process may boil down to one of risk avoidance — determining the best alternative within a set of buying criteria. When considering a bulk purchase of microcomputers, you can reduce the level of risk by probing vendors about the modularity, functionality, and processing power of their products.

In the next three sections, the modularity, functionality, and processing power of microcomputers are discussed in more detail than you may require. If you are not going to be personally involved in evaluating microcomputers prior to a volume purchase, you may want to skip this material and go right to Section 4.4.

4.3.1 Processing Power

Processing power determines what you can do with a microcomputer — and how fast. There are three ranges of microcomputer in common use today, distinguished by the type of processor they use. At the low end are the PC and XT classes of microcomputer based on Intel's 8088 chip, which was designed in the mid-1970s. In the middle range is the AT class of microcomputers based on Intel's 80286 processor, which runs three times faster than the 8088. At the high end is the 80386-based microcomputer that runs twice as fast as those in the AT class. The 80386 is also able to emulate multiple 8088 processors, making it better suited than the others for multitasking. Incidentally, when it comes to multitasking, an 80286 machine is the minimum requirement.

Within each class of microcomputer, there is another point of differentiation — clock speed. Expressed as millions of cycles per second, or megahertz (MHz), clock speed is an indicator of throughput, or the speed at which information can move through the system when it is working at saturation. The higher the clock speed, the faster bits (pulses or no pulses) are recognized and processed. Each class of microcomputer has its own range of clock speeds. The 8088 processor, for example, can operate within the range of 4.77 to 10 MHz, whereas the 80286 operates within the range of 8 to 16 MHz. The 80386, on the other hand, has an operating range of 12.5 to 25 MHz, which can put it into the performance category of some minicomputers, and even a few mainframes.

Another factor that affects the performance of microcomputers, especially those at the middle and high ranges, is the speed of the memory chips used in the system. The speed at which information is entered into and retrieved from memory is expressed in nanoseconds, or billionths of a second. An 8088 XT running at 8 MHz can get along fine with 150-nanosecond chips, but that kind of performance is not sufficient for an 80286 AT running at 12 MHz, which requires the faster 100-nanosecond or 120-nanosecond memory chips for reliable operation. At 150 nanoseconds, the 12-MHz AT would overrun the memory chips and introduce errors. This problem is especially pronounced in the 80386 machines, which may have to rely on "wait states" to avoid overrunning the memory chips. Memory chips perform three types of operations, consisting of one cycle each: they receive instructions, find the information, and send it to the processor. In making the processor wait one cycle before issuing a new instruction, the entire system is slowed down to permit more reliable operation. The processor may require two or three wait states when accessing system board memory, and 10 to 18 wait states when reading from memory expansion boards. One way to eliminate wait states is to use faster memory chips of 70 or 80 nanoseconds, but these chips are expensive and not yet available in large quantities. Therefore, the manufacturers of high-end microcomputers have had to compromise performance to make their products cost-competitive. This will change as the 8 to 9 million users of 8088-based microcomputers migrate to the 80286 and 80386 machines, which are virtually required for heavy-duty applications involving desktop publishing, computer-aided design, and database sharing.

Another distinguishing characteristic of the 80386 microprocessor is its ability to "pipeline," a feature that allows the various internal modules of the chip to operate concurrently. Pipelining means that the chip can address memory at the same time it retrieves or executes an instruction, for example. In performing these operations concurrently, the overall speed of the microcomputer is improved by 10 to 15 percent.

Among the other features that improve the performance of the 80386 is the

32-bit bus — or data path — linking the registers, which are also 32-bit. Registers are where computations take place; as such, they are the brains of any microprocessor. The more bits in a register, the bigger the portion of a problem it can handle within a processing cycle. The wider the data path between registers, the more information that can be moved to and fro at any one time. The 8088 and 80286 microprocessors only use 16-bit registers, which means that the 80386 can handle twice the information in the same amount of time, and move it from module to module two times faster. The registers can also operate as pairs of 16-bit registers, making the 80386 compatible with previous-generation instruction sets.

From a hardware standpoint, the 80386 processor has a feature called the "virtual 86" mode, which allows it to be partitioned into several virtual computers that can operate independently under a control program like Microsoft's Windows/386. You can put a spreadsheet program in one virtual computer, a data-base management system in another, and a word-processing program in still another. You can designate other virtual computers to handle communications, utilities, CAD, accounting, or any other applications programs. With the Windows/386 control program, the user can move in and out of applications quickly, transferring information from one program to another with unprecedented ease. This arrangement also permits the simultaneous execution of multiple tasks, and allows the 80386 to act as a host to dumb terminals.

The high performance of 80386 machines means that for the first time computers can take full advantage of the access time and data-transfer rate of the most sophisticated hard-disk drives. Because the 80386 can use huge amounts of data at a time, the speed at which the hard disk can locate any given byte becomes critically important. The AT class of microcomputers required hard disks with a minimum average access time of 40 milliseconds, whereas today's voracious 80386 machines require access speeds of under 28 milliseconds. The data-transfer rate is equally important with 80386 machines, but for another reason: they can handle data faster than conventional hard disks are capable of delivering it. Although the data-transfer rates of conventional drives are rated at 5 MHz, that is still too fast for PCs, XTs, and ATs. The 80386, however, is capable of handling data as fast as it can be delivered from the hard disk.

Despite the bargain-basement prices of even brand name 8088 machines, you would be buying yesterday's technology. A careful assessment of office needs is required before you decide to buy these older microcomputers. Although they may provide an acceptable level of performance as stand-alone devices, which perform relatively simple tasks like word processing, their performance deteriorates substantially in memory-intensive applications, making them awkward for desktop publishing, and quite unsuitable as network servers or for other multitasking applications.

Contributing to the gradual migration to 80386 machines in the office environment is the anticipated industry shift to OS/2, an advanced version of the MS-

DOS operating system announced in early 1987. This multitasking operating system, capable of unleashing the full power of the 80386 microprocessor, will encourage the development of a whole new generation of high-performance applications, which will not run on 8088 machines. A key advantage of OS/2 is its ability to overcome the 640-Kbyte memory limitation of DOS. This will become significant in the office environment upon the release of new software for desktop-publishing, accounting, presentation-graphics, and other high-performance programs, which will have the ability to address up to 16 million bytes of memory, allowing programs to become twenty times larger than those built around DOS.

4.3.2 Modularity

When selecting the hardware, modularity is a feature that can determine the success of integrating microcomputers into the office environment. Modularity means that you can purchase only what you need now, and add to it later, as your needs require. For example, to minimize your front-end costs, you may be attracted to the low-end bargain units that use Intel's 8088 processor. These microcomputers typically have a relatively slow clock speed of not more than 10 MHz. Nevertheless, you may find that this, plus the two 360-K disk drives, are sufficient to implement the word-processing package you have decided upon. After all, you can always add RAM (Random Access Memory) and an internal 20-Mbyte drive when the need to do spreadsheets becomes apparent, or when the next enhancement to the word-processing package requires more space than a single floppy disk can handle, but in eventually adding a drive card, you may find that it cannot be accessed. The problem is that the manufacturer allotted only enough device-address space to identify the two disk drives that came with the original unit. Despite the availability of expansion slots in the chassis, because the unit was not designed for modularity, you may not be able to add high-capacity storage, either internally or externally. When using a 20-Mbyte hard disk in a microcomputer that uses an 8088 processor running at 10 MHz, information handling can become quite inefficient, if only because it takes longer to search for files. This situation may worsen over time, because individual files tend to get scattered all over the disk. This simple example underscores the fact that microcomputer selection is not always so easy.

Before committing yourself to a vendor's hardware, find out exactly what brands of product it is compatible with, should you ever want to use the expansion slots to add things like memory, speed, graphics, color, and communications capabilities. Any vendor worth considering will have done performance tests on a variety of add-on products and will be able to recommend those that work best with their brand of microcomputers. Obtaining this information may take some digging. You may be able to get it from the manufacturers' customer-service staff; if not, they may be able to direct you to someone in their organization who can help. Another source of information is the mail-order firm which sells both microcomputers and add-ons. Since this type of firm hopes to get repeat business after the sale of a microcomputer, it must stock add-ons that perform best with the brands it sells. Use the toll free 800 number and ask to speak with a design engineer or product-evaluation specialist. More often than not, you will get the "how" and the "why" on products that go together best — insights you cannot rely on when they come from a sales representative who is under pressure to make quota.

4.3.3 Functionality

Closely related to modularity is functionality, which refers to how the microcomputer may be used to meet the requirements of different office environments with minimal modification. Since your office is not likely to remain static over the years, it is important to know if the equipment you are buying will still be useful as both work environment and job requirements change. For example, is the microcomputer capable of performing multiple tasks simultaneously? Can it perform such diverse functions as facsimile transmission, desktop publishing, and spreadsheets, should those needs arise? Can the microcomputers you buy today for stand-alone use be adapted to the networking environment? If not, what will it cost to modify them for networking?

The introduction of Intel's 80286 and 80386 microprocessors has turned microcomputers into powerful multifunctional devices capable of increasing office productivity with a minimum investment in additional hardware. Multifunctionality is of particular concern to small and medium-sized companies, which cannot afford to dedicate office equipment and personnel to specialized tasks. The multifunctional capabilities of today's powerful microcomputers may be implemented with the addition of plug-in boards.

The microcomputer can connect to the office PBX or key system *via* the RS-232C interface to record the details of all outgoing calls, such as the called number, the date and time of the call, the originating station, and the type of line used (e.g., WATS, FX, DDD). The telecom-management software arranges this laundry list of calls into a meaningful management report that can be used to control telephone costs, bill back customers, identify abuse, or monitor the performance of office personnel. Because the program works in the background mode, the microcomputer may be used for other applications. Such programs are now available for less than $1,000.

Voice-messaging systems also work in the background mode. For less than $350, any microcomputer with a hard disk can be turned into an office-message center. All that is required is a proprietary board, which plugs into a vacant slot of the microcomputer, and appropriate cabling to link the phones to the micro. Upon the third ring at an unattended station, the PBX or key system can "call forward" the incoming call to the designated message terminal, where callers may leave detailed voice messages.

Instead of this centralized approach, the cost of such messaging systems makes them economical for installation into many office micros. Callers can be greeted in a more personal way, and in the called party's own voice. Office managers can call in to retrieve their messages without wasting the operator's time. Some messaging systems have the intelligence to transfer calls to other extensions or call you when a message has been recorded. They even give callers the option of reaching the company operator by pressing 0, instead of leaving a message. Watch out, though; these boards usually require single-line telephone interfaces with the PBX–key system, many of which take full advantage of all six wires to implement advanced features, leaving none left with which to implement the answering service at the microcomputer. This means you will need a second single-line termination at your station. If you want to dial out through the microcomputer and it is not connected to a LAN, you will need a third single-line termination at your station, assuming that you still want full use of your phone for incoming calls.

One of the fastest moving developments in office technology is the integration of facsimile (FAX) with microcomputers. PC-based "virtual" FAX systems have all the features and reliability of conventional Group-III machines — and then some. Group III is the designation given to FAX units that are capable of sending and receiving a page at speeds up to 9.6 kb/s. Transmission directly from the PC eliminates hardcopies, unless you want them. Documents can be sent to multiple locations simultaneously and retransmitted if busy destinations are called; meanwhile, the FAX goes on to the next document. Documents can be accumulated from several office PCs *via* LAN or data switch and then FAXed after business hours to take advantage of lower calling rates. The next morning, a complete log of FAX activity is already waiting. Error-checking software insures copy quality by initiating retransmission requests. Automatic turnaround polling allows documents to be sent and received on the same call. FAX transmissions may be performed in the background mode, freeing the PC for other office tasks. The cost of equipping a PC with a FAX board and modem card is approximately $1,200 — less than the cost of most comparably featured Group-III FAX machines. Improvements in virtual FAX are coming fast and furious, leading some industry observers to predict the eventual demise of conventional FAX. For documents not already in computer form, however, an image scanner is required. Such devices connect to the microcomputer *via* special controller cards, and provide the means to "write" text and graphics to disk. The most popular desktop scanners cost between $900 and $3,500. If your FAX requirements are few and far between, the scanner can pull double duty for desktop publishing.

An emerging application of microcomputers is terminal emulation for packet assembly and disassembly (PAD). Instead of establishing a private network of their own, or even buying asynchronous terminals to access public packet networks like Tymnet and Telenet, companies may now turn office microcomputers into

PADs with the addition of a single plug-in board. These boards contain their own processors, memory, and software to handle PAD-function, protocol-conversion, and network-interfacing requirements. They also provide a variety of terminal emulations, including those of IBM, DEC, and Data General. You still require a dedicated X.25 link to the public or private packet network, either direct or *via* an X.25 gateway through a data switch or LAN.

Perhaps the ultimate in multifunctionality comes from U. S. Trade Research Inc. (Fountain Valley, Calif.), which has turned the microcomputer into a PBX, a feat not even AT&T has managed to achieve. A perfect example of the convergence of computer and communications technologies, the main PBX components, including receivers, transmitters, tone generators, and subscriber line interface, fit on a single printed circuit board, which uses only one expansion slot of the microcomputer. The software portion of the product operates in the background mode, allowing the microcomputer to be used for other tasks while it processes telephone calls. The program requires only 64 Kbytes of memory to support the standard-size add-in card, which provides for four trunks and twelve station lines. Each additional card provides 8 trunks and 16 lines. The PCBX is priced at about $300 per line, compared to a comparably featured PBX, which costs approximately $500 to $700 per line. The company plans to introduce an all-digital version of the product by early 1989, promising a total capacity of 1,000 lines, which will also be ISDN-compatible.

When purchasing microcomputers from value-added resellers (VARs), be especially discriminating when it comes to the functionality issue. You might be considering a VAR, if your company has a highly specialized need, which means that the VAR is supplying a highly customized hardware-software package, but can that product also handle the multitude of other office chores? Can it use the various applications packages you already have? Can it be integrated easily into the office network? If alternative operating systems like UNIX or Xenix are used, does the operator have the flexibility of switching over to DOS or OS/2 without time-consuming reinitialization procedures? Of course, you are buying a niche product to begin with, but there is no reason to invest in any computer with limited utility and connectivity potential.

There are a number of reasons why you might consider retail dealers for the purchase of microcomputers. Dealers like Businessland, Inc., and MicroAge, Inc., may be close to your location, allowing a more personalized business relationship to develop. Dealers also offer a range of services that your organization might not be in a position to handle, such as systems-requirements analysis and configuration assistance. After the sale, the dealer can install the necessary cabling, load the operating system and application software, develop custom menus, and perform media conversions. You can negotiate for other important services like after-hours support, on-premise training, equipment relocation assistance, as well as leasing and loaner arrangements. Many dealers also offer trade-in programs.

Buying from a dealer is not without risk, however. You cannot allow these convenience factors to obscure the technical-support capabilities of the dealer. You must satisfy yourself that the dealer has the resources required to handle both the AT and PS/2 hardware architectures, as well as the DOS, OS/2, and UNIX operating systems. Settling for anything less means that you may be limiting your flexibility to migrate to more sophisticated products as your needs change. You should also check into the dealer's systems-integration capabilities to insure that you can transport information across different operating systems and hardware architectures in a seamless fashion.

You will find that most small dealers are more sales-oriented than support-oriented. They simply do not have the resources to handle the growing level of technical complexity in the products they sell, nor can they support the broad range of offerings currently available. Consequently, they are increasingly turning to the manufacturers for resolving customer problems. The manufacturers claim that the dealers are ultimately responsible for support, because they interface with customers on a daily basis. Until this issue sorts itself out, you may be taking unnecessary risks when buying from your neighborhood dealer.

4.4 HARDWARE-SOFTWARE PLANNING

Given the fact that microcomputers are commodity items which differ widely in price, it is easy to be swayed by vendor claims of performance, or to become confused to the point of not being able to make a purchase decision. On the other hand, it is easy to play it safe by going with a top brand, and miss out on some really good bargains at little or no sacrifice in performance. Perhaps some final advice is in order.

There is no denying that purchasing higher priced name-brand microcomputers has some advantages. The manufacturer most likely became one of the top names in the industry because of its commitment to support, the quality and reliability of its product, and its leadership in implementing new advances in technology. Most likely, the company remained fairly stable throughout the industry's many boom and bust cycles during the early 1980s. You can be assured of the company's long-term success because hundreds of other vendors eke out a pretty good living by supplying add-ons and applications programs that are compatible with its products. When you add it all up, you know that the company will remain in business for a long time to come, and not leave you stuck with permanently disabled or obsolete hardware.

It is not necessary to buy the most powerful equipment, if you have no immediate use for it. You may want to consider the cost savings of adding on as your needs justify, being sure that the equipment you buy now can handle the range of possible upgrades without a noticeable drop in performance. In postponing the purchase of 80386 machines for as long as you can, you can reap additional cost savings as prices fall. You might even consider buying a moderately priced 80286 microcomputer, anticipating that low-cost add-ins will become available that provide 80386-level performance. Such reputable companies as Intel Corporation (Santa Clara, Calif.), AST Research Inc. (Irvine, Calif.), and Quadram Corporation (Norcross, Ga.) have already introduced motherboards that replace or supplement the 80286-based boards. AST even allows the user to toggle between 80286 and 80386 microprocessors.

If you are choosing a microcomputer that will be used as a network server, the 80386 is a must. Anticipate future growth by building in capacity up-front: heed Parkinson's Law by buying the fastest, largest hard disks and the maximum number of bus slots. After all, demand always expands to exceed available capacity.

Although the current version of OS/2 does not take full advantage of the 80386 processor, future versions will certainly do so. Until then, it is not really necessary to buy an 80386 microcomputer to take partial advantage of OS/2. The programs originally designed to run under DOS can run up to 20 percent faster on an 80286 machine under OS/2 which, incidentally, cannot run on 8088-based machines.

Keep in mind that when OS/2 becomes widely available, microcomputer manufacturers will offer their own versions of the operating system. As in the case of DOS, OS/2 may or may not run on each other's type of machines. This is because even slight differences in hardware can profoundly affect the performance of the operating system. Although OS/2 has features that allow manufacturers to make these differences transparent to applications and users, there may be cases when this type of patching will not be adequate to prevent problems. Therefore, if your organization maintains a mix of desktop computers from different vendors, it may be necessary to confiscate system disks from users after OS/2 is loaded into the hard disk. This will prevent mixups later, and save substantial man-hours in resolving this kind of problem.

One misconception about OS/2 is that it is merely a better DOS. Actually, moving from DOS to OS/2 is like moving from the XT to the AT — it not only feels different, but it opens up a wider range of tasks that may be performed. OS/2 and DOS are different systems, which means that users will have to be trained on OS/2. Training will consume organizational resources, at least initially, and users will need time to gain proficiency, but the start-up cost of moving to OS/2 will be justified by the increased functionality of the system and, consequently, the productivity gains of users.

If you want to tie microcomputers that use diverse operating systems together over a unified network, the means to accomplish that will be available in early 1989. Dell Computer Corporation (Austin, Tex.) intends to offer a triple boot feature with the ability to move between DOS, Xenix, and OS/2 as the application requires, and Microsoft in partnership with Hewlett-Packard will offer a similar product, LAN Manager/X (LM/X), which will provide a choice of MS-DOS, OS/

2, and UNIX from the same machine, allowing users to choose the most appropriate operating system based on the applications programs available to them.

In the purchase of microcomputers for the office environment, you must position yourself for the future. This means staying away from obsolete 8088 technology, which is already a decade old, no matter how attractive the price. New graphics displays, window-based applications, and operating systems require high performance microcomputers.

Before committing to a bulk purchase, use your applications programs on a few microcomputers, and then test them on those using OS/2. Then try new versions of your applications programs designed for use with OS/2. Compare the results. Until you do, you will not have the information you really need to make the right purchase decision.

As sophisticated applications programs designed for OS/2 continue to become available, there will also be parallel developments in imaging technology, optical storage media, and networking that will require the power of the 80386 microprocessor. Bargains in microcomputers abound, and they may serve you well in a few stand-alone applications, but you cannot position office operations for the challenges that lie ahead by investing in 8088-based microcomputers, which are essentially trailing-edge technology.

This is only a general discussion of some of the key features and capabilities that separate different classes of microcomputers. Knowing the basic factors affecting performance is sufficient for matching hardware to the applications programs you currently use, or plan to use in the future. When selecting microcomputers for office use, it pays to select hardware that will not be rendered obsolete as you add more functionality and migrate to more sophisticated applications programs.

One last thing — negotiate your best deal. Look for additional buying leverage among employees, who may jump at the chance to buy a reasonably priced microcomputer for home use. By including employees in your purchase plans, you can qualify for additional volume discounts, while enabling them to share in the savings.

4.5 LAPTOP COMPUTERS

Laptop computers constitute the newest element that must be tied into the overall strategy for office automation. Business people need not be out of touch with the office and suffer productivity losses just because they travel. With today's laptop microcomputers, all the office functionality they need can go with them, neatly packaged and weighing not more than 15 pounds. For the company, laptops are a relatively inexpensive way to support employees on the move with timely information and sophisticated productivity tools that can enhance competitive position.

With a laptop computer, any location can become a "virtual workplace." Equipped with from 20 to 40 Mbytes of hard-disk space, as well as battery pack and internal modem, laptops can perform a variety of tasks that extend the office environment to the car, the airport terminal, the hotel room — even right to the customer's desk.

Laptops are especially useful for salespeople who require the latest product-pricing and availability information. If a customer wants to compare the cost of your company's options with figures he already has from another firm, for example, the sales representative can call up that information instantly without fumbling through a voluminous product catalogue or making a hasty exit to place a long-distance call to headquarters.

When it comes to competitive advantage, the laptop can actually bolster the professional image of small businesses, allowing them to compete on a better footing with much larger firms. Instead of merely talking across a desk about a product's features, for example, the sales representative can take on the role of business consultant. In an interactive session with the customer, the cost savings associated with the product or its return on investment can be demonstrated with the laptop computer.

With a spreadsheet program already configured for that customer, the sales representative can perform a "what if" analysis to validate the product's cost savings under a variety of scenarios, or reveal the financial effect of using alternative offerings in the company's product line. Any customer variable can be entered into the financial equation, with the results even displayed in graphic form.

The laptop can also help close sales faster. In having a library of product information at their fingertips, there is no need for salespeople to waste time with follow-up correspondence or office visits to provide missing information — or worse yet, to retract stale information. In fact, when the salesman gets back to the hotel room, the order may be transmitted to headquarters *via* phone-line transmission using the laptop's internal modem. The company sends an acknowledgement of the order to the customer *via* regular mail. Not only can the laptop help get orders faster, it can make visiting customers more productive, substantially reducing the cost per sales call.

Managers, too, will find laptops useful for maintaining office-level productivity while on the go. Laptops have the speed, power, and memory capacity to handle any office task, including word processing, planning and scheduling, and data-base management. For added convenience, some laptops even offer resident "pop-up" programs that permit easy access to a notepad, calculator, appointment calendar, and telephone directory — all without having to close the file in progress.

Back at the office, idle laptops are pulling extra duty as replacements for desktop computers that are out for repairs. They also suffice for temporary use by new employees until desktops can be ordered and delivered. This allows new hires to become instantly productive. Managers are even unplugging their laptops from the LAN, rushing from meeting to meeting with their decision-making tools packed with them, and, instead of staying late at the office to finish up a project by morning, today's "knowledge workers" are taking laptops home to finish the job when it is convenient for them to do so.

4.5.1 Add-Ons and Add-Ins

Advances in technology have combined to make laptops lighter and cheaper. Their screens are easier to read, with users having a choice of gas plasma or backlit supertwist displays. The resulting market success of laptops has encouraged the proliferation of add-on devices to increase their functionality. For number-intensive applications, for example, you can plug in a separate numeric keypad. You can also plug-in a bar-code reader to facilitate inventory control. Special rechargers are available to put life back into dead batteries in as little as two hours, rather than the six to eight hours usually required to recharge the laptop while it uses an ac power source.

For extensive communications with the office, you might be better off choosing an internal modem for your laptop, rather than paying extra for the vendor's own unit. One feature worth looking for is "adaptive dialing," which allows you to handle the various pauses, access codes, and uncertain line quality of interexchange carriers. An "exclusion switching" feature prevents data-transmission errors, if someone inadvertently picks up the phone at the other end. If you travel abroad, look for a modem that adheres to the CCITT V.22 standard for 1,200 b/s communications.

If you prefer the added flexibility of external modems, you might like a unit with LED displays that indicate carrier detect, call in progress, speed, and weak battery. Some modems come with adapters that connect to acoustic couplers in situations where hotel phones do not have modular jacks. For laptops without internal modems, you can buy a battery-powered, 300 or 1,200-b/s modem that is only slightly larger than a cigarette pack.

Expansion cards are available to provide as much as ten megabytes of random access memory (RAM). One manufacturer has had the foresight to combine a 1,200-b/s modem and one megabyte of RAM onto a single board, so users are not forced to choose one over the other, just because their laptops may have only one expansion slot. Add-on chassis are available that clip onto the rear of the laptop, providing the means to use two extra expansion cards. Power users may even plug their laptops into a stand-alone five-slot chassis. Laptops with two full-sized expansion slots and multiple I/O ports are becoming more common, indicating manufacturers' awareness of the demand for still more functionality, and, to help combat eye fatigue, many laptops come equipped — or offer the option — of an external monitor port that allows you to use a desktop computer's full-sized monitor.

4.5.2 Data Transfer

In addition to light weight and full functionality, laptop users value the ability to transfer accumulated data and applications programs from desktops to laptops, and back again. You can plug a 5.25-inch external disk drive to the laptop to make conversions, but these units can cost $300 or more. You can replace a 5.25-inch disk drive in a desktop computer with a 3.5-inch drive for about $100, but you sacrifice portability. You can also transfer files between desktop and laptop computers by establishing the communications link *via* modems, but the slow speed of 300 to 1,200 b/s offered by most modems makes file transfers with this method a tedious chore, if you do not use the background mode.

The best method of data transfer involves the use of a combination cable-software package, which moves files between standard 5.25-inch and 3.5-inch disks at speeds exceeding 115,000 b/s. Some products even have the intelligence to check each block of data and retransmit automatically upon error detection. Transfers may be implemented by menu selections or DOS commands. The laptop's drives are operated by the desktop computer as if they were just two more drives installed under its own hood. These kits can also be used to send files between two laptops or two desktop computers. The cost for conversion kits like Lap-Link from Travelling Software Inc. (Bothell, Wash.) and The Brooklyn Bridge from White Crane Systems Inc. (Norcross, Ga.) are only $129.95.

Most data-transfer kits consist of software on a 5.25-inch disk and a null-modem serial cable. Some vendors thoughtfully provide their conversion software on both 5.25-inch and 3.5-inch disks, so the user can initiate transfers from either the laptop or the desktop computer.

Other products use a parallel cable that attaches from the laptop's external floppy-disk controller port *via* a 15-pin female D-shaped connector. The other end goes to the desktop's external floppy-disk controller port *via* the 37-pin male D-shaped connector found on the IBM/XT. (This port is not standardly available on IBM/ATs and XT compatibles.) Also, it is not always just a simple matter of adding the external floppy-disk controller port — some desktop computers have a fixed amount of address space, which is already allocated to disk drives, and so they will not be able to accommodate the address for another external device. In this case, serial cable-connectors may be used, which are not included in all transfer kits. You can always obtain a refund later, or buy the required cables, but you can save a lot of time and effort by making your first choice the right choice.

The growing popularity of data-transfer kits is leading to the introduction of some really nifty file-handling features. Some products, like Lap-Link, divide the screen. On the left side, it provides a listing of files in the local drive that can be flagged for transfer. When the transfer is complete, that file shows up on the right side of the screen as belonging to the remote drive. Other features include the ability to flag files for copying, deletion, or renaming.

In addition to protecting the user's investment in software, these kits give laptops access to a networked microcomputer's resources, including laser printers, magnetic tape drives, and optical-character readers. Users can even move between local and remote devices with a single keystroke — without performing a computer restart. A pop-up window lets users know which device has been accessed.

A transfer capability will become increasingly important as offices migrate from the 5.25-inch format to the 3.5-inch format. Before buying transfer kits, here are some things to be aware of:

- Do not be too impressed by the file-transfer speeds of these products. Some claim to implement file transfers at 225,200 b/s and higher. Such speeds are only possible by sacrificing error checking. Consequently, the higher the transmission rate, the greater the chance of errors.
- Be aware that modem-communications software is limited to 34-kb/s transmission — and that is with a proprietary data-compression scheme, but even that speed may be too slow for large data-base transfers. The special-utility programs and null-modem cables used in most transfer kits offer much faster transfer rates.
- Although transfer kits can be quite reliable, they are not very efficient for frequent use. Inserting cables and implementing the program can consume a lot of time. A better solution might be to use external disk drives, so that users can save files directly to the disk format of their choice.

4.5.3 On the Horizon

Laptops have helped legitimize IBM's decision to standardize its PS/2 line of microcomputers around 3.5-inch drives, which use 720-Kbyte floppy disks that hold double the amount of information on 5.25-inch 360-Kbyte floppies, but the increasing demand for more floppy-disk capacity will result in the widespread use of higher density 1.44-Mbyte disk drives by the end of 1988. Even though the physical size of 360-Kbyte, 720-Kbyte, and 1.44-Mbyte disks are the same, keep in mind that some computers do not allow them to be interchanged among the three types of drives and still perform reliably. For example, you can use a low-capacity disk in a high-capacity drive to write data to the hard disk, but you might not be able to move data from the hard disk to a low-capacity disk inserted in a high-capacity drive. Interestingly, the IBM PC/AT seems to be able to distinguish between the low-density, mid-density, and high-density disks very well.

Advances in data-storage technology continue at a break-neck pace. One development coming out of IBM's Almaden Research Center in San Jose, California, in late 1987 may find applications in future laptop computers. IBM demonstrated the feasibility of recording up to 1.25 trillion bytes of information on a single 3.5-inch hard disk. That is 50 times the capacity of today's highest density hard disk.

As more desktop-computer users continue to migrate from monochrome to color screens for graphics applications, the demand for laptops with color screens is increasing. Several Japanese firms are currently applying their expertise in hand-held televisions to bring color screens to laptop computers. Already, Seiko Epson Corporation has produced experimental flat panels measuring 4.7 inches with a dot density of 400 vertically by 640 horizontally. Although promising, this panel still falls short of the industry-standard 9-inch diagonal screen that would be required for the laptop computer market. Until flat-panel technology is perfected, users will have to make do with EGA and VGA screens, which will be introduced for laptops by early 1989.

Another innovation concerns modems. Currently, modems take the form of cards that can be inserted into an expansion slot of the desktop or laptop computer. Alternatively, stand-alone units may be plugged into the RS-232C port of the computer. Today's technology permits modems to be scaled down to only one or two chips mounted on a small board that plugs into the computer's main board. This development would allow the integration of communications with the computer's operating system, thereby greatly simplifying access to electronic mail and other dial-up services.

Laptops equipped with hard disks wear down the battery very quickly, limiting their usefulness when you are away from a power source for more than one hour. Conner Peripherals of San Jose, California, in February 1988 introduced a 20-Mbyte hard drive for laptops that uses only two watts of power. (A bedroom night-light consumes only seven watts.) The company expects to have production quantities available for laptops by the start of 1989.

Laptops are undergoing the same evolutionary process as desktops went through. Innovation will continue at a hyper pace, attracting more third-party vendors who will bring an innovative new twist to available offerings. Already, one manufacturer has introduced a 3270 emulation cartridge for its laptop, enabling it to mimic an IBM terminal for mainframe access, and another manufacturer provides a LAN interface to its high-end laptop. Oh, yes — the industry's first 80386-powered laptop was introduced by Grid Systems Corporation (Mountain View, Calif.) in January 1988.

Although laptops are not yet equivalent to desktop computers, the progress they have made over the last year in terms of power, functionality, and ease of use would suggest that that day is not too far away.

4.5.4 Where to Buy

Where to buy laptops is as important a consideration as which laptop to buy. The safest place to buy a laptop is a reputable computer store that has a national sales and service presence. These stores usually have several models of laptop to choose from and have demonstration areas where you can try out your favorite applications. During the demonstration, you will have a chance to compare several units by weight, operating speed, and screen legibility. A salesperson will be there to answer your questions, if they are not too technical, and, if your company has an account with the dealer, your purchase may qualify for a 10% corporate discount. Some dealers even like to keep their demonstration units moving, and will knock 10 to 15% off the store price.

If you know exactly what model laptop you want, you can buy one through a mail-order house for a whopping 25% savings. If the vendor is out of state, you can save even more, because you are exempt from paying state sales tax. When dealing with a mail-order firm, however, some caution is advised.

First, beware of discounts that are too good to be true. Some mail-order firms have been known to replace key components of brand-name laptops with "no-name" componentry, and then sell the removed parts as separate items. Swapping out disk drives, for example, can reduce the cost of a brand-name laptop by $200. To eliminate this concern, inquire as to whether the mail-order firm ships the product in its original, unopened factory carton.

If the firm offers 24-hour customer assistance, ask for the telephone number and try it out to determine its response time. Also, look for a mail-order firm that offers a "no questions asked" refund on returned items. Before buying, ask to speak with a technician so you can find out about the frequency of repairs for the laptop you are considering. Find out how long it takes to get the laptop back and if "loaners" are available until yours is fixed.

Another way to minimize risk is to buy only a brand-name laptop. If you have problems later, and the mail-order firm is no longer around, the manufacturer will honor the one-year standard warranty that came with the product. Some manufacturers will just replace your unit within 24 hours of receipt, rather than be bothered with a time-consuming repair, and you can also extend this kind of coverage by paying a little extra.

You should avoid buying laptops from so-called "liquidators." These firms sell brand-name laptops at outrageously low prices like $199, but these laptops are primitive, first-generation machines that are not suited for today's office applications. Usually, all sales are final and there is no technical support available, but if you know a computer hobbyist who likes to take things apart, these units make a great gift idea.

Although not intended to replace desktop computers, today's laptops offer excellent performance and functionality at a reasonable price. No office can long afford to be without them.

4.6 PRINTER TECHNOLOGY

Printers can be divided into three major categories. At the low end are 9-pin dot-matrix printers that are used mostly for drafts, but provide acceptable near-letter quality by performing an overstrike. Although inefficient, they perform reliably at up to 420 characters per second (cps) at a cost of only a few hundred dollars. Providing a transition from low-range to mid-range printers are those that use the 18-pin dot-matrix printhead, offering speeds of about 500 cps.

In terms of quality, the mid-range category is shared by 24-pin dot-matrix and daisy-wheel printers, which cost around $1,000. Twenty-four pin printers use print heads that fill in the gaps between dots by using more and finer pins in staggered rows so that one set fills in between the other. Since this is accomplished in one pass, print quality approaches that of a good typewriter, but without sacrificing speed. Daisy-wheel printers provide a continuous character pattern instead of using a pattern of dots to form characters, which limits their capability to handle graphics. Their print heads also strike the paper with enough impact to imprint multipart forms. In terms of speed, however, the two types of printer differ widely, with the 24-pin dot-matrix printer operating at greater than 650 cps, and daisy wheels at less than 120 cps.

At the high end are desktop laser printers that provide 300 × 300 dots per inch (dpi), making them better suited than the others for combining text and graphics. There is a wide variety of desktop lasers within the $1,500 and $6,000 range, offering incredible diversity when it comes to memory, speed, and features. At this writing, 600 × 600 machines are just being introduced to the marketplace, starting with Varityper's (East Hanover, N. J.) VT600. Although aimed at the office environment, these products are more plain paper phototypesetters than desktop laser printers. Prices begin at approximately $13,000.

Not only are laser printers getting smaller, smarter, and easier to use, but recent advances in controller technology, onboard memory capacity, encoding techniques, and optics have also contributed to their increased functionality, high performance, and low cost.

In using two microprocessors on the printer's controller — one for graphics processing, the other for I/O processing — the speed of printing documents that combine both graphics and text is dramatically increased over the plethora of printers still using one processor for both tasks.

More on-board memory has become available on laser printers, due in large part to the shift from bit-mapped to vector memory. Under the bit-mapped scheme, the entire page is represented as a mass of individual dots. On each dot, toner will either appear or not appear. In this way, an entire page may be printed at once. A letter-sized sheet measuring 8.5 × 11 inches contains 93.5 square inches. Since most desktop lasers perform at 300 × 300 dpi, they require a minimum of 1.0 Mbyte of memory to bit-map a single page, not including the bit-map control codes required to print it, which brings the minimum onboard memory requirement to 1.25 Mbytes. With vector graphics, only individual lines are encoded for printing. With substantially less memory allocated for mapping, much more is available to hold fonts, printer-emulation programs, and user-programmable features. Some laser printers are capable of interfacing with both bit-mapped and vector-graphic applications programs.

Innovative manufacturers are providing laser-printer add-ons that allow users to upgrade and tailor the printer for specific needs, eliminating the up-front cost for unneeded features. By inserting cards into the vacant slots of some printers, for example, users can add as many as 34 fonts and one megabyte of RAM, or add plotter emulation and 2.5 Mbytes of RAM. Other manufacturers offer plug-in cartridges, allowing users to select from hundreds of available typestyles. With first-generation laser printers, you could increase on-board memory to three megabytes and more. Second-generation laser printers, like Qume Corporation's (San Jose, Calif.) ScripTEN, come supercharged with three megabytes of RAM.

Some laser printers now have multiple-access ports to permit sharing among several office PCs, eliminating the need for LANs or data switches for this purpose.

Improved optical-imaging techniques are providing higher quality printing. Most laser printers rotate mirrors and lenses to reflect images onto a drum cylinder, which then transfers images onto paper. New laser printers use a stationary light source that displays characters onto a liquid-crystal "shutter," which changes from opaque to transparent. Light that is allowed to shine through the shutter is reflected onto a photoconductor drum that uses photocopy techniques to transfer the image onto a page. The result is sharper imaging and greater shade differentiation.

As many as 32 fonts may be printed on a single page. Some laser printers can even rotate a font for landscape or portrait printing, and print both on the same page. Lasers can handle labels and transparencies, and are able to handle envelopes with specialized add-on trays.

The life of laser printers is increasing. Most laser printers in use today boast a page life of 100,000 to 300,000. Newer models have pushed longevity to 600,000 pages. One manufacturer claims a life of 1.5 million pages. As technology improves and manufacturers seek to differentiate themselves from their competitors, we can expect even greater longevity.

Generally, laser printers are faster and more reliable than dot-matrix printers, and provide a lot more flexibility in terms of handling graphics and type styles; and they are much quieter than dot-matrix printers, making them ideally suited to the office environment. Laser printers, however, have their limitations. To begin with, toner cartridges cost about $60 each, and are good for about 6,000 pages. With second-generation laser printers, you not only have the recurring cost of toner cartridges, the drum must also be replaced every 30,000 sheets at an additional cost of about $600.

Despite all the hoopla about laser printers, innovation continues in dot-matrix printers. Laser printers are better than most dot-matrix printers in terms of quality printing. The 24-pin matrix makes printing in the "near letter–quality" mode virtually indistinguishable from that of the best daisy-wheel machines, providing a resolution of 360 × 360 dpi. This is better than the industry standard 300 × 300 for desktop laser printers, and in the draft mode some of these new-generation dot-matrix printers, like Output Technology Corporation's (Spokane, Wash.) TriMatrix PrintNet, can zip along at speeds greater than 800 cps.

Like their laser cousins, dot-matrix printers sport front-panel buttons for convenient selection of typestyles, pitch, line spacing, page length, and print quality. Switch settings on the rear of the printer even allow you to select from among a limited number of international character sets. Others may be selected through software command. Some dot-matrix printers even allow you to design your own character sets, and use them in combination with the printer's available type styles.

Paper handling has been enhanced through the use of a "paper-park" feature that allows users to switch from tractor-fed fanfold paper to single sheets without having to remove paper already in the tractor. A "tear-off" feature actually saves paper. After you tear off the last sheet printed on fan-fold paper, the printer reverses the fan-fold paper to begin printing at the top of the sheet.

Inexpensive "network" printers are making their way into the small office. Instead of three or more printers scattered around the office, one high-quality printer may be shared by as many as five users through the machine's serial ports. Any device capable of serial communications, including modems and servers, may be connected to such printers and exchange data in RS-232C or RS-422 formats at speeds of up to 19.2 kb/s. *Via* the serial ports, other printers may be connected, allowing jobs to be routed to the appropriate printer through the network printer.

There are also top-of-the-line laser printers that are built for heavy-duty use in shared arrangements. They offer duty cycles of 80,000 or more pages a month at 26 pages per minute. Desktop lasers, on the other hand, offer duty cycles of 5,000 pages a month at 8 to 12 pages per minute. Although top-of-the-line laser printers cost between $16,000 and $18,000, their networking capabilities make them very economical in high-usage office environments.

4.6.1 What to Consider in a Laser Printer

Given the popularity of laser printers and the confusing array of features that are becoming available, some product selection criteria are in order. Find out how much on-board memory the printer has; the more memory, the more fonts and software-controlled features the printer can store, and the faster text and graphics can be integrated on the same page. The amount of printer memory may seem impressive, but you must dig further to find out how much of that memory is available to the user for storing forms, frequently used graphics, and letterheads — all of which may be recalled with a simple command. Even if you have no immediate use for great amounts of memory, it is a good idea to find out how much may be added later, and at what cost. Remember, user needs have a tendency to expand with the available technology!

You should also look at typestyle diversity. For every standard typestyle, you will probably need bold and italic as well. Since cartridges or font disks may cost $200 to $400 each, you can minimize your front-end costs by determining your printing needs and evaluating the products that meet those needs with the fewest cartridges or font disks. If you plan to buy several printers from the same vendor, look into the possibility of negotiating free cartridges or disks. Do not forget, the salesman is on your side; if he knows his commission hinges on a few freebies, he will do what is necessary to please you.

There is a wide range of graphics capabilities among desktop lasers. Generally, the more types of pattern the printer is capable of producing, the more shade differentiation is possible. The range of the gray scale is particularly important in desktop-publishing applications where photos are integrated with text for newsletters and brochures. The range of the gray scale determines the quality of images, especially subtle facial features.

Keep in mind that many desktop lasers do not really print at their advertised speeds. Printing tasks that include graphics and high-density text may cut performance by as much as 50 percent.

Laser printers use toner cartridges, drum cylinders, and engines that rotate mirrors and lenses, making them more like photocopiers than other types of printers. This means that a laser printer will require servicing more frequently than most dot-matrix printers. Compare the service frequency of different models. Find out the nearest location of service personnel and service response times. Check into the local availability of spare parts and toner cartridges.

Keep in mind that toner cartridges will need replacement periodically and paper jams may occur at any time. With all this in mind, check for easy access to the cartridge and find out what is involved in clearing paper jams. The compactness of some laser printers can turn these relatively simple tasks into quite a nuisance, which can disrupt office workflow.

Printer emulation is important when choosing a laser printer. Since Hewlett-Packard's (Boise, Idaho) LaserJet printer has become the *de facto* industry standard, many sophisticated applications programs are designed to be used with the LaserJet. For this reason, other manufacturers let their printers accept and respond to LaserJet commands. If you buy a bargain-basement laser printer that is not capable of emulating the HP LaserJet, you run the risk of not being able to print out documents from your word-processing, spread-sheet, and desktop-publishing programs, because they do not include the driver program for that particular printer. It may cost extra to upgrade your applications package with the appropriate driver. On the other hand, if you invest in cheap laser printers made by an obscure offshore company, the cost and inconvenience of having a special driver program written may negate whatever savings you hoped to achieve in the first place.

If you plan to implement a shared arrangement, whereby several microcomputers may be connected to the printer, look into the paper-handling capabilities of the laser. The input trays of most desktop laser printers can only handle 100 to 150 sheets, which is unacceptable for multiuser configurations. Other laser printers can handle 250 to 500 sheets and have two bins for different sized sheets. They can even collate. Although such laser printers cost substantially more, they certainly can be justified if the cost is spread over many users. Alternatively, check into the possibility of using add-on paper-handling modules available from third-party vendors. Such modules connect easily to the most popular laser printers, adding 500-sheet to 1,000-sheet capacities, as well as the ability to handle envelopes.

Be aware that most laser printers still have a problem handling envelopes, but not for the reason that they lack the right feeding mechanism. The problem is that envelopes use low-temperature adhesives. When going through a laser printer, which operates at about 160 degrees or more, the envelopes either seal themselves, or they become attached to the laser's fuser roller. High-temperature adhesive envelopes are now available to remedy this problem, but this imposes another cost to running laser printers.

Laser printers require better quality paper than dot-matrix and daisy-wheel printers. To minimize contamination of toner and the wear to imaging components, paper with a uniform surface smoothness is required. This kind of paper has the added benefit of providing better image clarity, which is important to achieving good results in desktop publishing.

Despite the growing popularity of desktop laser printers, dot-matrix and daisy-wheel printers have a definite place in today's automated office. Dot-matrix printers are less expensive to buy and use than laser printers — when quality does not count. Daisy-wheel printers have a place in accounting and order-fulfillment applications, where raw striking power may be required to penetrate multipart forms. Laser printers, although more expensive, are the choice for desktop publishing and other applications requiring a combination of high-speed and high-quality output. Like dot-matrix printers, lasers may be shared to achieve greater economy.

Whether you choose dot-matrix or laser printers, be aware that their use on a local area network may pose some interesting management problems, which can be the source of considerable frustration to all who use them. For instance, some applications programs, like WordPerfect Corporation's (Orem, Utah) WordPerfect, require the use of DTR for hardware handshaking, while other types of programs use XON/XOFF. While the printer is in use under the DTR setting, print jobs using XON/XOFF will not be accepted into the queue, and *vice versa*. This problem can be remedied by printing files requiring the DTR protocol to a separate disk file, from where they can then be sent to the printer under the cloak of the XON/XOFF protocol. A cumbersome procedure, but it works.

Although possessing a few quirks, laser-printer technology is improving at an unrelenting pace. Not only have affordable four-color and 600×600–dpi printers arrived on the scene, but longevity has increased from 300,000 pages to more than a million pages, a development that occurred over the course of a single year, from 1987 to 1988.

4.7 DESKTOP PUBLISHING

Desktop-publishing systems provide the means to create typeset documents, combining text from word-processing programs with illustrations from graphics packages or tables from spreadsheet programs. Through optional scanning devices, photographs and artwork may be digitized and written onto disk, and then displayed, scaled, changed, cropped, and moved on the computer screen to suit any page-layout requirement.

If you are not using Apple's Macintosh for desktop publishing, you will require DOS version 2.10 or higher to run most programs on an IBM or compatible, as well as a minimum of 512 Kbytes of RAM. Keep in mind, though, that 640 Kbytes is recommended when creating chapters of more than 150 pages. Also required is a hard disk drive with one to three megabytes, depending on the type of printer used, to hold software and fonts. To accommodate illustrations, a graphics card is also necessary, either monochrome or color. A mouse is not required, but it can speed up the creative process. You may choose from among several types of printer, including laser, ink jet, or dot matrix. You may buy disks of clip art to illustrate your pages, or use a scanner to import images from other sources.

With desktop-publishing packages organizations can take control of their printed materials from start to finish, and exercise considerable creativity along the way, and if the variety or frequency of printed materials in the form of brochures, bulletins, newsletters, manuals, directories, and catalogues are enough, doing the job yourself with a desktop-publishing system can save substantial time and money over manual page-layout methods. This is because desktop-publishing programs mimic conventional pasteup methods, allowing the user to specify the basic information about the publication, such as its page size, number of pages, and whether the document will be printed on one side or two. In this way, the computerized approach introduces the elements of speed and flexibility that were not commonly found in manual methods of in-house publishing operations.

There are two basic types of desktop-publishing programs, interactive and code-based. Interactive programs allow the user to arrange text and graphic elements with a mouse, viewing on the screen exactly how the final page will appear when printed. This capability is referred to as WYSIWYG (what you see is what you get). Interactive programs are much easier to learn than code-oriented programs requiring the use of format codes, which replace the mouse and menus for arranging and formatting text. Using the bracket code [|R1.0], for example, indicates a right margin of one inch. Other bracket codes may be used to print lines, boxes, and shade patterns. Not all document-description languages are this arcane, however. Some are much more user-friendly, relying on mnemonic codes. Still, the numerous codes make such programs more difficult to use, because the learning curve is much greater than with interactive systems. The use of code-based programs requires operators who are strong in the ability to visualize, because they will not see the results of their labors as they perform page layout until they are ready to preview it on the screen before printing. Until operators become proficient in the use of such codes, you may anticipate considerable operator frustration and high costs due to wasted consumables.

Nevertheless, each type of program has its strong points for particular applications. Interactive programs are ideally suited for laying out publications of 20 pages or less that involve tightly coupled integration of text and graphics, and which involve a high degree of creativity with each revision. Long documents, like technical manuals, are best designed with code-oriented programs. Not only can this type of program create documents of virtually unlimited size, it can also automatically create tables of contents, footnotes, and indexes, making it ideal for long publications that require frequent revision. Some systems, like Ventura Publisher, can provide WYSIWYG *and* coding capabilities. Code-based systems offer more typographical control than interactive systems, and they are usually compatible with a broad range of high-end typesetting equipment. Like interactive systems, code-based products integrate graphics with text. Unlike interactive systems, however, the graphics are integrated into the text during the print stage — with many products there is no capability for on-screen WYSIWYG. For this reason, no graphics capability or unusually large memory requirement is necessary with code-based systems which, in turn, reduces the front-end cost of desktop publishing.

WYSIWYG requires a lot of memory, which is why such programs run slower than code-based programs. The benefit of WYSIWYG is that it gives the user instant feedback, which cuts down errors, and speeds up the editing process. One thing to understand when considering such products for purchase is that the screen image cannot match the clarity of laser output, even when using the highest resolution monitors. The more complex the fonts and graphics, the more profound the differences will become.

Not only are there great disparities between product categories, there are also many subtle differences within the same categories of products. This points to the necessity of performing a thorough needs analysis to be sure that the desktop-publishing system you buy fits your requirements and work environment. For example, a marketing group might be better off with the interactive approach, because most of its printed material is oriented to supporting the sales effort. As such, it will be relatively short and to the point, and executed with some creative flair to draw the attention of readers. The publications group would probably do well with the code-oriented approach, because most of its printed material is in the form of product manuals which may be quite long, each requiring a table of contents and an index. The content will be relatively uniform throughout, not requiring very much creativity from page to page.

Even within these two categories of desktop publishing, there are orientations to pick from: page-oriented products and document-oriented products. Page-oriented products, like Aldus Corporation's (Seattle, Wash.) PageMaker, give you control over every design element on a page-by-page basis for documents consisting of up to 128 pages. This orientation is ideally suited for marketing publications, which typically employ text and graphics on every page, but in a different format. This type of product is practical only for 24 pages or less; anything more can become a time-consuming burden, because of the program's page orientation.

With products that use a document orientation, you set the underlying page format, which will be adhered to throughout the entire document. This orientation is good for long documents like technical manuals and reports, which do not differ in format from one page to the next, and, anytime the text is revised, the entire document is updated as well, including pagination. By calling out a different style sheet, the whole document can be restructured at once.

At the risk of oversimplifying the buying decision, here is a rule of thumb to start you on your way. Use interactive, page-oriented programs for low-volume, low-frequency, page-layout work that will be performed by nonprofessionals. Use code-based, document-oriented programs for high-volume or high-frequency operations that can justify a dedicated operator. Also, code-based systems can be transmitted in ASCII form *via* electronic mail a lot easier than WYSIWYG systems that use bit-mapped formats.

Whether you choose the interactive or the code-oriented approach, you should be aware that taking responsibility for page layout is not as easy as it sounds. The person doing the layout must have an awareness of design and typography, must possess a sense of balance, and must appreciate the nuances that go into packaging words and images together in a way that will not only draw the attention of readers, but retain their interest as well. Also, editing images can be an extremely tedious undertaking for most people, because the process must be done pixel by pixel after the image has been digitized with a scanner. In other words, putting a desktop-publishing system in the wrong hands is like trying to douse a fire with gasoline — any way you look at it, the outcome is going to be a nasty business.

4.7.1 Hardware Requirements

Desktop-publishing products may entail significant front-end costs, depending on the nature and sophistication of their features. Some programs like PageMaker (Version 3.0) are sluggish, even on an AT-class microcomputer, because the use of Microsoft Windows places heavy demands on memory, processor time, and disk access. An 80386-based machine with 3 Mbytes of RAM would expedite work, especially when updating documents with a global search and replace. Others, like Xerox Corporation's (Rochester, N.Y.) Ventura Publisher, require a microcomputer with only 640 Kbytes of memory to produce documents of up to 150 pages. The low-end page-layout programs only require a microcomputer with 512 Kbytes of RAM and DOS 2.0 or later. With anything less than an AT, however, you will be limited in the number of pages you can create and the amount of graphics you can insert during any given session. Fortunately, the on-board memory of most printers is sufficient to provide a good selection of typestyles, thereby offloading the microcomputer for this requirement. You will also need enough hard-disk storage to hold page formats and text, as well as clip art and scanned images for future use. Afterall, there is no need to reinvent the wheel every time you want to issue the next installment of the corporate newsletter. The various programs required for word processing, spreadsheets, graphics, and scanning must also be accommodated. Simply stated, to take full advantage of desktop publishing, you will need at least one megabyte of RAM and a 20-Mbyte hard disk, and to avoid getting bogged down by machine speed, which can disrupt creative flow, an 80286-based microcomputer operating at 10 to 12 MHz is better, although an 80386 would be preferable.

4.7.2 Low-End Products

There are numerous interactive and code-based page-layout programs available for under $100, which may be used as a pop-up accessory in conjunction with your word-processing program, or as a stand-alone application program. These low-budget packages are quite limited in what they can do, however. You may only be able to use one font per document, for example, or integrate text and graphics by running the same sheet through the printer twice — once for the text, and again for the graphics. Some inexpensive packages offer a great deal more, such as the ability to create multiple column layouts, import ASCII text files up to 100 pages long, choose graphics from various libraries, and create custom fonts. Other packages are limited to supporting bit-mapped graphics, which means that your laser printer must be equipped with at least one megabyte of memory to print a full page.

For a slight increase in price, but still under $150, you can get desktop-publishing packages with predefined style sheets, which allows operators to become instantly productive by eliminating the need to start from scratch. You can pre-define such things as headers, footers, margins, and line spacing by embedding the format codes in the text created with a word-processing program. Then you can merge text and graphics created with other packages into a single document, and preview the results on the screen before printing, but if you want to change the text or graphics, you cannot do so while both are on the screen; you must do the editing within the respective application programs and check the result again in the preview mode.

If you are not already committed to desktop publishing and just want to try the concept before investing a lot of time, effort, and money, these budget packages may be what you are looking for. Do not buy them, however, with the expectation that you will achieve professional, camera-ready results. At best, these products are rudimentary design tools that are useful in conveying your ideas to print shops, which will generate the camera-ready pages for you.

4.7.3 Mid-Range Products

There are numerous desktop-publishing systems available within the $700 to $1,200 price range. At this level, vendors start offering powerful text-editing and graphics-integration capabilities with very precise control. If you are considering products at this level, you probably already appreciate the potential advantages of in-house publishing.

At this price range, many helpful features are available for the preparation of long documents. Through the use of embedded codes, a number of sophisticated features may be invoked, including automatic table of contents generation, automatic multilevel index generation, and automatic cross-reference generation. You can link text and graphics, which allows graphics to remain associated with a text reference like "See Figure 1.2 below," even after the reference has been moved to another page. You also have the choice of printing pages in landscape or portrait modes.

A frame anchor even keeps graphics in a specified position on the page relative to the text so that when the text is reformatted, the frames are automatically repositioned. You can reduce or expand images to fit the page layout. You can also rotate or copy images, or reproduce an image for multiple use on the same page. If you like starting page layouts with graphics, some programs will allow you to import text from another source — the text will automatically flow around the images. You can also create simple graphic elements to dress the page, such as bullets, ruled lines, picket fences, boxes, and gray screens. Janus Associates' (Boston, Mass.) SuperPrint, for instance, provides many built-in graphics capabilities, including the ability to draw circles, ovals, and rectangles. A choice of 72 different fill patterns is also included.

The most sophisticated publishing packages provide gateways that allow the user to import images or clip art from other popular programs, regardless of underlying format. Not only are bit-mapped or object-oriented graphics formats supported, but they can also be cropped with equal ease.

Some desktop-publishing packages allow you to create keyboard programs that replay frequently used keystroke sequences. This helps novices turn out work quickly and provides experienced users with a tool for expediting regularly used charts and graphs.

Other features include automatic hyphenation, widow and orphan control, and kerning. The kerning feature allows you to determine the spacing between characters, which adds substantially to the professional look of a page. One product, Lexisoft's (Davis, Calif.) Spellbinder Desktop Publisher, even controls leading in 1/300-inch increments. The leading feature allows the user to adjust the space between lines of text.

4.7.4 High-End Products

At the high end of desktop-publishing systems are microcomputer products costing $5,000 or more. At that price, they are not "desktop" systems anymore, but typesetting front ends. Although they support the most popular laser printers, they are really designed to drive phototypesetters. As such, they offer an economical alternative to typesetting terminals that can cost $20,000 or more apiece. To reap the full benefit of these publishing packages, you would have to buy a phototypesetter, at a minimum cost of $30,000. Such a purchase can only be justified for very high volume operations.

A good compromise between a laser printer and phototypesetter might be the previously mentioned VT600 from Varityper, a subsidiary of AM International (Chicago, Ill.). The VT600 is the industry's first plain paper phototypesetter, offering a resolution of 600 × 600 dpi at a cost of $13,000.

4.7.5 A Word About Documentation

Page-layout systems are slightly more complex than word-processing programs. For this reason, documentation should play a key role in your buying decision. Although the documentation for best-selling packages like PageMaker and Ventura Publisher is very good, novices may still find them difficult to understand. Fortunately, there are other sources of documentation available. Bookstore chains like B. Dalton and Waldenbooks carry a large selection of references on computer hardware and software, including easy-to-follow books that can help you get started with desktop publishing. Other books can show you how to get the most out of your word-processing program or spreadsheet. These volumes contain many useful tips that you will not usually find in the documentation that comes with the product.

4.7.6 What Next?

Publishing software continues to make substantial strides in sophistication, enabling users to produce documents of virtually any length and offering incredible flexibility in page layout. Documents of up to 128 chapters in length, each with up to 300 pages of text, are now easily handled, assuming that enough memory is available. Page layout is facilitated by on-screen rulers, automatic letter spacing, as well as improved hyphenation and cropping.

Software and hardware continue their symbiotic relationship: software takes advantage of hardware's memory capacity and operating speed to perform ever more complex tasks, while hardware makes use of software to deliver truly astounding performance. The end product of this relationship may soon blur the distinction between word processing and desktop publishing.

As advances in desktop publishing continue, making products more sophisticated and easier to use, a convergence between the two is inevitable in the not-too-distant future. It is only a matter of time before word processing includes many of the key features of desktop publishing, and the text-editing features of desktop publishing become as easy to use as today's most popular word-processing software. Word Perfect Version 5.0 from Word Perfect Corporation (Orem, Ut.) already integrates text and graphics on the same page. Continued innovation will blur the distinction between word processing and desktop publishing even further.

Aside from this convergence, the symbiotic relationship of software and hardware will manifest itself in increased integration of related functions. Desktop-publishing packages, for example, will not only provide support for more text and graphics standards, improve graphics handling, add more fonts, and support more laser printers, they will also provide more detailed on-line help, incorporate sophisticated windowing features, and improve text handling. Communications software will permit typed text, graphics, and scanned images to be transferred from any office microcomputer to a desktop-publishing station, which will automatically format the information according to user-defined parameters for output on a laser printer. All this without the need to move in and out of different programs or utilities!

Enhanced software has greatly improved the performance of optical character readers (OCRs), which read printed text into memory. Until recently, only the most expensive OCRs costing $30,000 or more had the capability of recognizing fonts and text with variable spacing. Now OCRs costing less than $3,000 have these capabilities, increasing their value for desktop publishing.

4.8 STORAGE MEDIA

A risk-free selection of storage media takes specialized knowledge. In lieu of reading this detailed section, you may want to seek the opinion of a qualified technician or consultant.

As today's "knowledge workers" become more dependent on computers for collecting and processing information, the cost of losing it also increases. It is not just the information that is lost, but the investment in labor that went into assembling and manipulating it, and the future benefits that could have accrued to the organization through its continuous use. As floppy disks gradually give way to high-capacity hard-disk drives in the workplace, the need has increased for convenient and reliable backup systems to guard against the accidental loss of voluminous information. With memory-intensive applications like desktop publishing and presentation graphics continuing their march into corporate offices, the need for high-capacity storage media becomes all the more important. Also, with the UNIX operating system moving into the office environment for its ability to support multiuser applications efficiently, the need for higher capacity disk drives is essential, and, along with it, the need for reliable backup systems.

Of course, floppy disks can be used to back up relatively small data bases, but swapping out 10 to 15 or more floppies becomes a tedious and time-consuming chore that most users "forget" to do, even if it is only once a week. Furthermore, continually backing up files could negate whatever productivity gains you hoped to achieve by computerizing office operations in the first place. Floppy disk backup, then, is neither convenient nor economical.

Recreating lost data stored on a damaged floppy disk is a mere inconvenience compared to the mind-boggling task of recreating data from a crashed 20-Mbyte hard disk — that is 10,000 pages of information lost! Even when microcomputers are linked *via* a local area network, the backup procedures used by the MIS-dp shop may not offer fail-safe protection. Any number of disasters are waiting to happen, like an operator improperly mounting a nine-track reel onto the tape drive, or using a tape drive with a misaligned head. Both can cause recording problems, which means that the files you thought were safely archived are still exposed to risk, because they were never really backed up, and if you forgot to leave your microcomputer powered up so the MIS-dp shop can poll it according to schedule, you may have missed your chance to save accumulated files — a disk crash can happen anytime, without warning. Although massive disk failure is relatively rare, the chances of losing or corrupting data are actually pretty good.

Even when everything does go right, there are myriad other uses for backup-storage systems, including archiving, data distribution, data transfers to other media formats, and secure storage of sensitive information. There are several types of backup-storage systems. Aside from nine-track reel-to-reel systems that are used most often in the mainframe environment, the most common is the one-quarter inch magnetic-tape cartridge-drive system, which adheres to the industry standard QIC recording format. The drive may be installed into the vacant front-panel compartment of the microcomputer. Some companies like Compaq Computer Corporation (Houston, Tex.) standardly equip their desktop computers with the vacant compartment, allowing the option of internal tape drives. A variety of stand-alone units are also available, which plug into the microcomputer *via* special controller boards that take up a single expansion slot. These units can even be shared among several workstations *via* the local area network, data switch, or A-B type switchbox.

Over the years, one-quarter inch magnetic-tape cartridge systems have evolved to meet the demands for increased capacity, higher speed, and higher reliability. Tape-cartridge units come in a variety of sizes and configurations to suit any office need, minimizing dependence on the MIS-dp group for information backup and archiving, and adding an element of convenience to office operations.

Unlike their nine-track reel-to-reel grandfathers, today's cartridge-drive systems use a combination of large scale integration (LSI) for scaled-down circuitry. The use of nonrotating heads greatly reduces the parts count. Combined, these developments not only provide fewer interconnections, but also lower power consumption and heat dissipation — important considerations when installing a cartridge-drive system directly into the microcomputer. All this, of course, translates into higher Mean Time Between Failures (MTBF), which is the indication of a product's long-term reliability.

When shopping for cartridge-drive systems, look for those that use direct-drive capstan motors, which eliminates the need for belts or pulleys to move the magnetic tape. Belts and pulleys have a tendency to stretch and skip with heavy usage. When this happens, speed variations in tape movement may be introduced, which can affect recording accuracy. This situation is analogous to belt or pulley-driven VCRs, which introduce speed variations during recording and cause distorted images to be displayed on the television during playback.

Direct-drive motors are also faster than belt or pulley-driven magnetic-tape systems, which means that less time is consumed for file backup, searches, and retrieval. Recent developments in technology have produced tape-cartridge systems that can back up a 20-Mbyte hard disk — the equivalent of about sixty 360-Kbyte floppies — in as little as four minutes, with only one error every trillion bits. Cartridge tapes can now hold as much as several hundred megabytes of information, with the drive system able to find any file within a few seconds.

Conventional reel-to-reel systems move the tape and the head together. Not only does this method of recording require a high level of mechanical complexity to implement, but movement tolerances between parts are also expanded, which may adversely affect data integrity. Further, when the head suddenly stops at the end of a cycle, the resulting shocks shorten the useful life of the head. Present-day tape-cartridge systems eliminate these problems through the use of a nonrotating head. Some units even feature dual read-write heads, one for writing the data to tape and the other for verifying the accuracy of the data as it is written.

Another aspect of cartridge-drive design that merits attention during product evaluation is the quality of the interface that comes between the head and the tape when the cartridge is inserted into the drive. The more pressure that is applied by the tape to the read-write head gap — where the actual transfer of information occurs — the more reliable its performance. This is because there will be less chance that dust, cigarette smoke, and debris will get lodged between the head and the tape to corrupt the data during reading or writing.

Beware of cartridge drives that still use two and three-gap heads. These designs provide only one-third to one-half the tape tension to provide the necessary pressure against any given head gap. Single-gap head designs, on the other hand, apply the greatest pressure, which will minimize data loss caused by tape-to-head separations during tape flying. Tape flying results when a thin layer of air holds the tape too far away from the head, which can significantly degrade performance.

The cost of tape-cartridge systems that install directly into the microcomputer vary between $350 and $1000, depending on type, quality, and capacity. This type of portable magnetic medium presents two common types from which to choose, the most popular of which is the one-quarter inch, or DC600-type, cartridge. Also available are mini–data cartridges (DC2000).

The cost of stand-alone units differs widely, from $2,000 to $8,000, or more. This might seem like a lot of money for a mere "safety net," but not when you consider the value of accumulated spreadsheets, mailing lists, sales orders, proposals, and other data that have been meticulously compiled by office staff over the course of many months, or the cost of reproducing them all.

If you do not want to spend money for tape-cartridge systems, the alternative will be more expensive over the long run. You will have to insure that office staff know how to use cryptic DOS or OS/2 "restore" procedures, which may not be adequate to recover all the lost data from a crashed hard disk. Even if some staff are proficient at using utility programs that greatly simplify the data-restoral process through the use of menus and windows, there is no guarantee that all the lost data will be recovered.

Some word-processing packages like MicroPro's WordStar (San Rafael, Calif.) can be optioned to back up files onto the hard disk automatically when they are closed. In the event one sector of the hard disk goes bad, you can access the backup copy, which typically resides on another sector. While effective, this method of backup will use up storage space fast, and it does nothing to protect against total disk failure.

With tape-cartridge drive systems, there is no need to fumble through manuals in search of obscure commands that may, or may not, prove effective in recovering lost data. Look for a cartridge-tape drive system that comes with a utility package that takes complete charge of the backup process, including formatting blank tape cartridges automatically upon insertion into the drive unit.

It is not necessary to image a whole disk; some systems can back up and restore a single file, a group of files, a directory, or a whole volume at a time. You may start, stop, or pause the backup operation at any time, and when it comes time to restore data from multiple cartridges, some systems will even prevent you from accidentally loading cartridges out of sequence.

Using the timer system within the software, you can schedule backups according to time of day and date. This handy feature permits file transfers between far-flung international locations at the most convenient times. You can also use the timer to schedule backups to occur at regular intervals. A screen message notifies the user when a file backup is about to occur. The user can take a coffee break or turn to other office chores. Many systems will operate in the background mode to permit use of the microcomputer for another application while designated files are being backed up. Some systems even have the intelligence to perform backups whenever an open file is closed.

The tape cartridges themselves are pretty tough — much more able to withstand harsh treatment than floppy disks. They are not affected by sunlight, moisture (just let them dry before inserting into the drive), or sudden changes in temperature. They are not sensitive to airport metal detectors that bombard objects with x rays. If the cartridge casing becomes damaged, the tape may be reloaded into a cartridge casing of the same type, and then retensioned by rewinding and fast-forwarding, before attempting to read from it or write to it. If the tape tears, it may be spliced, resulting in only minimal data loss. Most magnetic tapes can withstand enormous use — as many as 5,000 passes across the read-write head. Many cartridge systems, however, loop the tape in a serpentine fashion, which reduces the number of read-write cycles, thereby substantially extending tape life.

The plethora of vendor offerings makes buying tape-cartridge systems no easy matter. Here are some things to consider to minimize risk.

The speed at which files and entire hard disks can be backed up are very impressive, but you should also check the speed at which individual files may be located on the tape for writing to the hard disk. Look for a cartridge system that employs a bidirectional seek capability, which eliminates the need to scan the entire tape to locate a single file. Although this capability is now fairly standard among most vendors, there is still a lot of older equipment on warehouse shelves that does not include it. If you are buying tape drive systems at a discount through the mail, you should verify with the manufacturer that the model you are about to buy has the bidirectional seek capability. You cannot always count on mail-order salespeople for this kind of information. When in doubt, they always say "yes."

Compare the warranties of tape-cartridge drive systems. They vary between one and two years. One manufacturer goes so far as to offer a replacement within 48 hours, if you encounter a problem in the first six months. One manufacturer guarantees its cartridges "forever," and will replace without charge any cartridge that malfunctions during normal use, even if it happens years later. Of course, you will have to replace the lost data.

Cartridge tapes fall into three format categories developed by 3M Company (Minneapolis, Minn.). They differ by the size of the tape and the recording format they use. The DC600A is a large format, one-quarter inch tape. The DC2000 is a small format, one-quarter inch tape. The DC1000 is a small format, one-eighth inch tape. Since these categories are not interchangeable, one of the first decisions you will have to make before buying is what format you want to use. From a quality standpoint, there is no real difference between format categories. From a capacity standpoint, there is continuing progress within each category to pack ever more data onto the available tape. The real differences are to be found in the various drives available to operate the different cartridges, specifically their many features and the speed at which they can access individual files.

Look into the backup software that comes with the tape-drive system. If the software is not easy to use, the backup system will stay idle. Ideally, the backup system should be menu-driven and have help screens available in sufficient detail to help novices. At the same time, it should be command-driven for sophisticated users. The backup software should also be able to help you recover a magnetically damaged tape, or recover a tape when its directory has been lost.

Since part of your office-automation strategy is to position yourself to take advantage of new product offerings, you should select a cartridge-drive system that can be used with IBM's PS/2, even if your office currently uses only XTs and ATs. In doing so, you will not have to changeout tape drives when you finally switchover to the PS/2, and if you slowly migrate to the PS/2, the proper cartridge-drive system will become very handy for file transfers between the various disk formats.

This section has focused on the use of magnetic-tape cartridge systems as a reliable method of backing up high-capacity hard disks, only because such systems are currently very popular in the office environment. Instead of magnetic-tape cartridge systems, you may want to consider one of the many alternatives that are becoming widely available, such as data packs. Also called "Bernoulli" cartridges by Iomega Corporation (Roy, Utah), originators of the data-pack concept, these are cards that look much like 5.25-inch or 3.5-inch disks, only much thicker. Data-pack drives may be installed as an internal subsystem to the microcomputer, or used as stand-alone units. The technology is quite similar to that of hard-disk drives. Both use controlled air flow between a stationary plate and a flexible medium to stabilize the medium precisely for high-speed, high-density data recording. The Bernoulli drive includes a small computer-system interface (SCSI) that permits a 6.5-Mbit continuous-transfer rate. This, plus a closed-loop servo-mechanism and voice-coil actuator, enhances access time and offers data-transfer rates that rival the best high-speed Winchester disk drives. In this way, Bernoulli "boxes" offer speed advantages over magnetic-tape cartridge systems in that they can find any file within 25 to 50 milliseconds. Data-pack drives are capable of handling up to 60 Mbytes of data, far below the capacity of many tape-cartridge drives. Although both fall into the same price range, data packs generally offer the better access time and data-transfer rate. Of course, data packs offer the same level of portability as magnetic-tape cartridges.

If the cost of a backup system makes you hesitate about buying one, consider this: Would you drive around for long periods in your car without a spare tire? Of course not! Then why would you be willing to risk months of accumulated work by doing without a backup device to protect your data?

One-quarter inch tape-cartridge systems are very well entrenched in the workplace, and will not easily be displaced by emerging products based on newer technologies. Nevertheless, developments in tape-backup subsystems are continuing at a furious pace. New recording techniques, such as helical scan recording (HSR), which uses 8-mm videotape and 4-mm digital audio tape (DAT), are already showing market potential. HSR technology is based on rapidly rotating heads that write data diagonally across the face of a moving tape in very narrow tracks — the same way images are recorded by videocassette machines. Recording data in this manner allows much higher densities of data to be written onto the tape, while providing a total track length that is many times the physical length of the tape. As a result, it is possible to store several gigabytes of data, while allowing disk-to-tape backup speeds of 15 Mbytes per minute. This kind of performance makes HSR quite suited for use on local area networks and for storage-intensive applications like CAD-CAM. An 8-mm tape cartridge designed for HSR systems can hold in excess of two gigabytes of information — the equivalent of 6,000 floppy disks, or one million sheets of paper — at a cost of about $300 per cartridge.

A hint of things to come is offered by Nihon Digital Equipment Corporation (Tokyo, Japan). The company has announced plans to develop a "micro" tape-cartridge system that is half the size and uses half the power of the systems currently available. The company also wants to develop a hard disk that can hold ten times as much data as current technology allows — 200 megabits per square inch.

Two kinds of optical disc are available: CD-ROM (compact disc–read only memory) and WORM (write once–read many). Both are used for archival storage, because once information is written onto them, they cannot be changed, only read. Their enormous storage capacity makes them ideal for holding large reference works. Tri Star Publishing (Fort Washington, Penn.), for example, offers the gargantuan Oxford English Dictionary on two 600-Mbyte platters. Not only can users search for dictionary entries, some of which run longer than 60,000 words, but also for specific terms within the definitions, the etymological descriptions, or the many quotations that enhance the definitions. Storage Dimensions Inc. (Los Gatos, Calif.) offers 800 megabytes in one removable 5.25-inch optical-disc cartridge, and Toshiba America's Disk Products Division (Irvine, Calif.) offers a 12-inch write-once optical disc capable of holding four gigabytes.

Optical-disc technology is also making giant strides toward an erasable disk. Philips NV (Hilversum, the Netherlands) is experimenting with new surface materials which, if put into commercial production, would allow optical discs to be erased up to 1,000 times — roughly the number of times magnetic tape can be rerecorded before it wears out.

The new method of recording, however, entails a radical change in optical drives, because the laser beam heats the surface material on the disc to near its melting point. At this temperature, the material changes from crystalline to amorphous, which corresponds to the one and the zero required to code data into binary form. Until an erasable optical disc reaches the marketplace, users will have to make do with today's write once–read many (WORM) optical discs. Despite the limitation of only being able to write data onto them one time, optical discs boast incredible storage capacities, up to 200 megabytes on a single 5.25-inch disc. Theoretically, optical discs are capable of holding enough information to keep someone typing for 100 years at a rate of 60 words per minute! Until they become erasable, however, optical discs are only suitable for vast amounts of archival material, like finished circuit schematics and production drawings. For office use, optical discs may be useful for bulk references like dictionaries and directories, and for other frequently used resources like proposal and contract boilerplate. Considering the write-once limitation, today's optical discs are quite expensive, starting at $3,000. While other companies are merely speculating about how an erasable optical disc might be developed, Tandy Corporation (Ft. Worth, Tex.) beat everyone to the punch with an announcement of its breakthrough technology in April of 1988. At this writing, however, it is too soon to say anything about the long-term reliability of Tandy's product.

The microcomputer has enabled office managers to lessen their dependence on the MIS-dp group for information access and data handling. Today's add-on, magnetic-tape cartridge systems go a giant step toward severing that tie completely by allowing individuals, work-groups, and departments to assume full responsibility for archival storage and retrieval, as well as for the protection of routinely used files stored on hard disk. The benefits of this independence, of course, depend on the diligence of users with respect to backing up their files or, failing that, having in place systems that backup files automatically, according to user-defined parameters.

4.9 ELECTRONIC MAIL

Of all the communications technologies in use today, electronic mail is by far the oldest, tracing its roots to the primitive telegraph system of the early 1800s. TWX and Telex, too, are very much a part of the evolution leading to what is now called electronic mail, or simply E-mail. In its modern form, E-mail started out as a medium for message interexchange among government research facilities and the military. It quickly spread to the business world, where mainframes became the focal point for the swapping of messages. With their newfound processing power and memory capacity, minicomputers and micros now drive E-mail systems.

Under the E-mail concept, each user is issued his own electronic "mailbox," which is merely a unique identification number that permits access to a certain data base where messages are stored for retrieval by the addressee, or filed for future reference. The originator can send a message to one or more addressees on a selective basis, or broadcast the message to all E-mail users listed in the organizational directory.

Electronic mail may be implemented in a variety of ways, including:

- Mainframe, minicomputer, or microcomputer hosts;
- Local area network (LAN) operating systems;
- Wide area networks (WANs) like Arpanet, CSnet, NSFnet, and Bitnet;
- Service providers like AT&T, MCI, Telenet, Dialcom, and CompuServe.

Electronic mail is not intended to serve as the vehicle for remote file sharing, although it is capable of assuming that task. File sharing is already built into most LANs, and is available as a separate function over wide area networks (WANs). Electronic mail is better suited as an alternative to the kind of communication that formerly took place *via* memo stationary, post-it notes, short business letters, and phone calls — bursty, fast-paced communications. Users also take advantage of E-mail for the transfer of software, spreadsheets, and large documents.

Many new features have been added to E-mail to increase its functionality, making it more than just a vehicle for sending and receiving text messages, and, if not already incorporated into the operating system of a LAN, software modules may be purchased from third-party vendors. These packages usually offer a wider range of features than most LAN operating systems. Even if you already offer E-mail over your LAN, you may want to consider some of the many packages available that can enhance what you already have.

For LANs, one of the newest E-mail features is "registered" mail, whereby an urgent message greets the addressee upon powering up the terminal at the start of the business day. If the terminal is already on, an audio signal or on-screen message notifies the user to check his mailbox. Through electronic "folders," mail may be archived and retrieved by subject, date, and author. Not only do some systems reject mail to unknown users, they also offer the originator name suggestions when presented with only a fragment of a name.

It is no longer necessary to leave an application program to invoke the E-mail service. Numerous third-party suppliers now offer E-mail software that is designed to work with specific application programs. A "hot-key" feature allows the user to read or send E-mail without leaving work in progress. This eliminates the time-consuming steps of exiting one application program, calling up the mail program, sending the message, and then reversing the procedure to get back where you left off. 3X USA (Leonia, N.J.), for example, offers a product called Tel-E-Mail, a Lotus 1-2-3 add-in, that provides unattended background-communication, text-editing, and electronic-mail capabilities. The product allows users to capture and edit data from on-line information services and other remote computers, as well as create messages and modify text files, without leaving 1-2-3. Lotus Development Corporation offers LOTUS Express, which runs under Metro in the background mode for use with MCI Mail, a third-party E-mail service.

In addition to text, today's E-mail systems are equally adept at handling graphics and binary files, so graphs and tables, as well as spreadsheets, may be exchanged with other users. Moreover, E-mail works in the background, allowing other tasks to be performed in the foreground. You can be sending a spreadsheet or a report *via* E-mail, for example, while you start work on a new spreadsheet or report.

The major LAN suppliers, except IBM, offer electronic mail as part of their operating systems, or as an option. There are independent vendors who offer enhanced E-mail programs for LANs, providing features that LAN suppliers do not include in their programs. PCC Systems (Palo Alto, Calif.), for instance, offers a utility called "Snapshot" for its E-mail product called cc:Mail. Snapshot allows the user to take an electronic photocopy of anything that can be displayed on a screen, attach it to a message, and send it to another user, who can then view the entire message without using the original application program that created the captured image.

Most LAN operating systems provide various combinations of access levels, such as read-and-write and read-only, through user-names and password identification. Additional security is provided by some enhanced E-mail packages through encryption of text files or personal passwords that are not even accessible to the system administrator.

Soon there will be a choice between E-mail systems that comply with CCITT's X.400 messaging standards and those that offer proprietary gateways to the more popular host-based E-mail systems, such as IBM's PROFS and Digital Equipment Corporation's VAXmail. For organizations with remote locations which use different service providers, X.400 compliance will become important, because this standard facilitates the interconnection of various E-mail systems. The major electronic-mail service providers, including AT&T, Telenet, and Dialcom, demonstrated their X.400 connectivity in Geneva, Switzerland, at Telecom '87 in October 1987. The major E-mail service providers are expected to have X.400-based messaging services available by 1989.

You should not pick an E-mail system on price alone. You must select the features and performance you need, based on the requirements of your organization, as well as its various departments and work-groups. Make sure the E-mail system offers instant availability, message data-base management capabilities, and compatibility with a broad range of application programs.

If you discover that the requirements of your organization differ markedly from group to group, you may want to check into the possibility of using a UNIX-based mail system, if only because the multitasking power of UNIX makes implementation of E-mail extremely flexible. For example, you can offer no-frills E-mail to those who only have the need for internal messaging, while offering external mail capabilities *via* modems and gateways to those who must communicate with remote locations. Under UNIX, you may also set up basic E-mail services for inexperienced users, and then add more sophisticated features, as they become more familiar with the system. This prevents users from becoming so overwhelmed by implementation procedures that they are discouraged from using it.

In your needs assessment, you may find that users want to send elaborate spreadsheets and documents consisting of 100 or more pages to remote locations. This is really file sharing. E-mail is not designed for such transactions, if only because you would tie up the network for long periods and quickly run out of available "mailbox" memory. Using carrier-based services like MCI Mail, with their huge storage capacities, can be very expensive. An alternative would be to use file-transfer programs that allow users to communicate with each other through the public telephone network *via* modem. To minimize long-distance costs, these programs compress the files before transmission. Built-in error-detection schemes assure data integrity. Some file-transfer programs even offer encryption to safeguard sensitive information. As with E-mail, these products can implement transfers unattended.

Everyone has his or her own opinions about the benefits of E-mail. Supporters of the concept like the way it eliminates "telephone tag" and instills a certain discipline that increases personal efficiency and productivity. Others claim that E-mail leaves many people out of the information mainstream, just because they cannot type, or are too conscious of their inability to put their thoughts into writing. Like any other technology, E-mail is not for everyone. As a tool, its value must be first appreciated and then learned. With experience comes expertise, and with expertise comes productivity. Eventually, you will wonder how you ever got along without it.

4.10 THE FUTURE OF OFFICE AUTOMATION

The next milestone in office automation will involve the use of "expert systems" — computers and software that not only accumulate knowledge, but also emulate the reasoning processes people use to solve problems. These systems also attempt to mimic the consultation process, whereby an expert conveys information to the nonexpert in a way that can be readily understood.

Part of that reasoning process is pattern recognition. For example, expert systems are already being used to provide voice-recognition and voice-to-text capabilities to ease the drudgery inherent in manually transcribing, inputting, and manipulating text. These systems allow you to enter information into a file verbally, and then edit the text later. Even when words sound alike, expert systems will make the distinction based upon the context within a sentence to arrive at the correct spelling. Although such systems are good for getting raw data into text form, they still operate too slowly for most applications. Voice-to-text systems are already on the market for under $10,000. To safeguard sensitive company information, voice recognition will be added to computers as a security feature, with the operator "teaching" the computer to recognize his or her own voice — laryngitis notwithstanding — to prevent unauthorized data-base access. Voice recognition systems are available for under $5,000. As these types of system become more accepted in larger corporate environments, prices will fall dramatically, making them affordable even to small and medium-sized offices.

Procedure-based expert systems are already making their debut in the marketplace. Workhorse Systems (Dublin, Ireland), for example, has demonstrated a UNIX-based program that is capable of "recognizing" work-flow patterns and "learning" specific office tasks. The user "teaches" the system by providing it with a work-flow diagram of all the steps necessary to complete a project. Once the system "understands" the sequence of tasks and who is responsible for each, it assembles and prints documents without human intervention. The program can get information from files, ask for information from appropriate individuals, put reminders in their suspense files, offer data for review, chase down approvals, and check work in progress. Such products are ideally suited for monitoring complex projects involving many people to insure that the deadline is met. In this way, no single task is allowed to lapse without the supervisor being informed of it by the expert system.

Unlike other office-automation software, the Workhorse system does not include word processing, data-base management, or spreadsheet programs. Users can select their own applications programs and let Workhorse knit them together with its own functionality into a seamless environment that uses windows, pull-down menus, point-and-pick selection, and dedicated function keys.

In the not-too-distant future, a variety of technologies will be linked together through expert systems to provide unprecedented levels of office automation. It is not hard to imagine, for example, that someday managers will be able to request information needed for a corporate report verbally, and then have the expert system collect it from various distributed data bases, even if it must traverse different networks. If the information is not available internally, the expert system will initiate conversations with external computers that will grant access to commercial data bases, after obtaining appropriate billing information. Seconds later, the required information is integrated into the manager's report, repaginated, and printed in the preferred format.

Just when you thought you had purchased the most sophisticated productivity tools that would satisfy your office needs for years to come, along comes a new wave of products and improvements that leaves you begging for more. Continued innovation begets sophisticated office products with more features and functionality, and there is no sign that this trend will abate any time soon.

REFERENCES

[1]   Brown, Ronald O., "Office Automation Joins the Network," *Telephone Products and Technology,* November 1987, pp. 36–42.

[2]   Gartner Group (Stamford, Connecticut).

Chapter 5
Data Networks

5.1 LOCAL-AREA NETWORKS

The basic concept behind the LAN is quite simple: to facilitate the exchange of information between users of one or more networks, regardless of the equipment model, communication protocols, or transmission media that are being used at each end. The implementation of that concept is very close to being fulfilled, as vendors increasingly come to understand the value of coalescing around standards.

Although the LAN concept had been well articulated by the mid-1970s, there were no universally accepted standards to work from. Without such a blueprint, each vendor developed its LAN a little differently. Several LAN topologies emerged — bus, token-ring, and star. LANs differ in other ways, such as by access method, operating system, and transmission media. These differences affect ease of installation, network growth, and the up-front cost of implementation.

When such products were finally brought to market, users typically purchased all of the components of the LAN from a single vendor, who bundled everything from wiring to networking cards to the LAN operating system itself. Although convenient, this approach was not very flexible.

You could not customize your network by mixing and matching LAN components from other vendors to meet your unique requirements. Thus, if you wanted an Ethernet LAN, you had to install expensive coaxial cabling throughout your building, despite the immediate availability of twisted-pair wiring at each station. Now, of course, Ethernet may be run over twisted pairs.

Even when some vendors started providing both Ethernet and Token-Ring LANs, there was no way to interconnect them. Now there are a variety of devices — gateways, bridges, and routers — that can handle interconnection at various levels of efficiency and economy according to your needs and budget. Not only that, but the current trend among LAN providers is to unbundle their products to allow users to mix and match hardware, operating systems, and cabling.

In the simplest terms, a LAN is a conduit, or a family of conduits, linking various communications devices. The conduit, or conduits, may consist of ordinary telephone station wire from each device on the network. These connect in a wire closet to a backbone coaxial cable, and, possibly, a fiber link between buildings. For wide-area networking, connections to remote networks may be accomplished over leased lines.

Protocol conversion is accomplished at the network interface unit (NIU), enabling incompatible asynchronous and synchronous devices to communicate with each other over the same conduit. The NIU consists of a board that plugs into each device on the network, enabling it to communicate with other devices on the LAN that use the same communication protocol. These interface units may be pooled at a server to permit access on either a dedicated or contention basis. In addition to providing host access on a contention basis, data switches may be used as LAN servers and as gateways to other networks, provided that they are equipped with the appropriate network interfaces.

The operating system constitutes the heart of the LAN. It is the tool that the network administrator uses on a daily basis to help satisfy the needs of users. The operating system gives the LAN its features, including those for security and data protection, as well as those for network management and resource accounting. It determines the friendliness of the user interface and governs the number of users that share the network at the same time. The operating system, then, "drives" the network. It resides in one or more networking cards or dedicated network servers, which may vary in processing power and memory capacity.

Over the life of the communications network, LANs promise substantial cost savings in installation, maintenance, and expansion. Since a single backbone cable is used for a variety of applications, no additional backbone cabling would be necessary to accommodate new applications or changes in user requirements. Other benefits of LANs include:

- Rather than having to wire each terminal to a central switch, network expansion and terminal moves and changes are accomplished through plug-in connections to the LAN.

- Consistent throughput, despite the variety of applications or number of active terminals on the network.

- Higher transmission speeds and lower error rates that need not be limited by distance within a building or local building complex.

- Protocol conversion permits devices from different manufacturers to communicate with each other on the same network.

- Interconnectivity with other networks *via* gateways, bridges, or routers.

- Decentralized terminal distribution to the most convenient locations, rather than forced distribution within a centralized area.

- Enhanced network-management capabilities allow designated terminals to administer the network and monitor usage and performance for the purpose of relieving traffic bottlenecks or planning network expansion.

In evaluating a LAN, there are many discrete elements to consider, such as the topology, access method, operating system, and protocol — all of which are discussed in this section. You must also consider the types of wiring schemes that are available, and whether you will need gateways, bridges, and routers to provide connectivity with other networks. These topics will be discussed in separate sections. The final sections of this chapter explore the advantages and disadvantages of central-office LANs, and discuss the data switch as a LAN server, as well as the feasibility of using this device as a low-end LAN.

5.1.1 Topologies and Access Methods

In selecting a LAN, among the many decisions you will have to make is what kind of topology and access method you want to implement. There are three basic LAN topologies in common use today: bus, token-ring, and star. Although the principle of sharing is common to all LANs, the way sharing is done is determined by the access method, different types of which have strengths and weaknesses.

Ethernet was developed by Xerox in the mid-1970s and became a *de facto* standard with the backing of DEC and Intel. Ethernet is one of the best-known examples of a LAN based on the bus topology. The bus topology is contention-based, which means that the terminals vie with each other for access to the network. Each terminal "listens" to the network to determine if it is idle. Upon sensing that no traffic is currently on the line, the terminal is free to transmit. This access method is called "carrier sense multiple access" (CSMA). The trouble with this access method is that several terminals may try to transmit at the same time, causing the corruption of data. The more terminals that are connected to the network, the higher the probability that such collisions will occur.

Obviously, network throughput would suffer tremendously if this process were allowed to continue. To avoid this problem, each terminal sends out a packet with a preamble attached. The transceivers listen and send at the same time, comparing what is sent with what is heard. If these are not the same, a collision is detected. The transceiver sends a collision signal back to the attached device and sends bipolar violations on the Ethernet to make sure all terminations detect the collision. This preamble is called "carrier sense multiple access–collision avoidance," or CSMA-CA.

Because packets travel at about 50% of the speed of light over coaxial cable, one might think there is little chance of a collision, but during the very brief interval that it takes for a packet to traverse the network, terminals at the far end are completely oblivious to the fact that a transmission is about to begin. This "collision window" imposes a practical limit on the length of bus networks without repeaters. In its 802.3 standard for contention networks, the Institute of Electrical and Electronic Engineers (IEEE) recognizes 2500 meters — approximately 1.5 miles — as the maximum bus length, regardless of speed or cable type. (A bus network that has branches is called a tree, but it is essentially a bus network. For the sake of discussion, delta and mesh networks are also treated as variations of the bus structure.)

CSMA-CD (carrier sense multiple access–collision detect) is relatively simple to implement, which means that the bus network typically costs less to implement than other types of network. Since each device on the network is an independent unit, the failure of one will not disrupt the operation of the others. Terminals may be added or removed from the network without disrupting service. CSMA-CD is media-independent, which means that it works equally well over twisted-pair wire, coaxial cable, and fiber. The choice of medium is dependent only on the need for transmission speed, range, and immunity to interference from external sources of noise.

If the bus network remains fairly static in that few terminals will be added, traffic is uniform over time, and users can tolerate some delay from retransmissions caused by collisions of acquisition signals, the CSMA-CD access method may be an economical choice for a local network, but networks have a tendency to grow as organizational needs become more sophisticated. As the number of terminals increases and the volume of traffic grows, so does the possibility of collisions. That is when CSMA-CD starts to cause throughput problems. CSMA with "collision avoidance" (CSMA-CA) can help somewhat by avoiding collisions altogether, but this method may also deprive terminals of the chance to transmit in the first place. Many applications may not be able to tolerate the lengthy delays inherent with these access methods.

A more efficient topology is the token-ring. With its token-passing method of access, the token-ring insures that all terminals obtain an equal share of network time. The "ring" is essentially a closed loop, although its wiring scheme may cause it to resemble a star. The "token" is a byte that is circulated around the ring, giving each terminal in sequence a chance to put information on the network. This approach is commonly associated with IBM's Token-Ring Network.

Transmission is accomplished by the station seizing the token, replacing it with a packet, and then reinserting the token on the ring. Because each terminal regenerates data packets and the token, such networks are not limited by distance or speed, as are bus-type networks. Only the addressee retains the message, and only the station that put the message on the ring is able to remove it.

A further advantage of token-ring is that high-priority traffic takes precedence over lower priority traffic. Only if a station has traffic equal to or higher in priority than the priority indicator embedded in the token, is it allowed to transmit a packet.

Token-ring is not without liabilities, however. Adding terminals to the network must be accomplished without breaking the ring, and specific procedures must be used to insure that the new station is recognized by the others and is granted a proportionate share of network time. Failed terminals can break the ring, preventing all the other terminals from using the network. The integrity of the network can be protected by equipping each station with bypass circuitry. Thus, when a terminal fails, or a new one is added, there is no disruption to the rest of the network.

Token-rings are vulnerable to anomalies that can tie up the network for long periods, or at least until someone figures out what is going on. If a terminal fails before it has a chance to pass the token, the whole network is out of commission until a new one is inserted. The token may even be corrupted by noise, to the point of becoming unrecognizable to the stations. The network can also be disrupted by the occasional appearance of two tokens, or by the presence of a continuously circulating packet. The latter can happen when data are sent to a failed terminal, and the originating terminal gets knocked out before it can remove the packet from the ring.

To protect the token-ring from potential disaster, one terminal may be designated as the control station to supervise network operations and do the necessary housecleaning chores, such as reinserting lost tokens, taking extra tokens off the network, or disposing of "lost" packets. To guard against failure of the control station, every station is equipped with control circuitry, so that the first station detecting the failure of the control station assumes responsibility for network supervision. Such protective measures are achieved through greater network complexity, which adds to the cost of the network.

A variation of the token-passing scheme involves allowing devices on the network to send data only during designated time intervals. The ability to determine the time interval between messages is a major advantage over the CSMA access methods. Since each device transmits only during a small percentage of the total available time, no one station uses the full capacity of the network. Since the capacity required from station to station is neither uniform nor static over time, the overall efficiency of the network may gradually deteriorate. This time-slot approach does have one key advantage — it can support voice transmission. Time-slot access provides something that contention networks do not offer, which is access to the network that is consistent and predictable.

The star topology consists of a central hub to which all devices on the network are connected. This topology is familiar in the office environment, where each telephone is ultimately tied into the PBX. Another example of a star network entails several terminals sharing a single host. The star and the ring network share a key disadvantage, in that the failure of a single node can result in the failure of the entire network, unless provisions are made for hardware redundancy or bypass. In the star topology, the critical point, of course, is the central node.

An example of a LAN product that uses the star topology is AT&T's Datakit. In this case, all the network interface units and the interconnecting media are contained within a single cabinet to which the individual stations are connected *via* ordinary twisted-pair wiring. The system looks very much like a PBX, but it implements data communication. AT&T also offers StarLAN, which operates at 1 Mb/s, and StarLAN 10, which operates at 10 Mb/s. AT&T originally developed StarLAN to satisfy the need for a low-cost, easy-to-install local-area network that would offer more configuration flexibility than the Token-Ring.

Unlike the bus and ring networks, where intelligence is distributed throughout the network, the star network concentrates all of the intelligence required to run the network at a central hub. In the case of AT&T, the hub for StarLAN consists of a 3B2 minicomputer. Optional repeater units may be used to extend the reach of the network. The hub monitors network traffic to prevent collisions between workstations trying to access the LAN, and, since the hub knows the paths to all of the devices on the network as well as the address of each device, data routing is fast and efficient.

For interconnecting large numbers of microcomputers, Northern Telecom's (Richardson, Tex.) Meridian LANStar PC promises substantially higher performance. While many other microcomputer networks are limited to transporting data at only 19.2 kb/s, which is inadequate for many interactive applications, LANStar carries data thousands of feet at 2.56 Mb/s over twisted-pair telephone wire.

The high performance of the LANStar is attributable to the way it handles packets. LANStar is a true packet-switching system. Although StarLAN, Ethernet, and Token-Ring also use packets, the packets are not "switched." Instead of every station examining every packet over a common bus, each station on LANStar has a dedicated connection to the "packet-transport equipment" (PTE) at the hub. The PTE switches the packets and routes each one to only the addressee. This eliminates much of the overhead that plagues contention and token-passing networks.

Northern Telecom competes head-to-head with AT&T in the central-office digital-switch market. Like AT&T, Northern Telecom offers the digital-switch subsystems that implement ISDN, making LANStar a potentially viable entry point to ISDN, because its packets are designed to carry the 64-kb/s and 16-kb/s data channels that provide ISDN services.

Because LANs are essentially data-oriented, you might want to consider all of your options before choosing a LAN, including a CENTREX LAN, from a voice-oriented company. Despite the performance claims of Northern Telecom's LANStar, for example, the product has only recently caught on — this despite the company's demonstrated expertise in PBXs and central-office switches. This is significant, because a product's success has a lot to do with the level of continued support, and, despite all the media hype about ISDN, the concept is not exactly catching the world on fire. One reason is that AT&T and the other carriers simply have not demonstrated their expertise in handling data as efficiently as voice.

Moreover, there is a groundswell of opinion among large private-network users that the power and sophistication of today's packet networks already exceed what ISDN has to offer. (More about this in the chapter on ISDN.)

At any rate, the failure of one terminal on the star network does not affect the operation of the others, unless, of course, the faulty terminal happens to be the hub, but, since network intelligence is centralized at the hub, safeguards must be taken to protect it from catastrophic failure. Such measures may include an uninterruptible power supply, an alternate computer on hot standby, or using a fault-tolerant computer which has redundant subsystems built into it. Other steps may be taken to minimize the effects of a hub outage. For example, file servers in front of the hub may permit limited communication among the terminals connected to it. With such an arrangement, users would not be able to communicate with terminals not connected to the server, but they could at least continue to access files stored in the assigned disk area of the server.

Any way you look at it, protecting the hub against potential failure is going to add substantially to the cost of implementing the star network. Some users, however, may not feel burdened with such costs, since they already have the major hardware components in place for other applications. For them, buying into a star network is only a matter of adding an appropriate operating system and terminal-interface units. Today's data switches, for example, may be put to work as the hub for star networks, even providing gateways to other networks, including packet-switched networks. Another example of a low-cost star network is the central-office local area network, which is usually offered as part of the CENTREX package. Digital PBXs, too, may be used to support local networking. This would not have been possible two years ago, but is today thanks to the development of powerful microprocessors, like Intel Corporation's 80386, and recent advances in the way digital PBXs handle data.

For networking at the work-group level, the microcomputer may even serve as the hub for a star network. LAN kits, like QuadStar, available from Quadram (Norcross, Ga.), make this relatively easy and inexpensive to do. The QuadStarter Kit contains everything needed to get two stations up and running immediately. The $1,095 kit contains a half-sized board for one micro, a hub adapter that accommodates five more stations, 50 feet of twisted-pair wiring, and the Tapestry operating system from Torus Systems (Redwood City, Calif.). Additional hub adapters are priced at $375, and, using gateways, QuadStar can communicate with AT&T's StarLAN, as well as with Ethernet and IBM's Token-Ring. Another version of the QuadStar kit is the QuadStar Hub-12, which can tie 12 nodes into the hub.

So-called "LAN alternatives," such as the Zero Slot LAN offered by Avatar Technologies Inc. (Boston, Mass.), are becoming quite popular. So named because it does not require that proprietary adapter cards be inserted into every micro-computer, the Zero Slot LAN also dispenses with dedicated file servers. A single controller links as many as 20 micros and peripherals in a star configuration *via* twisted-pair wiring or RS-232C connections. Each device may be located as far away as 500 feet from the controller, which responds to requests by creating virtual circuits between nodes. Data transfers are performed at up to 115 kb/s.

Despite its apparent simplicity, the Zero Slot LAN offers some relatively sophisticated features. The network-management software, for example, allows the network manager to add or change connections, specify device types, set communications characteristics, name nodes, and determine password security levels using a menu-driven, fill-in-the-blanks scheme. In addition to supporting electronic mail, file transfers, printer spooling, and class of service, the Zero Slot LAN may be used to provide connectivity to IBM mainframes *via* an optional 3270 gateway device, which provides terminal-emulation and file-transfer capabilities. The cost is very reasonable at $100 per node for a 20-node network.

Another type of zero-slot LAN is Server Technology's (Sunnyvale, Calif.) EasyLAN. It is one step removed from Avatar's Zero Slot LAN in that it does not even use a controller — in fact, there is no hardware at all. Networking software is loaded into a microcomputer designated as the hub, which functions as a server. As many as 21 micros can be linked together *via* their serial ports.

Despite the impressive performance of microcomputer LANs, for the most part they do not provide the sophisticated management tools of general-purpose LANs. Although the lines separating microcomputer LANs from general-purpose LANs are rapidly blurring, the one thing that still separates them is the availability of management tools. This means that if you need to track network connections for security reasons, predict traffic bottlenecks, determine where problems are occurring on the network, or decide whether network usage is heavy enough to justify adding another server, you are not going to have the tools you need if you go with a microcomputer LAN.

In evaluating LANs, matters of topology and access method are among the considerations you will have to take into account. Together, they determine such things as throughput, the number of terminals that can be added to the network, the distance that the network can traverse, the ease with which moves and changes may be accomplished, and, of course, the cost of implementation. Committing yourself to a particular topology and access method before undertaking an extensive evaluation of current and foreseeable organizational needs may result in the premature obsolescence of your investment.

The complexity of network design and implementation has forced many organizations to turn to consultants for advice. There are consultants who specialize in putting together LANs and who pride themselves on being up-to-date on the latest products and emerging technologies. Many LAN vendors also provide consulting services, but you might be better off hiring a consultant who has nothing to gain from the specific recommendations made. Of course, there is no harm in listening to what a vendor has to say, verifying the vendor's claims with phone calls and visits to its customers, and then seeking a second opinion from a reputable consultant.

One final thought: instead of confining yourself to only one topology, you may want to consider using a combination of bus, ring, and star, even though they may use different access methods, protocols, and wiring schemes. Gateways, bridges, and routers make mixing networks in this way entirely feasible.

5.1.2 Network Operating Systems

The LAN operating system is simply networking software that gives the network its multiuser, multitasking capabilities. As such, it arbitrates service requests from many users. The operating system is composed of many modules, some of which reside in a microcomputer that acts as the server, while other modules reside in the printers and other network resources. These modules work together to provide the functions of the network, such as recognizing users, associating their identities with access privileges, and routing their requests to the appropriate server for action. Although the operating system is largely invisible to the user, it is nevertheless another aspect of the LAN purchase that requires attention.

There are two basic types of operating system: one grounded in DOS and the other in UNIX. DOS, of course, is the operating system used by most microcomputers. UNIX began as an operating system for minicomputers, although it is becoming increasingly popular for use on micros, mainframes, and supercomputers as well. Some microcomputers that have been modified for niche markets accommodate DOS and UNIX together, allowing the user to hot-key between them to run application programs written for either one or the other.

Since DOS was never intended to be used as the foundation for networking, vendors using it as a platform for their LAN operating systems have had to enhance it with shell programs that intercept multiple requests, buffer them, and divide processor time among tasks.

DCA subsidiary 10NET Communications (Dayton, Ohio) and TOPS (Alameda, Calif.), now a subsidiary of Sun Microsystems, are among the companies that have developed their own modifications of DOS to give it networking capabilities. Most DOS-derived operating systems originate from Microsoft's MS-Net. Other vendors like DEC, AT&T, 3Com, and IBM license pieces of MS-Net for use in their own LAN operating systems, but add their own menus and user interfaces.

DOS-based LAN operating systems share a number of characteristics, most notably, peer-to-peer resource sharing, which is the ability of any microcomputer on the network to use the resources of any other microcomputer, such as its printers and disk drives. The DOS add-on programs that offer multifunction capabilities work in background mode, so that anyone else on the network can use your disk drive or printer, for example, while you are busy with another application.

This peer-to-peer stuff sounds neat. It permits great flexibility in the use of available resources and it makes LANs economical for small work-groups, but peer-to-peer resource sharing has some serious drawbacks. It is hampered by slow response times, which limits network growth. Since many devices act as servers, the network can become difficult to administer. Furthermore, DOS-based systems can consume as much as 400 Kbytes of a terminal's random access memory (RAM), whether or not it is in use on the network. That is quite a lot, considering that the limit for DOS is only 640 Kbytes. In recognition of this problem, some LAN vendors put the bulk of their operating system modules on the networking cards. 3Com does this for its EthernetLink Plus and TokenLink Plus. Banyan Systems' VINES is capable of working off of expansion cards that may be added to the microcomputer to increase RAM. High-capacity RAM chips now provide 4 Mbytes of memory on generic expansion cards.

The standard features that come with DOS-based LAN operating systems vary according to the vendor. Most use menus, and include a print spooling and queuing capability. If you want electronic mail, group calendaring, job scheduling, or some other productivity tool, you can buy them as options, or get them from third-party vendors. There are some exceptions, however. Torus Systems includes a sophisticated electronic-mail capability in its Tapestry operating system, while 10NET offers electronic mail, chat, and network statistics as standard features.

UNIX is the other platform on which LAN operating systems are built. Unlike DOS, UNIX was designed for multiuser, multitasking operation, so no special modifications are required to deliver these capabilities. Although more efficient for a LAN operating system, UNIX must still emulate DOS to service requests from microcomputers. UNIX-based LANs, like Banyan's VINES, require a server. Under VINES, software modules that are functionally similar to the MS-Net LAN products run in each workstation. These modules communicate with the networking software in the server to pass along DOS service requests. The server accepts the requests, checks the identity and the class of service associated with the requester, translates the requests into messages that the server operating system can understand, and passes them to the server operating system. The server software sends back the requested data and issues appropriate error codes to the workstations when necessary.

Unlike MS-DOS, which takes care of mediating simultaneous user requests, it is the UNIX-based server that performs this task, as well as that of running multiple applications. This arrangement provides more efficient network performance and much faster response times. In some UNIX environments, however, there is no peer-to-peer communication; that is, the various workstations are not able to contribute their resources to the network for access by other workstations. Locus Computing's (Santa Monica, Calif.) PC Interface, however, does provide peer-to-peer sharing under UNIX. Generally, only one or a small number of dedicated computers perform the role of server to file, print, or run communications. It is worth noting, however, that Novell's NetWare is designed to allow the server in smaller installations to operate as a workstation, but a workstation cannot be used as a server.

The UNIX-based operating systems typically include such things as integral bridging, remote workstation support, and electronic mail as standard features. Add-on modules are usually available from the same vendor. Third-party application packages may also be used.

Given the differences between DOS-based and UNIX-based LAN operating systems, you should evaluate organizational needs prior to system selection. After all, nothing deflates the ego faster than thinking you have cut the best deal with a vendor, only to find out after installation that users cannot do half of what they thought a LAN should be able to do. What follows is a sampling of LAN features and capabilities that merit attention during the product-evaluation process:

- Dedicated servers *versus* resource sharing: Resource sharing slows down the operation of local programs. Although a computer dedicated to the server role provides faster network performance, it may not be cost-justifiable for small work-groups.

- Fault tolerance: When applications that are critical to a business are run on a network, the operating system can help improve survivability by automatically mirroring the contents of a disk drive or an entire server on a duplicate resource. If the first drive or server fails, the mirror image takes over. Some systems offer tape backup for subsequent reconstruction of the data files.

- Remote access: You can extend the reach of your network beyond the cable limit by providing users with dial-in access to the server or to a designated microcomputer. Some vendors offer password protection to prevent unauthorized access to the network. Find out how many passwords the system can accommodate. For an extra measure of protection against unauthorized dial-in access, look for a "lockout" feature that denies further log-in attempts after a defined number of unsuccessful attempts. This feature prevents break-in attempts by hackers using programs that automatically generate large numbers of log-in attempts with different passwords.

- Integral bridging: This feature allows you to mix and match LAN media and topologies. This is accomplished by putting more than one network interface card in the server, permitting communication from an Ethernet to a Token-Ring, for example. Some vendors like Novell and Banyan provide this as a standard feature, whereas 3Com offers integral bridging as an option. From two to eight different interface cards may be accommodated in the server, depending on the vendor.

- User interface: Most networking software provide menus as the means of selecting network resources and specific features. This can become very tedious for experienced users, especially the technical types. Check into whether or not the networking software permits the use of commands to select resources and functions, and whether requests may be strung together for batch delivery and processing. Also, find out if the networking software includes HELP screens, so users do not have to scurry for manuals, whenever they need assistance. Make sure these screens are easy to read and understand. This will cut down on the amount of time the network administrator must devote to hand-holding individual users. Some networking software even allows you to create on-line tutorials to facilitate user training.

- Network administration: Find out what tools are available to facilitate day-to-day administration of the network, such as account management, resource allocation, and billing. A user's account may include such items as his log-in name, log-in restrictions, password, full name, group affiliation, disk-space allocation, and other information deemed necessary for the daily administration of the network. Resources like disk space may be allocated to each user or work-group — a handy feature for networks plagued by users who do not believe in deleting unnecessary files or archiving infrequently used data. Network usage may be charged back to users, work-groups, or departments. Credit may be extended to those who exceed their monthly budgets. Find out what reports are available to track usage and costs, and if the reports can be customized.

- Performance reports: Find out what tools are available to facilitate network management. Usage statistics may include the amount of time a card has been active on the network, the number of packets it sent or received, the status of buffers, the number of collisions, and the number and kind of transmission errors. Such information can help you track down problems with cable or cards, and help you document the need for network expansion. High-end LANs offer some very powerful management tools like layer-by-layer tracing, which provides descriptions of network events relating to end-to-end connections.

- Diagnostics: Check into what diagnostic capabilities are offered to help technicians find problems on the network and configure the server for optimal performance. Such utilities might include reports of damaged packets or network errors. Some operating systems do both, while others only do one or the other.

- Utilities: Although of little value to the typical network user, utilities can provide network troubleshooters with such useful information as input-output activity, data-transfer locations, and displays of the memory locations used by different types of programs loaded in RAM, as well as the software interrupts they call and the links among application files and data files.

- Open programming interface: This capability allows you to develop distributed applications that can exchange data over OSI networks. The interface implements the OSI session layer and may be used with applications written in such languages as C, PL1, Assembler, FORTRAN, and COBOL.

- Productivity tools: Find out what productivity tools the network operating system provides. You can get pop-up message systems, calendars, notepads, and calculators. There are also pop-up menus that list network resources and device names. Users can establish connections right from the pop-up menu. Be careful, though; these programs have a tendency to gobble up RAM.

- Virtual links: Some networking systems can support two types of connection. In addition to permanent connections associated with each terminal, virtual links may be established according to the user log-in. Upon logging off, these connections are broken. This capability is nice to have, when users frequently access the network from different terminals.

- Transport negotiation: This feature enables both the sending and receiving computers to query each other about which transport class the other supports before any critical information is sent over the network. This feature can be disabled if the remote computer does not support negotiation.

- Security: Find out how many levels of access the system supports. In addition to the use of passwords to gain access to the network or particular resources on the network, you should have the ability to assign users levels of privileges, enabling them to read, write, modify, create, or erase files according to job responsibility or file ownership. Some networking software includes an "execute" privilege, which allows users to run programs, but not copy them onto floppy disk. Single files may be passworded. Whole directories may be hidden from certain categories of users. Some systems allow you to group users having similar privileges into a "class," which makes administration much easier.

- Encryption: If protecting sensitive information is important to your organization, find out what encryption options the vendor offers for its networking software. Some vendors only provide bit shifting or inversion as the means for encrypting data. Others use the Digital Encryption Standard (DES) supported by the National Security Agency (NSA) and the Department of Defense (DOD). Some vendors even encrypt user passwords.

- Disk format: The servers of non-DOS operating systems may use a special disk format. If so, this can prevent you from using the drive to run DOS-based applications. This may be something to watch out for, if you plan to use the server for other applications.

- Electronic mail: For some reason, users almost always expect that the LAN will include an electronic-mail system. They expect the system to store and forward messages, allow for direct replies and forwarding, and give the status of all messages. More often than not, the LAN does not include E-mail at all, and only a few LAN vendors offer it as an option. Fortunately, there are a number of third-party vendors who offer a wide range of E-mail packages to suit any level of user need.

- Barge-in messages: Some LAN operating-system vendors offer a messaging capability that is much simpler than E-mail. This allows users to exchange very brief messages that appear across the bottom of the screen, thus intruding on work in progress; hence the term "barge-in." These types of communication can be quite useful in coordinating network operations, but the capability for locking out such interruptions should also be included.

- Print spooling: Although this is a standard feature on most LANs, check for features that enable users to find out their place in queue or cancel jobs sent there by mistake. Look for capabilities that allow the network administrator to change the priorities of jobs in queue and to assign specific priorities to users.

- Disk caching: Performance may be improved by storing frequently requested data in RAM. This cuts down on the time wasted in frequently accessing the disk. Most networking software supports disk caching, differing only in the amount of RAM that the networking cards provide.

These are the principal features offered by today's LAN operating systems. You have to pick the ones that are most important to your work-group or organization, understanding full well, of course, that the number and sophistication of features will have a direct bearing on the cost of the product. If the features you want are not available from the LAN vendor, even as options, find out what third-party suppliers offer. Make sure that their products have been approved for use by your LAN vendor.

5.1.3 LAN Protocols

Protocols work behind the scenes to determine how data are to be transferred from one place to another, specifying the communication levels to be used and the format in which the data are to be sent. Protocols also provide the information needed by the machines on the network to coordinate data handling among each other. Finally, protocols perform a flow-control function to allow machines operating at different speeds to communicate with each other without losing or corrupting data. This includes arranging packets so that they arrive at the destination in their proper order. Some protocols are very crude and may only permit the transfer of data at very slow speed. Other protocols, like TCP-IP, can transfer data at high speed, even across different computer architectures and operating systems. Given such capabilities, attention to protocols is at least as important as having the right topology, operating system, and wiring plan.

Among the earliest interactive networks were those developed for use by the Department of Defense (DOD), where the network consisted, and still does consist, of many types of mainframe and minicomputer. The Transmission Control

Program (TCP) and Internet Program (IP) are the two protocols that were designed to integrate dissimilar computer systems. As such, they operate at the transport layer of the OSI model. Although the imminent demise of TCP-IP has been predicted many times over the years, these protocols are by no means obsolete. In fact, they enjoy wide usage on large, complex private networks where communicating, translating, and requesting data from diverse computer systems is done on a fairly routine basis.

Table 5.1 Most Commonly Used Network Protocols.

OSI Layer	Protocols	Supplier
Application-Presentation	NCP (NetWare Core Protocol)	Novell
	NFS (Network File System)	Sun Microsystems
	SMB (Server Message Block)	Microsoft Corporation
Session	APPC (Advanced Program-to-Program Communication)	IBM
	DNA Session Control	DEC
	NetBIOS	IBM
	RPC (Remote Procedure Call)	Sun Microsystems
Transport-Network	XNS (Xerox Network Systems)	Xerox Corporation
	TCP-IP (Transmission Control Protocol–Internet Protocol)	U.S. Department of Defense
Data Link–Physical	ARCNET	Datapoint
	Ethernet 802.3	IEEE
	Token Bus 802.4	IEEE
	Token Ring 802.5	IEEE
	StarLAN	AT&T

Vendors like Excelan sell TCP-IP software modules for specific computers and controller cards, which recognize each other on the network and communicate in a common format generated by the higher-level session layer and applications programs of OSI. Novell, 3Com, Banyan, and other networking software vendors package TCP-IP as plug-in boards to microcomputers, as stand-alone interface boxes, or as full-blown LAN gateways.

NetBIOS is another popular protocol, which was the result of a joint venture between IBM and Sytek Inc. (Mountain View, Calif.). This protocol was originally developed as an interface between the IBM PC Network Program (now called PC LAN) and network interface cards developed by Sytek. NetBIOS (Network Basic Input-Output System) includes a programmable entryway into the network, which allows microcomputers to communicate over network hardware without having to go through the networking software. This hardware independence is chiefly responsible for the emergence of NetBIOS as a *de facto* standard for local-area network software interfaces, and as more applications were developed to support NetBIOS, more LANs were developed to support the protocol. This symbiotic relationship promoted the growth of the LAN industry.

NetBIOS-driven devices simply establish virtual communications sessions with each other across the network, but this simplicity has its drawbacks. The naming scheme employed in NetBIOS does not work well between networks or with many LAN operating systems.

Although Novell, 3Com, and Banyan do not use NetBIOS to drive their interface cards, their operating systems do emulate NetBIOS, providing the same session-layer communications services that NetBIOS offers, but the combination of NetBIOS and TCP-IP can overcome the limitations inherent in NetBIOS. The Internet Protocol of TCP-IP can repackage the NetBIOS modules to permit their transmission intact through several layers of network names and addresses. Both protocols are so popular that a combination of NetBIOS and TCP-IP operating at the OSI session layer has been recommended by the Internet Activities Board, which oversees TCP-IP standards. LAN products that use both NetBIOS and TCP-IP are already in the works by such manufacturers as Ungermann-Bass Inc. (Santa Clara, Calif.) and Excelan Inc. (San Jose, Calif.).

What does all this mean within the context of evaluating LANs? It means that if you choose products that do not use TCP-IP or NetBIOS, or some combination of the two, your only other choices are to go with vendor-specific protocols, which can impose limits on your ability to communicate with other networks and limit your ability to integrate new services into your network. For example, SNA is IBM's answer to the OSI model. SNA is also a very obvious attempt to lock users into the IBM-mainframe environment. Within SNA is the Advanced Program-to-Program Communications (APPC) protocol, which corresponds to OSI's session layer in that it establishes the conditions that enable programs to communicate over the network. APPC is the foundation upon which IBM intends to build all of its future applications and systems products. DEC, too, has developed its own protocol stack to rival the OSI reference model. Unlike IBM, however, DECnet protocols will evolve toward compatibility with OSI. This will allow DEC to use such compatibility as a selling point against IBM.

If your network consists only of IBM PCs or compatibles, for example, and all you want to do is network them together and use a server for file sharing and IBM-mainframe access, you do not have to pay attention to what protocols the network operating system uses, since you are already locked into IBM by default, but if you want to move information across the network to DEC, HP, IBM, and other types of computers, and have those computers treat each other as peers, then paying attention to protocols becomes very important.

Another set of protocols has been developed by the Institute of Electrical and Electronics Engineers (IEEE) under the general heading of 802. IEEE 802.3 and 802.5 relate to microcomputer-based LANs. The former describes a standard for CSMA Ethernet–type networks on an electrical-bus topology, while the latter describes a standard for the token-ring architecture. Many vendors provide 802 interface cards, while IBM even offers an optional Ethernet port on its 9370 minimainframe computer. In fact, IBM provided much of the input and leadership on the general committee within IEEE that developed the 802 standard.

There are several distributed-file system network protocols, including SMB, RFS, NFS, and XNS. These protocols allow a computer to use the files of another networked computer as if they were local. The operating systems of the two networked computers link up so that a subdirectory made available on the host may be seen as a disk drive or as a separate subdirectory on the user's computer. In this way, application programs running on the user's computer can access the files and resources of the host without requiring special programming. Although these protocols operate similarly, they are not interchangeable. Computer vendors generally develop such protocols for use within specific product lines and then permit other vendors to license them to achieve compatibility.

Even though vendors make arrangements with each other to support each other's protocols in each other's products, you have to make sure that you match the specific protocols with those of the vendor whose equipment is used on your network.

There are subsets of the OSI reference model that govern networking within specialized environments. The Manufacturing Automation Protocol (MAP), for instance, draws upon OSI to implement a protocol tailored for the assembly-line and shop-floor environments. MAP specifies the use of a broadband cable network based on the IEEE 802.4 standard for token-passing media access control (MAC). It also includes the manufacturing message standard (MMS), an OSI application-layer protocol for formatting and transmitting commands between controlling programs and machines.

Likewise, Boeing Computer Services Co. (Seattle, Wash.) has developed the Technical and Office Protocol (TOP) for engineering, accounting, marketing, and other corporate functions that provide manufacturing support. TOP includes some OSI application-layer protocols that are not part of MAP, such as the X.400 messaging protocol and the virtual terminal protocol. This is because these capabilities were anticipated to be more relevant to the office environment than to the factory floor.

TOP specifies the use of IEEE 802.3, which includes a logical link control that allows users to specify devices to share Ethernet and token-ring adapter boards. The logical link control also defines interfacing for 802.3, 802.4, and 802.5 networks, so that users can build bridges between each at the data-link layer of OSI.

MAP and TOP are merging into a unified MAP-TOP protocol, which lays the foundation for a corporate-wide networking strategy that enables companies to exercise more leverage in the procurement of network components, thereby saving them more money. In eliminating dependence on proprietary solutions and ties to any one vendor, companies can shop around for products that offer the best price and performance, and when MAP-TOP is implemented throughout an organization, these benefits become more substantial, promising to reduce hardware and software costs by as much as 20%. Additional savings accrue through the reduction of the support staff needed to design, integrate, and implement network changes.

In determining which LAN provides the best fit for your organization, you must break out the crystal ball and try to determine what your network may look like a few years from now. If ISDN is a possibility for your network, then you might not want to invest too heavily in, or become overly dependent on, proprietary protocols. Positioning your network to take advantage of future services is a key consideration in any purchase decision, and protocols are no exception.

5.2 THE INS AND OUTS OF LAN CABLING

As LANs increase in popularity and more vendors enter the market, first-time buyers as well as veteran users must take care to select the best wiring plan for their requirements, but that is not as simple as merely deciding on one LAN over another. You must wade through seemingly countless cabling options and select the media that will deliver the best performance.

There are only three basic types of cabling used in LANs: coaxial, twisted pair, and fiber optic. Coax and twisted pair accommodate all of the requirements for an efficient LAN, including speed, data traffic, network growth, and distance. Fiber makes a good backbone facility because it can handle the high volumes of information passing between the floors of a multistory building, or between buildings in a campus-like environment, but even within these broad categories of cabling, there are enough differences to make selection difficult. As if that were not enough, each option may have a hidden cost associated with it.

Typically, user requirements are among the first things to be addressed during LAN selection. Cabling is usually the last item considered in the purchase of a LAN. Somehow that is supposed to take care of itself, down the road. Ultimately, however, it can cost at least as much as the terminals you are trying to interconnect, and sometimes more. Consequently, it is a mistake to underestimate the importance of proper cable selection. Once the LAN is purchased and the terminals are connected with the desired cabling, there is no quick fix available — at least not at a price you will be happy to pay.

You have a choice between thick and thin coaxial cable. Ethernet LANs, for example, traditionally use a 50-Ω, thick-baseband coaxial cable to transmit data at 10 Mb/s, but implementing the Ethernet LAN can be quite expensive, especially if you have not budgeted for it. Depending on volume and the vendor, the cost of such cabling can be $3 to $4 per foot. If you are a big user, once you have purchased the Ethernet LAN, there is not too much choice about the cabling you will need. Although it is now possible to use twisted-pair wiring for Ethernet, thick coax is still recommended to insure high performance, especially when a large number of devices will be using the network simultaneously, when bandwidth-intensive applications are being run, or long distances are involved between floors or between buildings.

There is thin-baseband coax cable available that is also 50-Ω, and it is cheaper and easier to install than regular coax. This makes it more suited for work-group microcomputer networks when relatively short distances are involved and slower data rates are acceptable. Opting for thin coax can shave as much as 25% off the cost of material and installation.

Although the industry is moving toward twisted-pair LANs, coax is still quite popular. It is a reliable, proven technology. In fact, IBM and DEC continue to recommend coax whenever possible for its reliability, even though both have wiring plans that include unshielded twisted pairs. Coax also has high bandwidth, which means that it can carry data at high speeds of 10 Mb/s and more over long distances, and it can support multiple channels. One thing to consider about coax cable is that while initial installation costs are high, the cost of expansion is incremental, permitting economical network growth, and, with broadband, more than one network may share the same cable, making it an excellent medium for video transmission, whether it be for security, video conferencing between buildings, or quality-control applications in manufacturing.

Coax, however, also has a dark side. Physical-link problems with coax are the cause of most LAN problems. Future maintenance problems can be minimized by specifying quality connnectors, particularly the baluns (*bal*anced-to-*un*balanced). These are interfaces that enable installers quickly to link cables with different electrical impedances — such as twisted-pair and coaxial cable — to allow different vendors' equipment to share the cabling system. You should also make sure that only experienced installers are involved in setting up your LAN. Still, the medium is inflexible and cumbersome to install. It cannot be bent too sharply and may even break, if it is not handled with care. Installation costs tend to be quite high, and coax does not lend itself to terminal equipment moves and changes, unless it is already installed in every nook and cranny of the company, which would entail even higher up-front costs.

For these reasons, coax as the primary LAN-wiring medium may be on its way out in favor of twisted pair — copper wires twisted around each other, which may, or may not, be encased in a jacket to provide a shield against electromagnetic interference. The twists help cut down on noise interference; since noise will be added to both wires equally and in phase, the twists produce a canceling effect.

Some vendors wrap additional shielding around individual pairs within multipair wires, to cut down further on noise interference.

Twisted pair will probably become the dominant LAN terminal wiring medium for several reasons.

- It is much easier to install than coax. It requires fewer connectors, and its connectors attach more quickly.
- The bend radius is smaller than coax, making it easier to install in a wider range of environments.
- Unlike coax, twisted pair is versatile in that it can handle voice as well as data.
- Token-Ring LANs can run at 4 Mb/s on unshielded twisted pair. The emerging Ethernet twisted-pair standard will boost this speed to 10 Mb/s. Shielded twisted pair will allow even higher speeds.
- Businesses that have outgrown CENTREX phone systems in favor of digital PBXs have extra twisted pairs already in place at every station, which can be put to use as LAN connections.

There are many types of twisted-pair wiring, each available in different gauges, insulation and shielding materials, and bundle sizes. The most common gauges are 26 American Wire Gauge (AWG), 24 AWG, and 22 AWG. The lower the gauge, the thicker the wire. Although most in-house wiring is 24 gauge, 22 AWG is sometimes used to extend the transmission distance of in-house wiring. Shielding improves the resistance of the wire to external noise and electronic radiation, but it also decreases the transmission distance due to signal loss. Signal loss results from the tendency of the signal frequency to migrate from the wire to the shielding, where it bleeds off into the ground surrounding it. Some shielding materials are better at limiting signal loss than others. In addition to differences in gauge and shielding, twisted pairs vary in terms of bundle size. Bundles may consist of as few as two twisted pairs to as many as 25 twisted pairs, depending on the application. Wiring between standard single-line phones to the PBX may consist of two, three, or four twisted pairs, for example, while an attendant station (a secretary's multiline phone) may require 12 to 25 twisted-wire pairs.

Both Ethernet and IBM's Token-Ring Network can use shielded twisted pair and deliver reliable data rates of up to 16 Mb/s over relatively short distances. Under IBM's Cabling System, based on a star topology, there are nine types of voice and data-grade twisted pair offerings. Types 1, 2, and 3 are the most widely used.

Types 1 and 2, 150-Ω shielded twisted pair, are typically used for higher speed LAN applications. Type 1 consists of two data-grade twisted pairs. Type 2 consists of two data-grade and four voice-grade twisted pairs. Type 3 is made up of four voice-grade unshielded twisted pairs similar to those providing plain old telephone service (POTS). Types 1 and 2 are better for reducing noise interference and the bit-error rate of transmissions than Type 3, but they may cost as much as $1 more per foot.

Both shielded and unshielded twisted-pair wiring have well-defined limits on how many devices they can support on the token-ring network. The limit for shielded twisted pair is 260, while the limit on the unshielded wire is 72.

A token-ring using IBM's Type 3 unshielded twisted pair is considerably less expensive than either of the shielded types, since it is the unshielded type that already exists behind office walls, but the unshielded wire carries some risk. That medium is limited to 1-Mb/s data transmission, and the bit-error rate on unshielded twisted pair can have a significant effect on performance. IBM claims that its Type 3 wiring can support 4-Mb/s transmission, but only up to 150 feet. That distance does not present a problem for most users, but what about large office buildings, like the World Trade Center in New York City, where each floor is an acre square? Fortunately, it is standard installation practice to locate an Intermediate Distribution Frame (IDF), or wiring closet, within close proximity to the telephones being served. In fact, an AT&T study revealed that 97% of all office phones are connected within 180 feet of a wiring closet. There, a patch panel can provide the means to connect station wiring to a coax backbone.

Type 3 wire is more susceptible to electromechanical interference caused by fork lifts, elevator motors, or power generators, for example. Even analog-telephone ringing signals can cause enough interference to disrupt data transmissions. There are also a variety of line impairments that can break down the data signal, like attenuation, crosstalk, and jitter. Attenuation is signal loss due to resistance in the cable. Crosstalk is the unintentional coupling of two separate signals running in opposite directions. The twists in the nonshielded wire are supposed to cancel out the effects of crosstalk, but it may not do the job completely. Jitter is the instability of a signal from one bit to another. This impairment is especially troublesome on large networks, because jitter accumulates linearly as the number of devices on the network increases, and this occurs more rapidly over unshielded than over shielded wire. Reflections of energy, stray capacitance, or a change in impedance add to jitter — and get worse at higher speeds. The only incentives to using unshielded wire are its low cost and its immediate availability at each station.

AT&T's StarLAN adheres to the same standard as Ethernet (IEEE 802.3), but, unlike Ethernet, StarLAN offers data speeds at 1 Mb/s over twisted 4-pair 100-Ω unshielded wire. StarLAN is much less expensive because existing wiring can be used. The StarLAN wiring scheme works in tandem with an installed PBX. StarLAN uses only two pairs of wires to transmit and receive, leaving the other two pairs for the telephone connection. A data rate of 1 Mb/s may seem reasonable, but for high-volume data transfers, StarLAN may not fit the bill. For high-bandwidth applications like CAD-CAM or desktop publishing, StarLAN 10 would be required, which provides 10-Mb/s transmission.

AT&T has its own blueprint for cabling, which it calls the Premises Distribution System (PDS). Whereas IBM and DEC continue to recommend shielded twisted-pair wiring or coaxial cable, AT&T's PDS specifies unshielded twisted-pair wiring for error-free transmission at 6.3 Mb/s for distances up to 1,000 feet. PDS is an eight-pair scheme designed to provide telephone, Token-Ring, and 3270 terminal connections.

The use of unshielded twisted-pair wiring may be risky in a number of ways. Passive line conditioning–impedance matching equipment is required at each end of the twisted pair. Regardless of how well this equipment works, the quality of the initial installation is unknown and, even if excellent, may have deteriorated over time as a result of modifications, expansions, reroutings, and shoddy workmanship. Also, even new unshielded twisted pairs require additional equipment in the form of filters to eliminate crosstalk interference and minimize the effects of external electromechanical interference. The cost of the filters, plus the cost of the line conditioning–impedance matching equipment required to operate over unshielded cable, is comparable to the cost of installing all new shielded cable. Finally, there is still some question as to whether unshielded cable represents a viable platform on which to build future networks.

Unshielded twisted pair simply cannot handle large network demands. Maximum signal distance is only about 150 feet from workstation to patch panel, and if a ring topology is employed, a ring can contain only about 72 nodes.

In contrast, the IBM Cabling System — which consists of double-shielded twisted-pair wire — has a transmission distance of 600 feet from patch panel to workstation in an Ethernet network. For token ring, it can support up to 260 terminals on a single ring, with an average signal distance of 300 feet. In addition, the IBM Cabling System can support multiple rings; unshielded twisted pair cannot.

Whether shielded or not, the quality of the twisted pair itself is important. High-grade twisted pair maintains a consistent 150-Ω impedance across the frequency spectrum. That is important, because the 150-Ω impedance is likely to become an industry standard. Not only is the Electronics Industry Association (EIA) considering this, but IBM has also indicated that all of its future equipment will operate at 150 Ω. Increasing data rates result in increased sensitivity to outside electrical interference, making shielded twisted-pair wiring mandatory. Networks with shielded data cables serve high data rates more reliably.

Fiber-optic cable, made of hair-thin glass or plastic that conducts light pulses, can carry network data over many miles without repeaters at speeds exceeding 100 Mb/s. This makes fiber appropriate for baseband backbones that link remote LANs. Although the maximum speed of fiber is constantly being pushed higher in research labs, the American National Standards Institute (ANSI) has specified the data rate of 125 Mb/s in its Fiber Distributed Data Interface (FDDI). Proteon Inc. (Westborough, Mass.) offers an 80-Mb/s fiber LAN called Pronet-80, while Scientific Computer Systems Inc. (San Diego, Calif.) announced a 178-Mb/s token-ring in June 1988. Token-rings capable of delivering 500-Mb/s performance between supercomputers are in the development stage.

In addition to providing greater bandwidth capacity and increasing the distance between network nodes, fiber offers many other advantages over copper wire. Its immunity to electromagnetic interference makes it the preferred medium for riser cable in tall office buildings and for extending networks to factory areas where heavy machinery is in constant use. Fiber is more resistent to moisture, which can seriously impair the performance of copper wiring. It is also more secure than other types of wiring, the radiated signals of which can be tapped by hackers.

Fiber is not without its drawbacks, however. The cost of the terminating equipment is still too high for most LAN users, and installation costs are higher than for other cable types, because expensive equipment must be used to cut, align, test, and seal the fiber during splicing. There are also fewer experienced technicians to maintain the terminal equipment and do the splicing. All this will change, of course. As innovation continues, and as fiber becomes more commonplace, more qualified technicians will be available and prices for terminal equipment will fall.

Many networks being built today will evolve to operate at a much higher data rate, perhaps well above 10 Mb/s. IBM plans to introduce technology for a 16-Mb/s Token-Ring, and advises users to plan for 20-Mb/s transmission in the future. All things considered, it makes little sense to be penny-wise and pound-foolish by selecting low-performance cable, and if you already have enough justification for installing fiber, it makes sense to plan ahead for 125 to 178 Mb/s.

Now that you know some of the factors that should enter into your decision-making about LAN cabling, who should you get to install it? You could consider doing it yourself, assuming that you have the manpower available and that they know how to isolate unmarked, unterminated, or unusable pairs. Also, if you lack proper cabling documentation, in-house staff must know how to use such test devices as a volt-ohm meter (VOM), or something more exotic like the Time Domain Reflectometer (TDR). Wiring sounds like a simple enough task, but it is definitely a job best left to experienced work crews.

Data-communications service companies and telephone-interconnect vendors may have the installation experience you are looking for. Although they may be very knowledgeable in their respective areas, they usually do not have experience in both. A detailed reference check is your best source of information about whether such firms can really do the job. Ask for references that use the same LAN that you plan to install, have the same topology as you plan to implement, and have roughly the same number of terminals you have. Because LAN problems may not be traceable to the cabling until long after the installation, make sure the references have had the LAN up and running for at least three months. That is usually sufficient time to track down a problem with cabling and have the installer back on site to correct the problem. Then you can ask the reference about the vendor's response time and helpfulness in resolving such problems. If the vendor hands you a list of references that have only recently used its installation services, you will not get the opportunity to delve into these matters, because your questions will probably not be relevant.

Many PBX and LAN vendors offer network-design and installation services, but they may be biased toward specific wiring schemes, which may or may not be what you need. The representatives of these types of firm might talk very authoritatively about which wiring scheme you should use, and may even try to convince you that they know what is best, since they perform such services day in and day out. Never forget, though, that many of these firms are aligned with other manufacturers whose equipment works best with theirs under a particular wiring plan. This may make future expansion more costly, limit the number of terminals you can have on your network, or make moves and changes difficult.

If you are setting up a LAN for the first time, it might be best to hire an outside consulting firm to design the wiring plan and supervise installation. Consultants typically will have a broad range of experience to draw upon, which you may find preferable to narrowly focused equipment vendors or specialized contractors and interconnect companies.

Cabling systems are not easy to install, and they can be very expensive to troubleshoot, so it is important that installation be done right the first time. No matter whom you choose to design and implement the wiring plan, you should have an in-house technician work side-by-side with the installation crew. The in-house technician can perform an oversight function to insure that cabling is not laid too close to sources of electromechanical interference, fluorescent lighting, or electrical power. He can make sure cable splices and connections are performed properly, that specified wire types and components are indeed being used, and that cables and patch panels are properly labeled, and he can see to it that the cable system is being mapped while the work is in progress, and not left to the installer's memory after the work is done. This kind of supervision will insure a quality installation and facilitate network expansion at the lowest cost.

5.3 GATEWAYS, BRIDGES, ROUTERS

LANs were originally intended for intraorganization networking; within a campus environment, for example, or even within a building. After Xerox developed Ethernet, it was used within its Palo Alto Research Center (PARC) in California. Only after the Department of Defense developed TCP-IP did Xerox start building gateways to link its remote sites. Initially, LANs were also designed to link a common set of hardware, but once other organizations like DEC started to use them and TCP-IP became more widely used, LANs started to be used for linking diverse hardware and operating systems.

In recent years, a variety of devices have become available to facilitate networking, while relieving host computers of the processing-intensive tasks of protocol conversion and routing information to appropriate destinations. Specialized devices like gateways, bridges, and routers are available to handle these chores, and they can do them more efficiently than host computers. Furthermore, these devices may be shared among many users, which substantially lowers the cost of networking.

Each type of device is designed to operate in conjunction with a different layer of the Open Systems Interconnection (OSI) reference model (Table 5.2), which provides specific network functionality. In addition to reducing operating costs, the use of such devices can extend the reach of network users and improve LAN performance.

Gateways and bridges connect different LANs. A gateway interconnects networks or media of different architectures by processing protocols to enable a device on one type of LAN to communicate with a device on another type of LAN. Thus, an SNA gateway, for example, may be used to interconnect a microcomputer network to an IBM-SNA mainframe. The gateway, then, acts as both a conduit over which computers "speak" and as a translator between the various protocol layers.

Table 5.2 Open Systems Interconnection (OSI), Reference Model.

Number	Layer	Function
7	Application	Selects appropriate service for application
6	Presentation	Provides code conversion data reformatting
5	Session	Coordinated interaction between end-application processes
4	Transport	Provides end-to-end data integrity and quality of service
3	Network	Switches and routes information
2	Datalink	Transfers units of information to other end of the physical link
1	Physical	Transmission onto the network

The seven-layer OSI model provides a common base for all LAN protocols, which are sets of messages with specific formats and rules for exchanging information. Each layer of the OSI model constitutes a process that must take place for a message to get from one place to another. Each layer works with the layers above and below it.

A bridge operates at the data-link layer of OSI to interconnect LANs having the same type of operating system. Since both LANs have the same operating system, the bridge does not need the capability to perform protocol conversion. In this case, bridges merely look at the LAN-level source and destination address to see where the packet is going. This means that a bridge can interconnect DECnet, TCP-IP, or XNS networks, for example, but it cannot insure that users on one can talk with users on another. For that, a router is needed.

A router has more intelligence than a bridge in that it can handle up to three levels of addressing (see summary diagram in Figure 5.1). It keeps a map of the entire network, including all the other devices operating at or below its own protocol level. Whereas a bridge only checks the address of a packet to see if it is bound for another network, a router looks deeper into the packet. Then, referring to its internetwork map, it examines the status of the different paths to the destination and decides how best to get the packet to the addressee. Since routers are protocol-specific, there is no protocol conversion involved in this process.

These are only basic categories of LAN products, however; not all products fit neatly into these categories. Many vendors have developed hybrid products that include functions traditionally associated with one or another of these categories. Thus, an "intelligent" gateway device may include some attributes commonly associated with bridges and routers. Another device may operate as a gateway, but default to a bridge under certain circumstances. Single-unit bridges and routers — "brouters" — are really bridges that include some router capabilities. To complicate matters, some vendors of simple controllers and front-end processors are calling their products gateways or bridges, and what the Department of Defense calls a gateway on its Internet is merely a router in OSI terminology, while Novell's NetWare bridge is the equivalent of a router in OSI.

5.3.1 More About Gateways

Because organizations are comprised of specialized work-groups, different networks may be in place to meet the requirements of their users. For example, the research and development division of a large company may have chosen an Ethernet LAN to support its applications, while the technical documentation group may have chosen AppleTalk as the means of linking Macintosh microcomputers to support its publishing operation. The two groups discover that they can eliminate duplication of effort by drawing upon the resources of the other. A server equipped with both Ethernet and AppleTalk circuit boards performs the necessary protocol conversions that allow users on both networks to exchange files. When a server performs protocol conversions that allow information to be exchanged between two different networks, it is called a "gateway."

As organizations become more complex, the need to share files and communicate all kinds of information across diverse networks becomes necessary to improve efficiency and productivity. The need to connect dissimilar networks may also come about as the result of corporate-merger or acquisition activities, or it may stem from the desire to interconnect with packet-switched networks for economical long-haul data transport. Whatever the justification for linking dissimilar networks, gateways are designed to do the job.

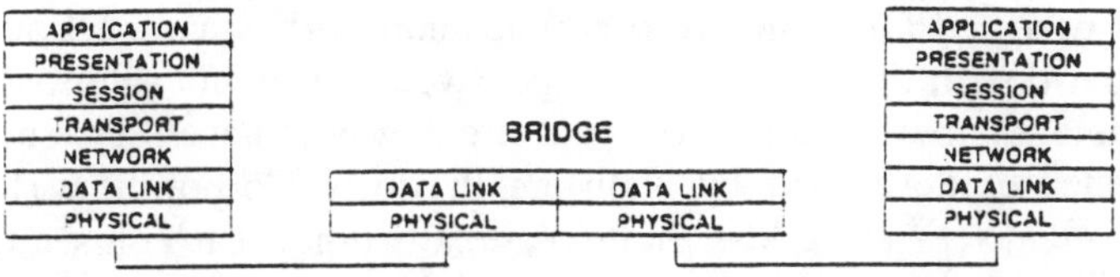

Bridge — *Bridges and routers are similar in function and therefore often confused. Bridges connect LANs with compatible protocols at the Data Link layer and above. Only the Physical layers differ. Because the connection is just under the three layers which define the LAN protocol (such as TCP/IP, DECnet or XNS), bridges are protocol-independent.*

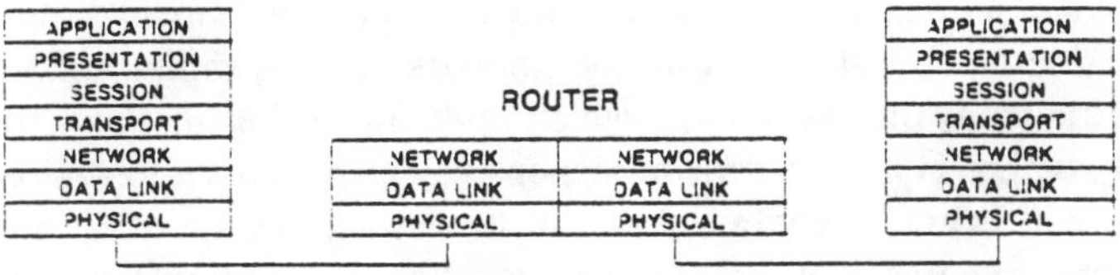

Router — *Routers connect LANs with protocols in common at the Network layer and above. Since they do connect at the Network layer, they are protocol-sensitive and thus can link TCP/IP, DECnet or XNS-based LANs. A multiprotocol router is needed to connect their combinations. Protocols without a Network layer (such as DEC's terminal protocol LAT) cannot be routed; they must be bridged.*

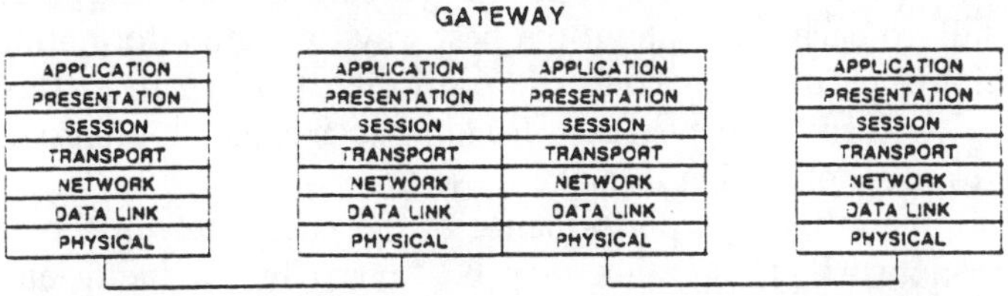

Gateway — *Routers and gateways are often confused because what the Internet calls a gateway is a router in OSI terminology. In the OSI model, gateways provide a communication path between two LANs using entirely different LAN types and protocols. They perform protocol conversion for all seven layers of the OSI model. Gateways are application-specific. An X.25 gateway can connect a PC LAN to a public data network, for example.*

Figure 5.1 Summary of gateway, bridge, and router functionality. (Courtesy *LAN Magazine*, June 1988.)

Before gateways became available in the early 1980s, users had to purchase special boards, like Digital Communications Associates Inc.'s (Alpharetta, Ga.) IRMA or Micro Tempus Inc.'s (Montreal, Can.) Tempus-Share, one for each microcomputer, to permit occassional access to the host. Such products did the job they were designed for, which is to permit micro-to-host connections, but at $400 to $800 per board, they were a very expensive connectivity solution. To trim operating costs and streamline the network, a gateway is necessary, such as the X.25-based gateway developed by Network Products Corporation (Pasadena, Calif.) for Novell's NetWare LANs. In this case, a two-port card is used that plugs into an expansion slot of a microcomputer, which is designated as the server. One port on the board provides a 64-kb/s connection, while the other provides a 19.2-kb/s connection. Together, they support up to 32 concurrent sessions. The gateway includes a packet assembler-disassembler (PAD) function, eliminating the need for separate PADs.

Such arrangements also simplify network management. Instead of having to monitor the traffic from 100 micros on the network, for example, you only have to monitor traffic from the file server, which appears to the host as a single peripheral device. There is no need for an expensive cluster controller, because the gateway replaces it. When a separate server is used as a gateway, other advantages become apparent, such as reduced cabling costs and installation time, as well as easier moves and changes. In fact, users can change the physical location of their equipment, but their logical address on the network stays the same, and with communications functions offloaded from the host in this way, valuable processing resources are freed up to do more important tasks.

Some vendors are developing so-called "intelligent gateways." These are gateways that talk to each other about the best way to route information, taking into consideration such things as congestion, priority, performance (throughput, delay, error rate), security, and even cost. Building such capabilities into intelligent gateways relieves users of having to make these decisions.

Congestion may affect the performance of the entire network, or only one gateway of the network. Congestion may be caused by an inefficient routing scheme, causing more traffic to stay on the primary data link or links longer than necessary, thus slowing down the entire network. Alternatively, congestion may actually be in the gateway itself, which may occur when it is presented with too many packets to filter. Finding out the cause of the congestion can make a difference in determining whether there is any point in trying to reroute through other gateways attached to the network. If the entire network is congested, then all of the alternate gateways that are located on the other side of the network can be bypassed entirely in favor of a hop through an entirely different network.

When congestion is detected, the intelligent gateway can even consider the priority of information that is to be routed; whether local or internetwork traffic should be given preferential treatment, for example. Prioritizing information is also important in the management of the network. Intelligent gateways will have the ability to let diagnostic information pass through or around congested areas, providing real-time status reports on each link.

In hopping through to another network to avoid congestion, security becomes a concern. This concern is addressed by the intelligent gateway's ability to distinguish between routine and sensitive information during the routing decision. This would entail that the user flag the information as "confidential" in the address.

The gateway occupies a strategic position on the network, which can be used to enhance overall management. The gateway can extract detailed information about the data traffic that is passing through it, as well as the status of the data links it interfaces with. The gateway can insure that the links are handling data reliably, without exceeding user-defined error-rate thresholds. It can also check on the various protocols being used, making sure that there is enough protocol-conversion processing power available for any given application.

There are a lot of factors that deserve consideration when evaluating gateways. Grilling vendors before the purchase will help you avoid potential problems later. Start by finding out if the gateway is certified to operate on *your* LAN. It is not enough to accept vendor assurances that its gateway runs on any NetBIOS-compatible network, for example. You must find out if the vendor supports your particular NetBIOS-compatible LAN. Slight differences in the way vendors implement NetBIOS may prevent you from running certain applications over your LAN.

Speed may play an important role in your selection of a gateway. Some gateways operate at only 9.6 kb/s, which may cause traffic bottlenecks when connected directly to a 10-Mb/s Ethernet LAN. Most LAN gateways are capable of operating at 19.2 kb/s and 56 kb/s, which can also cause traffic bottlenecks. In such cases, the protocol must be able to do flow control through the gateway and respond to gateway-congestion indicators. Other gateways, like DCA's IRMALAN DFT and IBM's 3270 DFT, offer variable operating speeds. Several vendors, including Network Products Corporation and DCA, plan to deliver 64-kb/s and 72-kb/s gateways respectively by the start of 1989. (For wide-area networking, several low-speed channels may be multiplexed over a T1 link to the remote gateway.)

In evaluating gateways, look at the level of transparency embedded in the product. This is especially important if the gateway will be used for wide-area networking. The actual path that connects the user to any device on the network should be chosen by the gateway to balance load levels, bypass failed links, and find the most economical route. All of this should occur automatically as the result of a single connect request, regardless of the equipment location or the protocols involved.

With large networks, the ability to detect, isolate, and diagnose problems becomes very important. You should find out what network-management tools are available with the gateway that will enable you to correct problems quickly. Look for graphic displays that allow you to monitor network performance, determine buffer levels, and check network control-signal status. Make sure that the management system allows you to configure channels, links, and other network access devices like bridges and routers. It should also have the ability to bring gateway ports on-line or off-line as required.

When choosing a gateway, assume that your network will experience growth. As your network grows in size and complexity, the ability to plan becomes very important, but you cannot plan without having relevant information at hand. For this reason, you should find out what kinds of report are available, as well as what formats the information is presented in. Statistics should become available automatically by time of day, or on demand, by issuing the appropriate command. Make sure the gateway management system can archive network statistics for trend analysis. This can assist you in predicting network growth.

If security is important, you will want a gateway that implements controlled access. This involves assigning specific ports to certain microcomputers. When the microcomputer requests access to the gateway, it is given the port reserved for it. Since no other microcomputer can access that port, security is enforced. Under this scheme, each port may have access privileges associated with it. One port may provide access to all mainframe applications, for example, while another port may be limited to only one. The trouble with dedicated access is that idle ports cannot be used by anyone else, which means that you must sacrifice efficiency for security.

If security is not one of your concerns, gateway access on a first-come, first-served basis may provide more opportunities for users to link with the mainframe or other network resources, since this arrangement does not tie users to specific microcomputers. Some gateway products permit both shared and dedicated access, allowing some ports to be reserved for specific microcomputers and the rest pooled for general use.

Some gateways now support windowing, which allows all multiple sessions to be displayed on-screen at the same time. The user can even cut-and-paste from window to window to make use of any and all information. This is a handy feature in publishing environments, where the user must draw upon many sources of information to develop corporate reports or reams of technical documentation.

Be aware that file-transfer software may not be included with the vendor's gateway and terminal-emulation package. Transfer software allows files from the mainframe to be downloaded to the microcomputer, not an easy task, since the mainframe is treating the micro as a terminal without local disk storage. If no file-transfer software is included, do not dismiss the product yet. You might be able to get the gateway at a bargain-basement price, and choose the file-transfer software you really want from a third-party vendor. Of course, you have to make sure that the third-party software works on the gateway you want, but the extra effort you put into this evaluation process may be well worth it.

Since dedicated gateway devices are usually more expensive than a microcomputer, if your resources are limited you may want to consider using a micro-computer as a gateway device. This allows the micro to emulate a terminal at the same time that it provides a gateway to other microcomputers. Dedicated gateways, however, are faster than microcomputer gateways, and since gateway software, terminal-emulation software, and DOS software will be vying for space, you must know how much RAM you will need. Most microcomputers can be upgraded with at least 2.5 Mbytes of RAM, which can handle all of these requirements fairly easily. Problems occur when you try to use processing-intensive applications on top of all that, like large spreadsheets or CAD-CAM.

Gateways are in wide use on LANs of all types. You should have no trouble obtaining references from vendors and getting performance information from customers. Additionally, LAN topics are covered virtually continuously in the trade press. You can call the authors of technical articles for additional information. Especially be alert to articles authored by consultants. These people write such articles in the hope that someone will call for further information, and possibly be in need of their consulting services.

Finally, ask vendors what they mean when they use nebulous terms like "universally approved gateway hardware," "superior design," and "proven compatibility." These phrases sound nice, but they beg amplification. Some vendors deliberately use such language to disguise the fact that their products are still under development.

5.3.2 More About Bridges

Although the terms "gateway" and "bridge" are often used interchangeably, there is a subtle difference: a gateway connects dissimilar networks, whereas a bridge connects similar networks. A bridge may connect two or more LANs within the same building, or geographically separate LANs. Local bridges are capable of operating at 10 Mb/s, whereas remote bridges typically operate at from 4.8 kb/s to 1.544 Mb/s, depending on the type of leased line used.

The bridge knows enough about the data-link protocol to do some minor routing based on which network provides access to which address groups. A hierarchical routing feature filters local data traffic so that local network performance is not affected. In this way, the bridge receives a packet of data, scans only to the network address, and passes it on to the appropriate network, where it is ultimately routed to the appropriate addressee.

Bridges are ideally suited to interconnecting like networks, where no protocol conversion is required, security concerns are minimal, and rudimentary routing is all that is required. For example, in a campus-like environment, you might want to connect each building's local network to the fiber-optic backbone. The bridge can be used to keep local traffic isolated to a building, or to a cluster of buildings, and off the superhighway. Backbone traffic does not enter a building's local traffic unless it is addressed to a node there. The use of bridges at this level also provides an effective way of expanding the capacity and physical reach of computer resources, while minimizing the performance costs associated with interconnection at higher levels.

Bridges come in handy when trying to interconnect multiple versions of the same LAN product. For example, AT&T's original version of StarLAN may be bridged to StarLAN 10, the company's more sophisticated product for high-bandwidth users. AT&T implements bridging through either a dedicated bridge unit, its Information Systems Network (ISN), or a router.

AT&T offers two versions of its dedicated bridge unit. The StarLAN 10 Network 1:10 Bridge links StarLAN 10 with its slower 1-Mb/s StarLAN system at a cost of $4,500. The StarLAN Network 10:10 Bridge connects StarLAN 10 networks to each other at a cost of about $7,000. ISN is AT&T's backbone network that supports synchronous and asynchronous data communication on both local and wide-area networks. It is optimized to support multiplexed host access, host-to-host, terminal-to-host, and terminal-to-terminal connections using what it calls a "perfect-scheduling" access method, which is a variation of the time-slot method of access. Software version 5.0 of ISN can connect up to nine StarLAN 10 networks together, while providing bridges to StarLAN. Bridging StarLANs may also be accomplished with AT&T's router, but you cannot really justify buying the more expensive router, if you just want to bridge similar networks.

Some vendors offer bridges to link their various LAN offerings. 3Com subsidiary Bridge Communications Inc. (Mountain View, Calif.), for example, provides a bridge to link its Ethernet and Token-Ring.

Bridges differ from gateways in that they do not do protocol conversion. Although gateways perform sophisticated routing functions, bridges perform a rudimentary form of routing that can only be described as "filtering." The bridge accomplishes this by monitoring the traffic on the networks connected to it and "learning" the addresses that are associated with each network. In this way, the bridge filters the traffic destined to remain on the local segment of the network, and broadcasts the rest to the other networks.

After determining that a bridge offers all of the interconnectivity you need, you have some more choices to make. You can either buy a stand-alone bridge, one that can be plugged into a multiplexer, or a hybrid bridge-router. The advantages of stand-alone products include portability; as complete, self-contained units, you can put them wherever you need them with minimal disruption to other network components. The trouble with stand-alone products of any kind is that they tend to be bulky, consume a lot of power, and take up the lion's share of available space in equipment rooms. Finally, stand-alone units are not usually expandable.

When interconnecting remote LANs, stand-alone bridges have a serious drawback: they require a dedicated connection to a remote bridge, regardless of how much the line is used. Aside from the obvious savings in cost, power consumption, and space, a bridge that is an integral part of a T1 multiplexer allows you to allocate bandwidth for either LAN-to-LAN communications, or to voice or data communications — whenever such needs arise. Micom, for example, offers card-mounted bridges for its T1 multiplexer. The multiplexer can allocate between 9.6 kb/s and 1.5 Mb/s for the point-to-point LAN connection. The remaining bandwidth may be allocated to voice and data. Even though stand-alone bridge units may cost less than plug-in card-mounted bridges, the monthly leased-line charges can quickly override the cost savings on the stand-alone hardware.

Another advantage of the integral bridge is that the multiplexer's existing management system can monitor and collect error and usage statistics from the bridge. Furthermore, the bridge's filtering capabilities allow the network manager to restrict the types of packet that go out over the bridge, thus alleviating traffic bottlenecks, or limiting access to certain types of network resources.

Depending on the complexity of the network and the protocols used, you may want to consider a hybrid bridge-router unit. The device houses two main boards; one makes intelligent routing decisions, while the other performs the filtering function characteristic of a bridge. If the user wants to select a pure bridging operation to achieve maximum packet throughput, for example, the routing function of the hybrid device can be supressed entirely by sending the appropriate command.

In large organizations with multiple networks, a bridge-router might be a better choice than separate bridges and routers. A bridge-router can be configured to allow bridging between VAX clusters running DECnet, for example.

5.3.3 More About Routers

In terms of complexity, a router falls between a gateway and a bridge. The function of a router is to join LANs at the network layer of the OSI reference model, which has two levels: Internet and Subnetwork. DECnet, on the other hand, does not have the Internet layer, so its routers work at the network level only, and DEC's terminal protocol, LAT, cannot be routed because it does not conform to the specifications of the network layer. In other words, to be routed an application must use a protocol that performs the functions associated with the network layer.

Each network protocol has a routing protocol built into it. Through the routing protocol, the router discovers the addressing information and shares it with other routers and hosts on the network. The information the router needs to route data is built into the packet itself. To send packets to their destinations, a router must perform several functions.

When a packet arrives at the router, it is held in queue until the router is finished handling the previous packet. Then, it scans the destination address, and looks it up on the routing table. The routing table lists the various nodes on the network, as well as the paths between these nodes and how much these paths cost. If there is more than one path to a particular node, the router will select the most economical path. If the packet is too large for the destination node to accept, the router breaks it down into a manageable frame size. This capability is especially important in wide-area networking, where telephone lines provide the link between LANs. With smaller packets, there is less chance that the data will be corrupted by a spike on the line, but if that should occur anyway, and a retransmission is necessary, having smaller packets results in less information to be retransmitted.

There are two types of routing: static and dynamic. In static routing, the network manager must configure the routing table himself. Once the routing table is set, the cost of paths on the network will never change. This might be all right for a LAN confined to a small geographical area, but not for wide-area networking. Although a static router will recognize that a link has gone down, it will not reconfigure the routing table to determine the next lowest cost, nor will it rebalance the traffic load.

A dynamic router reconfigures the routing table automatically and recalculates the lowest cost path. Some routers, like those offered by DEC, Cisco Systems (Menlo Park, Calif.), Ungermann-Bass, Proteon, and Wellfleet Communications (Bedford, Mass.), even rebalance the traffic load. Of these, Proteon's 4200 has been the most popular router.

Because routers are protocol-specific, you may need more than one router to support all of your organization's networking needs. There are some multiprotocol routers becoming available, however, which will let you route several protocols at the same time. Proteon's (Westboro, Mass.) routers, for example, can handle IP, DECnet, and XNS. Wellfleet routers can handle DECnet and IP, while Cisco Systems' routers can handle all of these plus DDN and ISO. All three vendors support speeds of up to 1.5 Mb/s for wide-area networking (WAN). If you require this much functionality, though, you might be better off with a gateway.

Routers are very good at bypassing link failures and congested nodes, which is critical for networks that cannot afford outages. This is because routers share information with each other through the OSI network layer. Bridges cannot do this, because they do not have access to the network level through a routing protocol. Thus, when one bridge gets overloaded, the others will never know.

Whether you finally select a gateway, bridge, or router — or put them together to link your networks in various ways — knowing what these devices are designed to do (Table 5.2) will minimize the risk of making a poor purchase decision. Asking vendors the right questions, talking with references, and obtaining written assurances from the vendor about the cost and availability of future product upgrades will reduce your level of risk even further.

5.4 CO-LAN

Among the latest offerings of telephone companies are central office–based local-area networks (CO-LANs), which are intended to stave off the continued onslaught of digital PBXs, while making local-loop bypass opportunities less attractive. For CENTREX users with fairly limited needs for data switching, central office–based local-area networking may be worth looking into. Instead of coughing up a lot of up-front money for on-premise hardware and cabling, users can tie into the CENTREX LAN *via* inexpensive data-over-voice (DOV) multiplexers at each station (Figure 5.2). This arrangement allows both data and voice to be carried over existing CENTREX wiring. Voice and data are then separated at the central office. Voice signals are routed to the voice switch, while data signals are sent to the adjunct digital data switch which routes them to another terminal device on the CENTREX LAN.

CENTREX LANs take the risk out of choosing products and vendors; this has already been done by the telephone company as extensively as possible. The CENTREX option also minimizes the cost and responsibility for day-to-day network management and hardware maintenance; the telephone company provides 24-hour service and support. Users do not have to worry about investments in LAN technology becoming obsolete; the telephone company will integrate new technology to protect its customer base from potential erosion. From a customer's point of view, there is no requirement for additional floorspace, heating and cooling plant, or extensive building rewiring.

These were pretty much the same reasons for turning to CENTREX in the first place, instead of opting for PBX ownership. Turning to CENTREX for data networking, then, would seem to be a logical decision, and, as an additional incentive to go with the CENTREX LAN, some telephone companies, like the subsidiaries of NYNEX, hold out the promise of ISDN connectivity. This is a clever way for some telephone companies to presell ISDN, but only if they use a digital switch for LAN services. Although LANs and ISDN are not the same, they are complementary. This notion is not so farfetched as it may seem. Simultaneous usage of the telephone and the terminal over the same pair of wires is an ISDN-like feature, which is already in widespread use through DOV systems.

As telephone companies continue to provide digital end-to-end connectivity, the infrastructure put into place by CENTREX LANs may be one way to migrate to basic rate ISDN (2B + D) once the appropriate interfaces are installed at both the central-office and customer premises. For customers, it might be well worth the investment of $800 to $1,000 for such interfaces, since ISDN offers much more

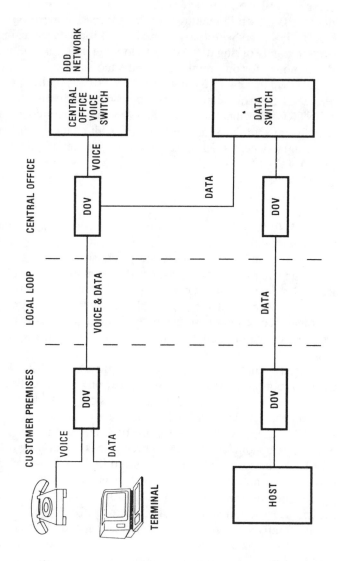

Figure 5.2 Typical CO-LAN configuration.

bandwidth and functionality over the same pair of wires than is currently available *via* the CENTREX LAN. If this price seems high, do not fret yet; count on an enterprising firm to come up with an economical "ISDN server" that will allow many terminals to share the same ISDN interface, once standards become more stable. The cost of such interfaces will plummet as ISDN becomes more widely available, and as more companies come to appreciate its value.

Until ISDN comes into widespread use, however, and becomes more affordable, DOV may be the best option for small-to-medium sized companies, at least for the foreseeable future. From a performance standpoint, there is little risk, because DOV performance on existing loop plant has been amply demonstrated. By controlling the frequencies used by the frequency-shift keying (FSK) circuitry in the DOV units, performance over existing nonloaded cable can be accurately predicted.

In some regions, such as the one served by NYNEX, regulators permit only CENTREX users to take advantage of the LAN service. In other regions, such as the one served by Ameritech subsidiary Ohio Bell Telephone Company, even non-CENTREX users are eligible for the LAN service, although they would have to assume the cost of additional trunks to take advantage of the data-switching capabilities of the central office.

Although not offering the same level of functionality as on-premise LANs, CO-LANs provide an array of features that appear to be most in demand.

- Password protection restricts access to the CENTREX LAN to authorized users only.
- Closed user groups limit a terminal's access to specific terminals and resources on the CENTREX LAN.
- A variety of applications are available, including word processing, electronic mail, remote job entry, remote file access, and database sharing. Users may select applications based on current needs and add others as requirements justify.
- Automatic data-speed matching (autobaud) adjusts the receiver's speed to equal that of the caller.
- Autoconnect allows specified terminals to connect to selected hosts automatically upon logging on to the LAN.
- On-line directory assistance provides screen displays of the names and numbers of other CENTREX LAN users to facilitate communication.
- Session hold allows communication to be suspended temporarily without having to reestablish the connection to continue.
- Synchronous as well as asynchronous data may be transmitted at speeds of up to 19.2 kb/s (64 kb/s is currently possible, but not tariffed at this writing).
- A variety of protocols is also supported, including 3270, 2780/3780, SDLC, and X.25 to facilitate communications between terminals, microcomputers, workstations, and hosts.

A number of options are available through the CO-LAN, including modem pooling, which allows users to make and receive data calls *via* the public network. Another option, internodal trunking, permits users to interconnect multiple CO-LAN nodes or data switches. Queuing lets the user wait for access to a selected destination until a previous data call to that destination is completed.

The cost of CENTREX LAN service varies from region-to-region and according to the applications the user chooses. For users in the Ohio Bell serving area, the average monthly charge is $20 per connection, while the cost of on-premise DOV multiplexers are $250 each.

5.4.1 Data-Over-Voice

DOV technology has been available since 1979. It was originally developed to solve organizational cabling problems. As organizations expanded and more terminals were added, it became impractical and prohibitively expensive to run separate cables from the host to each terminal. When DOV emerged and demonstrated its reliability, it was eagerly embraced by the user community.

DOV involves multiplexing data onto voice lines at frequencies above the voice band. This way, data do not interfere with voice. Although both are entirely independent, the two different signals may coexist on the same line. This arrangement offers several advantages, such as eliminating the need for separate cabling for voice and data, and since ordinary twisted-pair telephone wiring is involved, equipment moves and changes are greatly simplified. Instead of paying $1,000 or more to pull or reroute wiring to a new location, users just plug into the phone outlet at the new location.

The customer-premise unit is generally placed on the user's desk, under the telephone, since it is similar in size to a desktop modem. The telephone plugs into the DOV unit *via* an RJ-11 connector, while the terminal plugs into the DOV unit *via* an RS-232C connector. Componentry internal to the DOV unit converts the binary digits to one of two FSK frequencies. At the other end, the FSK frequencies are converted back to binary form. Data are carried over two carrier tones, one of which is applied at 36 to 40 kHz and the other at 72 to 80 kHz — well above the frequency range of 0 to 4 kHz used for voice. In this way, data are overlayed on single two-wire twisted pairs so that the cable carries both full-duplex data and undigitized voice traffic simultaneously.

Because FSK is potentially susceptible to crosstalk interference, other schemes are now in use to implement DOV. Applied Spectrum Technologies Inc. (Minneapolis, Minn.), for example, uses a "spread-spectrum" technique whereby data are carried over a broad band with low energy density. Not only does this technique minimize the effect of crosstalk interference, it also permits forward error correction for more reliable data transmission.

At one time Gandalf Technologies used FSK to achieve DOV, but the company now uses time-compression multiplexing, a technique that increases the data rate to 64 kb/s.

Over the years, DOV improved in terms of speed, range, and diagnostic features to the point where it is now appropriate for implementing central office–based LANs. As such, DOV has become a viable interim step toward ISDN. As with customer-premise DOV systems that take advantage of existing in-house wiring, central office–based DOV systems take advantage of the existing local loop. The vast majority of telephone subscribers are located within three miles of a central office, well within the 18,000-foot cable range of DOV equipment. (Some DOV equipment developed for the European market has a range of nine miles.) At the central office, another DOV device demultiplexes the two types of signals from the customer's unit before they reach the CENTREX switch.

Because DOV uses transmitting frequencies below 50 kHz and receiving frequencies below 100 kHz to achieve full-duplex operation, crosstalk is minimized. In not having to add power to overcome attenuation, the distance and the number of adjacent pairs carrying data within a cable may be increased. This means that properly designed units can perform at extremely low bit-error rates when used over standard nonloaded cables.

The telephone company implements diagnostics to insure the proper operation of the CO-LAN. Through an ASCII terminal, thousands of DOV channels may be monitored for test status and alarms. Oftentimes, problems are spotted and corrected before users even notice that a problem has occurred.

The DOV link makes an attractive CENTREX service enhancement, not only to provide LANs, but gateways to packet networks as well. In fact, some telephone companies are considering incorporating DOV into their packet networks to augment their antibypass strategies. The customer-premise DOV unit provides asynchronous to X.25 Level 3, single-channel protocol conversion, making it possible to connect to packet switched networks *via* the CO-LAN. Packet networks provide efficient, error-free data transmission between host computers and remote terminals. Properly equipped DOV units make using packet networks easy and affordable.

Most DOV systems are analog, although newly installed systems combine voice and data onto single 64-kb/s digital channels. Before plunking down several hundred dollars per DOV unit, you should find out what plans the telephone company has for upgrading its CO-LAN from 19.2 kb/s to 64 kb/s, and what it will cost you in terms of upgrading or changing out on-premise DOV units.

Another way that telephone companies implement CO-LANs is through AT&T's DataKit Virtual Circuit Switch (VCS), which performs as a central switch in local and wide-area network environments. It combines virtual-circuit capacity with an 8-Mb/s centralized short bus to provide high throughput and virtually instant network access by many users. The principal advantage of DataKit VCS is its compatibility with a wide range of communication-equipment types, as well as the throughput, productivity, and economy that are achievable with the system's packet-sized internal switching techniques. The Datakit VCS can also be accessed by dial-up modems. Through an X.25 gateway, the user can access remote computers *via* public packet networks like Telenet and Tymnet.

Companies that appreciate LANs, but do not want to be bothered with setting up and maintaining their own systems, may find CO-LANs the next best thing to owning one, but before committing to the LAN services offered by the telephone company, check into the telco's expansion plans. This includes finding out what application packages they plan to offer in the future, the time frames for their availability, and their cost. The time to bargain over the cost of future services and upgrades is before you sign a contract. After you become a customer and have paid for the on-premise DOV units, you have lost your leverage.

Even though CO-LANs sound like a great deal, never forget that the service is being operated by the telephone company. For many telcos, CO-LANs are a relatively new service. Although the technology that makes the service possible is reliable, the administrative procedures used by the telco to service customers may still have some bugs. You have to satisfy yourself that the telco will respond to your needs in a timely manner. Check with present CO-LAN users to find out what problems they have experienced, if any, and how fast the telco responded. Ask users what promises the telco made before the start of service, and if those promises were honored after they became customers. Find out if the users would opt for the CO-LAN, if they had it to do all over again.

Finally, check into the telephone company's disaster-recovery plan. Since you are using the same lines for voice and data, even to communicate internally, any outage at the central office will have a profound impact on your organization's ability to continue even routine operations. Make sure the contract specifies penalties for nonperformance in case the telco cannot provide service. The fire that swept through one of Illinois Bell Telephone Company's largest hub offices in May 1988 only points to the need to be concerned about such disaster-recovery plans. That fire destroyed or seriously damaged most of the 50,000 circuits tied into the Hinsdale facility, including a number of fiber-optic lines. The AT&T 1AESS switch was damaged beyond repair. Although traffic was diverted to other hubs in stages, corporate networks operated in a degraded mode for about a week — all because the telco had not backed up its circuits.

5.5 DATA SWITCHES

Despite all the vendor-generated hoopla about the imminence of fourth-generation PBXs and LANs capable of integrating voice and data to handle a multitude of diverse communications needs, comprehensive, economical products have failed to materialize. Perhaps integrated, cost-justifiable connectivity solutions will yet evolve from humbler beginnings and eventually prove their value in the user community, but the patience of communications managers is already wearing thin. Rather than wait any longer, many communications managers are rediscovering the basic data switch — and finding a few surprises in the process.

In addition to offering greatly improved port-contention and restriction facilities, as well as fail-safe security options, today's leading-edge data switches deliver true nonblocking data-transmission speeds of up to 19.2 kb/s for asynchronous data and 64 kb/s for synchronous data. Advances in technology allow total modular design, which permits economical growth from as few as two computer lines locally to thousands of computer ports worldwide and, just as importantly, are breaking the $100-per-port barrier at only 30 lines.

Data switches — also known as port-selection or port-contention devices — have been in mainframe computer environments since their invention by Gandalf Technologies (Wheaton, Ill.) in 1972. They permit a number of users to share a relatively limited number of computer ports from one or more hosts and associated peripherals. Prior to that time, terminals were permanently connected to computer ports. Users who required access to other computers or peripherals had to go through different terminals, or have the connection manually changed at a patch panel. Over the years, data switches have evolved from relatively simple port-selection and contention devices to sophisticated communications controllers that augment LANs and WANs.

Unlike matrix switches, which are designed for permanent connections under the control of the network manager, data switches are designed to establish connections dynamically under the control of each user as his needs require. In recent years, data-switch manufacturers have made giant strides in turning the basic data switch into a powerful multifunctional network-management tool. Such sophistication offers an economical migration path to LANs. For those with a LAN, the data switch makes an efficient LAN server or gateway to packet and T1 networks.

Terminals may be connected to the data switch in a variety of ways. To insure data integrity, direct connections may be accomplished with RS-232C cabling at distances of up to 50 feet. For longer distances of one to two miles, between buildings for example, terminals may be connected to the data switch *via* line drivers or local multiplexers. Some vendors have integrated these devices into their data switches, which are offered as optional plug-in cards. Terminal-to-switch connections may also be accomplished *via* the extra twisted pairs that are already in place in most office environments. Connections to remote data switches may also be accomplished through dial-up phone lines.

Some data switches can carry voice traffic with the addition of a special card that digitizes analog signals for transmission at 64 kb/s (PCM) or 32 kb/s (ADPCM). This capability makes the data switch a viable, low-cost alternative to integrated voice-data PBXs, which were designed for voice and later modified for data. Beyond that, this capability enhances the value of the data switch as a gateway to the T1 network.

Expanding the one-chassis configuration beyond the maximum line capacity usually requires a networking card, which plugs into each of several chassis to achieve a daisy-chained arrangement. Chassis may also be linked together with statistical-multiplexer cards to allow the chassis to be distributed throughout the network and connected together over high-speed digital facilities. This kind of arrangement can result in total port capacity of 3,000 or more lines. In fact, the maximum number of lines in a data-switch network is theoretically limitless.

Some data switches, like the #1-ISS from Netrix Corporation (Herndon, Vir.), have combined X.25 packet switching and transparent circuit switching into a single box for wide-area networking, even providing a graphics interface to facilitate management.

5.5.1 Redundancy

Sequel Data Communications, Equinox, Gandalf, and Micom are among the innovative manufacturers who have introduced products with enough redundancy virtually to eliminate concern that the data switch constitutes a single point of network failure.

Not only is control logic concentrated on a single processor card, but optional redundant logic allows some data switches to activate an optional secondary processor card automatically if the first one fails. Configuration instructions are automatically copied into the standby card upon cutover, eliminating the need for manual reentry. Some data-switch makers claim to offer redundant logic, but require that the processor card be pulled and replaced upon failure. Not only does this type of arrangement stretch the commonly accepted definition of "redundancy," it also is really not practical, especially at unattended installations.

Some manufacturers, like Sequel Data Communications (Raleigh, North Carolina), offer redundant power supplies, which plug into the switch and activate automatically upon failure of the primary power-supply unit. The faulty power-supply module may even be safely replaced with another plug-in unit with the data switch in operation to prevent unnecessary downtime.

The redundant backplane or "split backplane" protects the switch from damage that may occur from the failure of components connected to the bus. In the event of a failure on the bus, the switch automatically cuts over to the spare backplane to maintain uninterrupted operation.

These levels of redundancy are especially important for data switches, because they are usually configured in the center of star-type networks. Without such protection, the data switch is potentially the only single point of failure that could bring down the entire network. Together, redundant logic, redundant power supply, and redundant backplane offer communications managers the ultimate fail-safe system for uninterrupted service.

The emergence of universal backplanes (distinct from "redundant back-planes" mentioned earlier) allows vacant card slots to be filled with optional boards that greatly enhance the functionality of today's data switches. One such board provides a statistical-multiplexing capability which delivers up to 36 channels per line for long-distance networking between data switches *via* modems or digital data services. Other options, like security callback, may be added *via* separate boards that also insert into any vacant slot.

Gandalf's (Wheaton, Illinois) PACX 200 and Sequel's SDC series of data switches are among those which perform continuous background diagnostics to provide the capability of automatically disabling faulty channels. If a requested port is out of service, a message notifies the user that the port has been busied out. When the appropriate channel board is replaced, the data switch automatically reenables it. Of course, the communications manager still retains the option of using a terminal keyboard to disable any port from any terminal for any reason.

5.5.2 Contention

Port contention today may be determined by the communications manager in keeping with organizational needs. For example, the communications manager may decide to group ports with similar characteristics and capabilities under a "class" name to permit a more efficient utilization of shared resources. Even with today's smaller switches, at least 20 password-protected classes are possible to restrict access and enhance system security. The class of service may designate particular ports of a host computer. Other classes may designate a high-speed printer, a modem pool, or a gateway to a packet network.

When a user enters the connect command and personal password, the switch will attempt to complete the requested connection. If the requested port is busy, the user is put into queue and notified of changes in position with a screen message. Some users may be assigned a higher priority in the queue than others. When a high-priority user attempts to access a busy port, he bumps other waiting users by automatically assuming first place in the queue. All other contenders for that port are then notified of their new status.

The "single sign-on" feature available with some data switches allows users to reach anywhere on the multinode network without performing cumbersome manual reroutes if the primary route is busy. Also referred to as "dynamic re-routing," single sign-on works much like the direct-dial telephone network — when a call is dialed, it is automatically routed over the most direct route, but, if that route is blocked, the "call" is rerouted to reach its destination. Today's data switches work much the same way to transport data over the multinode network, a process that is entirely transparent to the user.

Not only do modular switches like Sequel's SDC series permit two different configurations per chassis, they can also implement one or the other automatically *via* a time-of-day clock. A late-night crew of software people can plod along with cumbersome design work, for example, and be restricted to accessing the main computer from 6:00 P.M. to 8:00 A.M. The class-of-service feature keeps these users from being able to access sensitive financial information that may be stored in the host. When the workday begins at 8:00 A.M., the time-of-day clock changes the data switch back to the primary configuration for general access — without interrupting existing connections. This enhances overall network performance in that available resources may be reallocated during the day for optimal usage based on the varying needs of different classes of users. Various alternative configurations may be stored on floppy disk and implemented by a microcomputer with only a few keystrokes.

The session "toggling" feature offered by Sequel enhances operator productivity by permitting two connections — one designated primary and the other designated secondary. The operator can toggle back and forth between them to perform multiple tasks simultaneously. For example, a batch file transfer can be in progress over the secondary link while a real-time data-base search is being performed over the primary link. When the batch file transfer is completed, the primary link can be put on hold while another file transfer is initiated over the secondary link.

If one of the destinations is busy during the attempt to connect with the toggling feature, the user is placed in queue and may return to the primary connection. An audible signal or screen message notifies the user that the connection has been made to the alternate port. When the second connection is made, the port number and name are displayed on the user's terminal. A message indicating that the other destination is on hold is also displayed.

Gandalf and Sequel are among the few data-switch makers who offer a "third-party connect" feature, which allows the communications manager to connect two called ports together from a terminal using still another port. The use of a terminal keyboard to establish a connection between two called channels, rather than manually jumping them together at a patch panel in the computer room, greatly simplifies the task of network management.

5.5.3 Flow Control

With a data-rate conversion capability, the communications manager does not have to match terminals with computer ports; each computer port may be set at its highest rate. The data switch uses a buffer to perform the rate conversion for any device that communicates with the computer at a faster or slower rate. This means that users do not have to be concerned about speed at all, and network managers do not have to waste time changing the transmission speeds of computer ports to accommodate lower speed devices. Thus, a computer port set at 19.2 kb/s may send data to a much slower printer, but for reliable data-rate conversion, the connecting devices must be capable of flow control; if not, there is a risk of losing data.

When XON/XOFF is used for flow control, the switch buffer will be prevented from overflowing during data-rate conversion. When the buffer is in danger of overflowing, an XOFF signal is sent to the computer, telling it to suspend transmission. When the buffer clears, an XON signal is sent to the computer, telling it to resume transmission. These settings are also used for reformatting character structures, enabling devices of different manufacturers to communicate with each other through the data switch.

5.5.4 Management Features

Many data switches allow the communications manager, instead of being confined to one terminal, to log on to the computer from any terminal operating at any speed to reconfigure ports conveniently while making rounds. Once connected, there are a variety of other functions that can be invoked to enhance operating efficiency.

A "broadcast" feature allows the communications manager to transmit messages to individual users, with delivery controlled by the time-of-day clock. The same message may be sent every hour, or a different message may be sent to different users all at once.

A special link feature lets the communications manager make permanent connections between a terminal and a port. Sometimes called "nail-up," this feature allows the continuous connection of devices like printers, which have no keyboards from which they can be logged on or off the system.

With a "force disconnect" feature, the communications manager can disconnect any port, any time, for any reason. Open files are even closed automatically with the proper disconnect sequences, including any required control characters. This handy capability is also found in a "time-out" feature, which enhances system efficiency by automatically disconnecting idle ports after a predefined period.

The communications manager can assist inexperienced users by watching what individual operators are doing, character-by-character, through the "monitor mode" feature, and, when there is a need to know who is connected where and who is waiting for a port, the communications manager can call up the network-status display for the answer.

With today's switches, configuration instructions need only be entered once; primary and secondary configurations are stored in battery-maintained RAM or nonvolatile EAROM (electronically alterable read only memory). As mentioned earlier, configuration data may also be developed and stored in a PC for later downloading to the data switch *via* an RS-232C interface.

When the data switch is equipped with a "logging port," the communications manager can obtain a complete record of port connects and disconnects to aid in maintaining network security. When used with the optional security call back, this feature provides a precise audit trail of connections, as well as the users who made them. In addition, all alarms may be logged for output to a microcomputer or designated port on the host to aid in network analysis and management.

5.5.5 Security Features

The security of computer data continues to be a legitimate concern among communications managers. Some data-switch manufacturers have made quantum strides in addressing security issues and now offer comprehensive security features that cannot be disabled by users for the sake of operational expedience.

Although data-switch makers have made security user-friendly, this does not mean penetration by intruders has been made easier. Aside from the security-related features already mentioned — passwords, class names, alternate configurations, and logging port — there are now security options that specifically address dial-in access. Communications managers may now choose from among several levels of protection to thwart unauthorized dial-in access effectively. The most basic security feature offered by some vendors entails the use of passwords. A user who calls into the switch from a remote location is required to enter a valid password. The data switch checks its directory for the password and, upon finding it, completes the connection. Thousands of passwords may be supported. Computer hackers, however, have been known to seek out valid passwords by combining number-generating algorithms with automatic dialing. With this technique, it is only a matter of time before any host system is breached. What is needed is a protective security system designed to guard against unauthorized dial-in access, while confounding the most tenacious hackers. Here are some of the security features currently available:

- A bare-bones security option permits a connection to the network when a user dials in and inputs a valid password. If the password is invalid, the switch will hang up and pause for 15 or more seconds before allowing the user to try again. A maximum of three tries is allowed before the switch hangs up.
- A more sophisticated security option allows a user to dial in over an autodial modem and input a valid password. The switch will hang up and call the user back over that modem. With an invalid password, the data switch simply hangs up and never calls back.
- Still another security option allows a user to dial in and input a valid password. The switch hangs up, then calls the user back on a different line or line group. If the password is invalid, the switch does not call back.

5.5.6 Local-Area Networks

When CXC and Ztel announced plans to introduce integrated voice-data PBXs a few years ago, communications industry pundits and media literati praised the coming of such products as the quintessential networking solution, but the promise of the so-called "integrated" approach to networking failed to live up to expectations. Few were prepared to pay the hefty price to find out for themselves if such products could pass data as reliably as voice. Even today, the performance of some products would suggest that the data capability was added to the voice PBX more as an afterthought than as a result of design. Enter the LAN.

As recently as 1986 industry gurus were touting the LAN as *the* connectivity solution. Data switches were slighted as "outmoded" technology that would have to be quickly replaced if companies hoped to increase white-collar productivity and compete effectively in the "Information Age." Many of the promises of LAN technology, however, failed to materialize, and for many companies the cost of implementation still outweighs the benefits LANs have to offer.

Because so much publicity had been bestowed upon LANs, many users have come to believe that such products are now the only viable solution to local-area networking problems, but today's data switches outperform LANs in many respects. Aside from offering better security features than LANs, data switches offer more consistent performance and connectivity potential — and at a much lower cost than most LANs.

Data switches generally use circuit switching, whereas LANs use packet switching. A LAN provides an instantaneous data rate of up to 10 Mb/s, and some, like Proteon's Pronet 80, operate routinely at 80 Mb/s. The connecting devices must contend for the available bandwidth. If several devices are executing file transfers at the same time, other users will be limited to significantly less bandwidth. In this case, CAD-CAM users might find response times stretched to intolerable levels. Beyond that, LANs can get bogged down when a number of users are on the facility simultaneously. With LANs based on token-passing schemes, throughput can be seriously degraded as each user waits for a turn to transmit or receive data. Although data switches are limited by the number of circuits they provide, they offer users a more consistent level of performance than LANs in that a constant 64 kb/s of bandwidth may be provided at all times. When more bandwidth is required for a particular application, it may be provided by using several channels arranged in parallel.

There is no denying that LANs have a distinct advantage over data switches in terms of speed. LANs, like Ethernet, transmit data at speeds of 10 Mb/s and greater, while the most innovative data switches like Gandalf's PACX 2000 are capable of delivering nonblocking speeds of only 64 kb/s for synchronous data. The data switch, however, will be able to switch both the primary-rate and the basic-rate access lines of ISDN, since those services are built around 64-kb/s DS0 channels. In choosing a data switch that supports 64-kb/s transmission, you can expect a smoother migration path to ISDN when such services become available in your area.

This brings up the connectivity issue. Data switches offer much more connectivity potential than LANs. In wide-area networking, for example, data switches may perform as well as LANs — but at much less cost. Gateway functions may be added to data switches economically through plug-in cards that provide the appropriate protocol converters. This permits asynchronous or synchronous terminals from one vendor to communicate with the mainframes and network nodes provided by other vendors. It also permits the data switch to connect to specific types of LANS, like IBM's Token-Ring or AT&T's StarLAN, and routing data between LANs is not transparent to the user, as is the case when routing data between data switches. Data switches may also be equipped with interfaces that allow communication over digital facilities like DDS or T1.

Even though LANs have demonstrated their capability of linking different types of computer equipment into a single, cohesive network, the proliferation of standards has made choosing a LAN quite complicated. Because data switches use the RS-232C standard for data interchange, compatibility with virtually all vendors' products is assured. With the protocol-conversion, flow-control, and data-rate conversion capabilities of today's leading-edge data switches, questions of compatibility may not arise, if equipment selection can be geared around the data switch.

Data switches offer many other advantages over LANs. On a homogeneous network, port selection among a small number of terminals or PCs can be more efficiently handled by a data switch than by a LAN. Even on a heterogeneous network of incompatible devices, the data switch may be the best choice. Data switches offer stable interfaces, time-tested maintenance and control procedures, and proven technology.

Although network-configuration considerations should be hammered out before the purchase of a LAN (number of users to be supported, types of applications, *et cetera*), configuring a local network for optimum performance still involves trial and error. It may take weeks or months to fine tune the LAN, whereas the same might not be true of some data switches.

Configuring the LAN oftentimes requires extensive vendor support or the services of outside consultants, because it entails specific knowledge of operating systems to determine such things as cache-buffer sizes, time-slicing parameters, and the amount of memory to allocate for caching files. With data switches, the task of configuration is relatively simple, because the communications manager merely steps through a series of menus and supplies the required information to configure the network properly. Some data switches also allow the use of abbreviated commands for configuring the network so that experienced managers can avoid waiting for several levels of prompting.

Although the low cost, high performance, and functionality of data switches are certainly attractive, LAN users might well question the need for even a leading-edge data switch. Likewise, those considering the purchase of a leading-edge data switch might wonder about the wisdom of eventually investing in a LAN.

Actually, neither type of user has reason to worry; local nets and data switches are not mutually exclusive, but complementary. For example, recent improvements in packet assembler-disassembler (PAD) products make leading-edge data switches especially useful and cost-effective as servers to bridges or gateways between local nets, or to broadband, proprietary, and public data networks.

PADs format data into packets for transmission over high-speed links. Each packet contains addressing, error-detection, and control information which determines its route and assures the integrity of the data. The X.25 protocol and other industry-accepted protocols specify how this data is packaged for acceptance by the packet data network. Terminals linked to the data switch may be multiplexed to the PAD to send or receive data from the packet data network. Of course, data switches that are optionally equipped with the X.25 software do not require PADs to communicate with packet data networks. In this case, the PAD function is carried out internally by the switch and made accessible to users through designated ports.

As a company's LAN expands to meet the increasing need for data communications, modular data switches can be used to add management precision and control to operations, while saving time and money.

At the same time, the protocol-conversion, flow-control, nonblocking-throughput, and T1-networking capabilities of today's data switches make them a viable, low-cost alternative for those who have not yet purchased a LAN. When the need for a LAN is fully justifiable, these data switches can play a subordinate, but important, role as servers. Thus, the capital expenditure for a LAN is postponed as prices continue to drop, and, once the expenditure is made, the original investment in the data switch or switches is totally protected. Also, because of their port-contention capabilities, data switches may be used to allow microcomputer users to share peripherals, thus offloading the LAN. Although LANs perform this function well, data switches do it for about $100 per port *versus* about $500 per LAN connection.

Even the purchase of a small data switch will not be wasted if the organization quickly outgrows it. If the organization cannot be adequately served by adding incrementally to modular switches, but must graduate immediately to a high-end data switch of much greater capacity, the smaller unit can be put to use as a terminal server to the larger switch.

Equinox Systems offers further evidence that data switches are not an outmoded technology. In May 1988, it started shipping a "zero-slot" microcomputer network controlled by communications software and a data switch. The product, SwitchLAN, allows users to transfer files, share printers, and access host systems

— all without network cards. The software runs on the microcomputers, which are linked *via* serial port and twisted-pair wiring to the Equinox data switch. The switch is capable of networking 1,000 microcomputers, and the software is capable of running many tasks in the background mode, including file transfers, electronic mail, terminal emulation, and printer spooling and sharing. As innovation continues in data-switch technology, the distinction between these types of product and LANs may become increasingly fuzzy.

5.5.7 Adding It All Up

LAN vendors are quick to point out that the value of their products must be seen in terms of useful life — specifically, their ability to absorb more terminals of ever greater complexity, while continuing to offer consistent response times, high reliability, and rapid recovery from failure. Today's feature-rich and affordable data switches offer acceptable nonblocking throughput for nonexotic applications, as well as modularity and a degree of configuration flexibility to meet virtually any data-networking need effectively.

Finally, leading-edge data switches offer "self-healing" circuitry that comes from numerous levels of redundancy. These data switches also use existing copper wire and, with completely connectorized cabling, may be unpacked, tested, and configured for operation in a single afternoon. Although many hours may be consumed in naming the nodes, assigning passwords, determining classes of service, and devising alternate configurations, keep in mind that configuring a LAN is much more involved.

Advances in technology have surely made the leading-edge data switch a viable alternative to LANs and integrated voice-data PBXs. Moreover, these contemporary data switches are flexible enough to manage differing technologies, fit into various network topologies, and handle the increasing levels of complexity that are so much a part of today's data networks.

Communications managers contemplating connectivity solutions would find that today's innovative data switches may be a simple and economical alternative to LANs, when networking needs are relatively simple, and would make a suitable LAN server when networking needs grew more complex.

Chapter 6
Computers

6.1 WORKSTATIONS

Workstations were developed in the early 1980s to fill the void between low-powered microcomputers and high-cost minicomputers. Technical professionals with heavy research and design needs, such as product developers, architects, engineers, and scientists, needed more processing power, memory, and display resolution than microcomputers were then capable of delivering. Although minicomputer systems delivered such performance, they were costly and inefficient when it came to interactive applications and, consequently, did not have the "feel" of personal systems.

Sun Microsystems and Apollo Computer were among the first to recognize the need for general-purpose workstations. Not only did both companies introduce products with high-speed processors, large amounts of on-board memory, disk storage, and high-resolution displays, but networking capabilities as well. This meant that the general-purpose workstation could share expensive resources like file servers, printers, and plotters.

Workstations are usually sold to original-equipment manufacturers (OEMs), who bundle them with specialized hardware and software and then sell the entire system as a stand-alone product. A manufacturer of data-communication products, for example, might order a customized version of the workstation to interface with its network-controller board. The network controller, in turn, would interface with such devices as modems, multiplexers, and data sets to collect performance information from various points along the network. Special software translates the various inputs and arranges the data in a meaningful format, which is sent to the workstation for graphic display or to a printer for hardcopy reports. The most sophisticated network-management systems also take advantage of the processing power, memory capacity, and high-resolution display capabilities of the workstation to depict an entire communications network on the screen. Through a system of icons, windows, and pull-down menus, specific information about any aspect of the network may be called to the screen in any level of detail required. For example, the operator can use the "mouse" to point to and zero in on a specific segment of the network. When the screen displays the configuration diagram associated with that node, he can again use the mouse, this time to zero in on a multiplexer, for example, whereupon the entire cabinet configuration is displayed, showing all shelves and boards. Using windows, specific-performance data may be displayed about any of the boards in the shelf. From a single high-performance workstation, an operator can receive alarms, identify faulty components, and initiate restoral actions. Among vendors of network-management systems, Sun, Apollo, and Hewlett-Packard workstations are most used at the high end, while IBM PC-AT or compatible microcomputers are most used for the low-end network-management systems.

6.1.1 Workstation Characteristics

Workstations are typically evaluated according to the following features:

- Processing power, as measured in "millions of instructions per second" (MIPS). The MIPS rating of today's workstations varies from just under 1 MIPS to 60 MIPS for superworkstations. Industry observers are predicting 100 MIPS machines by 1990.
- Main memory (RAM) of from 4 to 8 Mbytes for low-end workstations; up to 32 Mbytes for high-end workstations, which may be expanded to as much as 128 Mbytes.
- Cache memory for low-end workstations may run as high as 32 Mbytes; for high-end workstations, up to 64 Mbytes.
- Hard-disk storage for low-end workstations is typically from 40 to 80 Mbytes; for high-end workstations, from 140 to 560 Mbytes, and climbing. Both low-end and high-end workstations may be diskless, although this is a feature mostly associated with low-end workstations.
- The system bus for low-end workstations is 16-bit; for high-end workstations it is almost always 32-bit.
- Operating systems for low-end workstations may be DOS, UNIX, or related Xenix; high-end workstations may run DEC's VMS, or DOS and UNIX concurrently.
- In terms of networking, most workstations were designed for resource sharing in the multivendor environment, and do it quite well *via* Ethernet. IBM uses the Token-Ring, of course, while Apple uses its AppleTalk.
- Screen size varies from 12 to 16 inches at the low end and from 17 to 20 inches at the high end. The larger the screen, the more practical it becomes

to use several windows at once. Both low-end and high-end workstations come in a choice of monochrome, grayscale, and color displays, which differ in pixel density.

- The pixel density (resolution) of low-end workstation displays may be as low as 640 × 480, whereas high-end workstations offer megapixel displays as high as 1600 × 1280 for monochrome and 1280 × 1024 for color.
- Color displays differ in terms of color planes, with 4-, 6-, 8-, and 16-bit color planes being the most common. Third-party vendors are providing 24-bit displays that allow 16.8 million possible colors, of which 256 may be displayed on the screen at any given time.
- Workstations support many of the ease-of-use features that originated with microcomputers, such as windows, icon-based desktop-file organizers, HELP screens, pop-up menus, and mouse-driven interfaces for point-and-click operation. On high-performance workstations, however, the power of these features is greatly amplified.

There are two window environments that are in contention as the standard for workstations. Sun Microsystems is promoting its NeWS (network-extensible windowing system), while Apollo, DEC, Hewlett-Packard, and others are supporting the X Window System developed at MIT. Both windowing systems are network-based, client-server systems that allow tasks to share displays in a network of like machines.

Some low-end and many high-end workstations come with an Ethernet interface, which enables them to tie into a local-area network and access a file server. This is important if you want to exchange complex graphics and files between workstations; share expensive plotters, printers, optical storage media, and high-capacity tape drives; send electronic mail from one workstation to another; and upload CPU-intensive tasks to a minicomputer or mainframe. Ethernet is the most widely used local-area network for workstations. The IBM RT PC, does not support the Ethernet interface, but does provide one for token-ring. The Macintosh II, which Apple Computer has positioned as a low-end workstation, may be networked *via* AppleTalk, which is much slower than Ethernet and token-ring.

With regard to networking software, Sun Microsystems has positioned its network file system (NFS) protocol into an industry standard. NFS lets the workstations and servers of different manufacturers on the same network share files and swap messages. The UNIX that most workstations use offers remote file sharing (RFS), an AT&T network protocol. Apollo offers its network computing system (NCS), which is a set of tools that helps support data and program distribution across the network. DEC, of course, offers its proprietary DECnet. TCP-IP is the most commonly used nonproprietary network protocol.

Until recently, many workstations were built around the Motorola 68000 family of microprocessors, including those of Apollo, Sun, and Hewlett-Packard.

Others, like Intergraph (Huntsville, Ala.), use the 32-bit Clipper chip set developed by Fairchild Semiconductor Corporation's Advanced Processor Division, now a wholly owned subsidiary of Intergraph. Some low-end workstations like Compaq's Deskpro 386, IBM's PS/2 Model 80, and TeleVideo Systems' TeleSTAR/386 are really high-end microcomputers that use Intel's 80386 microprocessor. Overall, though, they do not yet have enough power really to qualify as workstations, although the IBM PS/2 Model 80 will probably evolve to that point in the near future. Of the three, the TeleVideo's TeleSTAR comes closest to performing like a low-end workstation. It comes with 4 Mbytes of RAM, expandable to 16 Mbytes. Both its monochrome and color displays offer a pixel density of 1280 × 1024. It uses Microport's DOS Merge-386 and runs UNIX and DOS concurrently. It may be networked over Ethernet and it supports TCP-IP, RFS, and NCS network protocols.

Recognizing a large market for workstations among high-end microcomputer users, Sun Microsystems introduced the Sun386i family of workstations in early 1988. The 386i/250 boasts a 25-MHz 80386 CPU with an 80387 coprocessor that delivers 5-MIPS performance. Like other 80386-based microcomputers, Sun's 386i may be networked *via* Ethernet across a multivendor environment, permitting all types of computing resources from microcomputers to mainframes to be networked together. The 386i runs both UNIX and DOS applications. In addition to running multiple DOS applications simultaneously, users may even cut and paste text between UNIX and DOS windows, and share files and data between UNIX and DOS applications. The Sun386i comes with 4 Mbytes of RAM, expandable to 16 Mbytes, and a wide choice of monochrome and color displays. The 386i is not just another beefed-up micro trying to pose as a workstation; its functionality and performance put it firmly into the category of "workstation."

DEC's VAXstation and IBM's RT PC are built around proprietary chips. In recent years, workstations have begun to move away from standardized chips in favor of proprietary designs that surmount the performance limitations of standardized chips.

In the race to squeeze higher levels of performance out of the CPU, workstation developers at one time simply replaced one microprocessor with another in the same family which offered a faster clock. Thus, a 14-MHz 68020 system gave way to a 16-MHz 68020 system, which eventually gave way to a 20-MHz 68020 system. While moving to the next more powerful member of a chip family increased the power of the CPU in the short term, this method of boosting performance had its limits over the long term. Technical professionals continually demand more cycles per second, so they can perform increasingly complex tasks and do them much faster. Since there are performance limits to standardized microprocessor architectures, workstation developers had to look to proprietary chips designed around higher-level architectures.

One of those architectures is parallel processing, which appears to be quite promising over the long term, but would require massive revisions to software, negating the huge investments of users. Another architecture, which a number of companies appear to have adapted in one form or another, is based on reduced instruction set computer (RISC) principles.

RISC processors perform faster, because their simplicity allows much easier system implementation. The vast majority of all computation requires a small percentage of any CPU's instruction set. Yet, conventional processors include unnecessary instructions that consume clock cycles during program execution. This translates into the use of multiple cycles per instruction, which holds back performance. RISC processors, however, execute most instructions in a single cycle of the CPU clock, raising overall performance exponentially. RISC processors are from three to seven times faster than CPUs based on conventional architectures.

The performance of RISC processors is enhanced through the use of "pipelining," a process that optimizes the execution of instructions. Sun Microsystems, for example, uses a four-stage pipeline operation in its 32-bit RISC microprocessor architecture, which it has labeled SPARC, short for Scalable Processor ARChitecture. The four operations that must be performed to process an instruction are: fetch, decode, execute, and write. When execution control transfers from one stage to the next, a new instruction is delivered. The rapid movement of instructions through the various stages of the pipeline results in higher workstation performance.

RISC-based processors also use fewer transistors. Not only does this result in faster speed, it also means that newer technologies can be integrated into RISC-based systems more easily, because there are fewer elements with which they must interact. Some companies, like Sun Microsystems, have fully capitalized on this new architecture by developing compiler-technology and sophisticated optimization algorithms that make application development easier and program execution much faster. Moreover, programs that currently run on DEC's proprietary MicroVAX processor or Motorola's 68000-based systems may be recompiled to run much faster on machines based on the RISC architecture. IBM's RT PC and Hewlett-Packard's HP 9000 Series 800 are among other workstations that use RISC-based processors.

Aside from developing new processors to boost performance, the move to proprietary systems helps workstation makers differentiate themselves from microcomputers which, it seems, are still on a collision course with workstations. In moving away from the chips that drive the mass-production engines of IBM, Compaq, and Apple, workstation makers hope to protect their competitive edge. While the performance of microcomputers is usually measured by the speed of the system clock, the performance of workstations is measured in millions of instructions per second (MIPS). The first workstations offered only 1-MIPS performance. Advances in chip technology led to the introduction of workstations with 4-MIPS performance in 1987 and 10-MIPS performance by 1988, as represented by Sun's 4/200 Series. In mid-1988, however, MIPS Computer Systems of Sunnyvale, California, introduced a 12-MIPS workstation family, the M/120, but the M/120 was intended as a multiuser network server, rather than a powerful single-user workstation.

One measure that further distinguishes workstations from microcomputers is floating-point processing performance. Most workstations come with special hardware for increasing the speed of floating-point mathematics, which are calculations that are used mostly in scientific problem-solving, as well as in complex graphics manipulations. The less expensive workstations offer a floating-point processor, which is essentially a coprocessor that works in conjunction with the main CPU to boost processing speeds by as much as 300%. More expensive workstations even offer floating-point accelerators. The Sun-4/200 Series CPU is among the most powerful for calculation-intensive applications, featuring an integer-unit chip, a floating-point controller chip, two floating-point arithmetic processors, and high-speed cache memory, which are all connected *via* a 32-bit bus. Floating-point performance is measured in MFLOPS — millions of floating-point operations per second. The Sun-4/200 offers 1.6 MFLOPS, compared to the 50 KFLOPS of most microcomputer coprocessors. Interestingly, MegaMath Corporation (Sherman Oaks, Calif.) introduced a PC-AT add-in board in mid-1988 that endows 80286 and 80386 microcomputers with MFLOP performance. The company's $1,995 Floating-Point Engine boosts performance to 12 MFLOPS, or 240 times more than was previously available, thus bridging another gap between micros and workstations.

Other distinctions between microcomputers and workstations include multitasking and configuration flexibility. While microcomputers perform many tasks in background mode, workstations deliver true multitasking performance. Not only can multiple windows be displayed on the screen, but each window can also run a different application.

Although microcomputers are quickly closing the gap, workstations have more configuration flexibility than microcomputers. A workstation may be configured as a stand-alone system, diskless node, terminal server, file server, or communications gateway. Some workstations may be configured as terminal servers to which as many as 50 terminals may be connected. High-speed mass storage up to 3.6 Gbytes is available for file sharing and access by diskless nodes when a high-end workstation is configured as a file server. File servers are customized workstations, dedicated to serving clusters of terminals.

Such workstations are configured with larger disks and larger memory, as well as with larger cache memories, than are standard workstations. Apple offers software that enables one of its systems to act as a server, but it does not perform up to the level of a server configured in hardware. Apollo, Sun, and Hewlett-Packard offer servers that are part of their workstation families. IBM does not yet have a server. Also, when configured as a communications gateway, the workstation can offload communications-related tasks from other systems.

As far as cost is concerned, there is a workstation to suit any budget. At the low end are TeleVideo Systems' TeleSTAR/386 Model 15DL and Apple Computer's Macintosh II, priced at $3,995 and $4,796 respectively. The diskless TeleSTAR/386 comes with a monochrome screen, as well as a 16-MHz 80386 CPU and a 10-MHz coprocessor which delivers 2.2-MIPS performance. The Macintosh II uses an 800-Kbyte floppy disk drive and a monochrome screen, as well as a 16-MHz 68020 CPU and a 16-MHz coprocessor which delivers 2-MIPS performance.

The big names in the workstation market — Apollo, Sun, and DEC — also offer low-end workstations. Sun offers the diskless 3/50M with a monochrome screen for $4,995. This model comes with a 15-MHz 68020 CPU, but a 16.67-MHz 68881 floating-point unit is available as an option. The 3/50M claims an MIPS rating of 1.5. Apollo offers the diskless DN 3000 with a monochrome screen for $4,490. This unit comes with a 12-MHz 68020 CPU and a 12-MHz 68881 FPU to deliver 1.3-MIPS performance. DEC's entry into the low-end workstation market is the diskless VAXstation 2000 with a monochrome screen, which sells for $4,600. It comes with a 20-MHz VAXstation II processor and a proprietary FPU to deliver 0.9 MIPS.

The mid-range is the largest category of workstations with prices spanning $7,000 to $30,000. These machines operate at from 3 to 4 MIPS and typically offer a choice between color and monochrome displays. Minimum hard-disk capacity varies from 70 Mbytes to about 160 Mbytes, but may be substantially higher when ordered as an option. The Sun 386i/250 falls into the mid-range category with a price tag of about $14,000.

At the high end of workstations are the RISC-based machines; of those, Sun Microsystems' 4/260 stands virtually alone. The 4/260 comes with a 16.67-MHz proprietary CPU and a Weitek 1164/1165 FPU to deliver 10-MIPS performance. Disk-drive capacity is 560 Mbytes, and the display is 19-inch color with 1152 × 900 pixel density. Total price: $85,500.

Arranging workstations into price categories does not always lend itself to a neat separation of products by performance and functionality. In fact, comparing products on the basis of price alone may not give you the slightest clue as to what to expect in the way of a workstation's performance and functionality. Selecting a workstation is a matter of matching applications to a specific product and then comparing the performance of different products when running those applications.

6.1.2 Buying Considerations

Assuming that your applications are not very complex, one of the first decisions you might have to make is whether to go for a high-end microcomputer or a low-end workstation. After all, both cost about the same and are very close in terms of performance. A quick comparison of microcomputers and workstations, however, reveals some fundamental differences between the two that should go into your decision-making:

- The performance levels of microcomputers and workstations may start out on an even footing, but the proprietary chip architectures of workstations overcome the inherent performance limitations of microcomputers. While the clock speed of Intel's 80XXX chip family may have already peaked, the proprietary chip architectures of workstations are boosting performance in quantum leaps, with no signs of slowing down.
- Expansion is limited with microcomputers, whereas it is basically unlimited with workstations. A microcomputer, for example, will quickly reach its expansion limit when the vacant card slots are filled up, and, as more cards are added, overall performance may diminish. Adding more functionality requires that you abandon that investment and move into the realm of workstations. With workstations, expansion requires the addition of hardware, but the operating system and application programs can migrate across the entire product line.
- Memory capacity is usually limited on microcomputers, which may be configured with as much as 1 Mbyte of RAM, but with heavy graphics applications this amount of memory can be consumed quickly. You can expand RAM to as much as 4 Mbytes, but even that pales in comparison with low-end workstations, which may be configured with as much as 4 Mbytes, and expanded to 16 Mbytes.
- In terms of connectivity, microcomputers were designed for stand-alone use; networking capabilities were added later as an afterthought. A workstation is designed for resource sharing and networking across the multivendor environment. Therefore, it can more seamlessly support work-group networking objectives through easy data-sharing and distributed-computing applications. Although microcomputers are adept at networking text *via* token-ring and Ethernet LANs, workstations more readily bridge graphics and connectivity.
- The distributed-computing orientation of workstations further sets them apart from microcomputers, along with the applications-development tools and common-file systems that these integrated environments usually possess.
- Operating systems of microcomputers are still oriented around DOS, which is hampered by the 640-Kbyte memory limit. Most workstations are oriented around the more powerful UNIX, which is a virtual requirement for technical applications. While some microcomputers can run UNIX and DOS, they usually do not allow the user to run both together, let alone with the interactive ease that some workstations allow.
- Graphics, especially those requiring subtle color shading and three-dimensional rendering, are better handled by workstations than by micros, which do not yet have the processing power to keep megapixel displays adequately refreshed. Micros also do not manipulate a lot of pixels very well. While microcomputers are adept at text handling, graphics handling is still in its formative stages by workstation standards. While single-user microcomputers

are just beginning to deploy windowing systems, for example, workstations already offer sophisticated networked windowing. With some workstations, each DOS window may access up to 640 Kbytes of RAM, plus up to 2 Mbytes of Lotus/Intel compatible EMS RAM. As if that were not enough, such systems even allow the user to run the popular microcomputer windowing systems, such as Microsoft Windows and DesqView, within a workstation window!

Another factor that may influence your choice between a workstation and a microcomputer is the environment in which the machines will ultimately work. If the users require close integration with the installed base of microcomputers, then the choice of a workstation, with its proprietary, non-DOS networking software may pose a problem, and if your applications for the new machine consist mainly of text and spreadsheets, choosing a workstation over a microcomputer may be construed as overkill. If the machine is intended for technical users, however, you are going to want something that can do all that plus detailed graphics, in which case a workstation is more appropriate.

This brings up another point on which to evaluate workstations: many workstation makers have been so preoccupied with developing high-performance design tools that they are just getting around to embedding data-base management, report writer, forms generator, spreadsheet, and business graphics into their workstation software products. Evidently, they forgot that technical people have the need for these capabilities as well.

In comparing workstations, you may want to check for expansion slots that allow the use of inexpensive IBM PC-AT or compatible option cards. Some workstations, particularly those at the low-end, accommodate several expansion slots, so that the workstation can take advantage of such things as communications adapters, internal modems, and graphics adapters. These slots can also be used for data-image acquisition cards such as those used by optical scanning devices for text or image input to disk.

You should also check that the workstation comes equipped with Ethernet or Token-Ring ports, as well as serial and parallel ports for connections to local printers, modems, and other peripherals.

Because workstations are used extensively for networking, check into the connectivity potential of specific workstations. Some vendors like IBM and DEC are quite adept at networking among their proprietary products and systems, but fall far short in the multivendor environment. In this regard, DEC has only recently offered emulation packages that provide a direct link with devices operating under IBM's SNA, and only after bowing to customer pressure. In contrast, Sun Microsystems prides itself on being able to smooth the integration of its workstations with other vendors' systems. The company's SunLink wide-area networking product offers workstation connectivity in either IBM or DEC environments. Connecting to IBM equipment may be accomplished locally or remotely using SNA or BSC protocols, which may be used to support batch or interactive processing. Through SunLink, the workstation can even emulate an IBM cluster controller under SNA, or an IBM-3179G graphics display terminal when connected to an IBM mainframe. In DEC environments, Sun workstations use DECnet protocols for integration.

Even if your company is solidly committed to using IBM or DEC systems, you should consider a workstation that can at least be inexpensively upgraded to connect with other systems, just in case new requirements emerge. This also applies to X.25 public data networks like Telenet or Tymnet, or defense networks like Milnet and DRINET (Defense Research Internet), which replaces Arpanet. Also determine the vendor's commitment to emerging international networking-protocol standards like OSI, MAP, and TOP. After all, positioning yourself to take advantage of future opportunities at minimal cost constitutes one of your principal objectives under the strategy for risk avoidance.

If your workstations will be used for software engineering, you should determine their suitability as platforms for computer-aided software engineering (CASE) for both native software development and for cross-development to other computers.

Look for features that automate system administration. For example, there are workstations with the capability of scheduling and automating file-backup operations. There are tools that enable you to add, remove, or reconfigure network resources, control access to application programs, and restructure user accounts with only a few clicks of the mouse. When evaluating high-end workstations, however, make sure that the vendor has not sacrificed ease of use just to achieve performance gains over the products of its rivals.

A good indicator of a workstation's long-term viability in the marketplace is the number of third-party vendors that support it with such things as application programs, add-on equipment, interfaces, and display enhancements. Since workstations are OEM-ed quite often, you can check the list of big-name companies that resell the workstation as part of their products. After all, they must be convinced of the company's future position in the marketplace, or they would not have put the success of their own products on the line. Look at the number of units sold *versus* the competition. Solid market share indicates organizational stability and provides some assurance that future growth will be adequately financed. Another way to determine the credibility of the workstation manufacturer is to find out what joint-development agreements it has with other reputable firms. Also look at how many companies have licensed the firm's proprietary technology. This may affect the chances for its future standardization throughout the industry.

Although vendors have typically assisted users struggling with networking concerns, only recently have they started providing such services under a formal program, sometimes in the form of joint ventures with other companies that provide complementary products. For example, Apollo, Hewlett-Packard, and Northern Telecom launched a joint venture in 1988 called Corporate Networks Operation, which provides network-consulting, design, and custom-integration services. The only problem with such arrangements is that the solution the company proposes invariably involves the products of the joint-venture partners. If you are looking for objective opinions, you are better off seeking the services of a consultant, or a *bona fide* integration-services firm.

Another thing to consider about workstations — or any other fast moving product — is not to buy too early. Newly introduced workstations that are fundamentally different from the vendor's other offerings may not have enough application programs available to make an early purchase worthwhile. By the time the packages you want are ready, the price of the workstation will have gone down. If you purchased early, you did not get the most out of your initial investment. This advice especially applies to the new breed of superworkstations that are coming down the pike. If the $100,000 price tag for a single 60-MIPS unit does not dissuade you, perhaps the lack of application programs will, unless, of course, you want to use such machines as a file server to other workstations.

As workstations and microcomputers increasingly look alike and act alike, the ensuing competition will exert downward price pressures on both. Any way you look at it, the customer will come out the winner.

6.2 SUPERMICROS

Scarcely a week goes by without some hot-shot vendor announcing the performance rating of a new computer that soars past those of existing products, whether they be supermicrocomputers, minis, mainframes, minisupercomputers, or workstations. In fact, the number-crunching power of some supermicrocomputers, minis, and workstations surpasses that of mainframes and approaches that of minisupercomputers, blurring the distinction between each. As a result, it is getting increasingly difficult to categorize these products by any widely accepted criteria such as cost, processing power, architecture, or application. Even the technologies used to achieve new heights of performance are vastly different, and usually involve the use of proprietary architectures.

More often than not, the industry abides by the designation placed on a product by its manufacturer. Sometimes the manufacturer decides how to classify its product according to what competitor it is trying to position itself against. Some manufacturers try to set themselves apart from the rest of the industry by inventing new classifications for their products, such as "Personal Supercomputer," trademarked by Apollo Computer Inc. That term, strictly a marketing device, was deliberately selected to evoke images of a multimillion-dollar mainframe scaled down to a relatively inexpensive, single-user, desktop machine.

Typically, microcomputers are brought into corporate offices to fill the information-processing void left open by an unresponsive or overloaded centralized data-processing organization. While a few microcomputers may prove quite useful, as their numbers increase they tend to create new problems. Originally conceived as stand-alone machines, microcomputers do not lend themselves to information sharing. A few microcomputers hoarding some information for personal use are not really a problem, but many microcomputers hoarding vast amounts of corporate data may indeed present a serious problem. Moreover, expensive peripherals like hard disks, tape drives, laser printers, and networking interfaces are usually needed for each user, which increases the costs associated with decentralized information handling.

The new generation of high-powered microcomputers alleviates many of these problems. The real value of such machines is not in their use as single-user, or even multitasking, machines, but in their multiuser information-sharing capabilities that provide the functionality microcomputer users want, but at a cost organizations can afford. Such machines are generally referred to as "supermicrocomputers."

The supermicrocomputer is distinguished from the microcomputer by its ability to support multiple users for sharing information and resources. This makes supermicrocomputers ideally suited for small businesses and the work-groups of larger companies. To determine the need for a supermicrocomputer, you must first find out what users are doing with microcomputers. You will typically find that word processing, spreadsheets, business graphics, and communications make up the majority of office applications. The next thing you have to determine is whether these common applications are more cost-efficient when done on a dedicated basis or on a multiuser basis. More often than not, the multiuser system will win out as the most cost-justifiable solution.

A multiuser system allows information to be shared among many users, rather than isolated on the hard disk of someone's microcomputer. Data bases, word processing, and spreadsheets are accessible by many users, providing substantial productivity and cost benefits. Trying to share information using stand-alone microcomputers involves time-consuming procedures and, consequently, becomes very costly.

The availability of micro-to-host communication packages has resulted in a less efficient utilization of host resources, because many times users request the same data, and, with on-line data-base updates, users risk downloading different versions of the same files to their microcomputers. In the multiuser environment, a request is made to the host only once, so everyone can access it through the supermicro, which performs the function of a file server.

When it comes to software, rather than equipping each user with a copy of the application programs, only one copy is needed for the multiuser system. Even under a software licensing agreement, this is much more economical than buying multiple software products. The multiuser system also makes it more efficient to maintain the software; since it resides at only one location, installing the latest software version is a lot easier and less expensive than if it has to be done for multiple microcomputers.

The multiuser approach also cuts down on the cost of hardware. Instead of equipping each microcomputer with its own hard disk, for example, only one high-capacity hard disk is required for the supermicrocomputer, which may be shared by all users. Also, through the multiuser system only one high-performance laser printer is required for shared use, instead of having a printer for each microcomputer. The same economies apply to other high-cost peripherals like tape-backup systems, optical-disk storage, and plotters. The cost savings inherent in the multiuser approach mean that you can afford to purchase higher quality peripherals, since you do not have to squeeze the most hardware out of a limited budget. Higher quality peripherals are easier to justify for shared use, since the productivity of many users may be affected by cheap equipment that may accrue a lot of downtime.

Do not forget, though, there is generally a hefty price to be paid for multiuser systems in the form of contention for resources, congestion, reduced throughput, and limited expansion.

6.2.1 Buying Considerations

There are two approaches to implementing the multiuser environment. One involves software, which enables a designated microcomputer to mimic a supermicrocomputer in a small configuration. The other approach is hardware-based, involving the use of high-powered industry-standard chips, the performance of which may, or may not, be enhanced through proprietary techniques.

With multiuser operating systems, the performance of an application on a remote terminal depends on the power of the host system and the number of terminals connected to it, as well as on the efficiency of the operating system.

Of the software approaches, that offered by Virtual Systems, Inc. (Walnut Creek, Calif.), for example, partitions the serving microcomputer's central processing unit to support multiple tasks for as many as 32 terminals. The firm's Quick Connect software uses a special scheduling algorithm to allocate available CPU processing time remote to terminals, which are connected to the host in a star configuration *via* RS-232C links. Terminals may log on to the serving micro to access application programs like spreadsheet, word-processing, and data base–management systems. A warm reboot may be performed at the terminal without disrupting other users sharing the host microcomputer, and, if the terminal keyboard locks up because of an application-program error, the user may reboot the terminal from the serving microcomputer.

Many vendors of multiuser software packages assume that all the terminals will not be in use at the same time, and, therefore, claims of supporting 50 or more terminals go unchallenged. When all the terminals are in use, however, disk performance degrades considerably. Depending on the applications being run from the various terminals, disk performance may drop by as much as 20 percent with only a handful of active terminals in the configuration, if frequent disk I/O (input/output) is involved. This limitation may not become apparent until after the system is installed, in which case the vendor will try to convince you of the need to purchase caching software and high-capacity RAM, but this is nothing more than a variation on the old "bait-and-switch" scam, whereby a vendor deliberately underprices a product to get your business, and then bills for "extras" to get the equipment functioning properly. After the initial purchase, you do not have much choice — you have to buy the extras to protect your original investment.

Some operating systems support only ASCII terminals and micros, while others offer as many as eight terminal emulations. Multitasking capabilities also differ. Some support four multitasking applications at the server and two at the remote terminals. At a remote terminal, however, the application running in background mode is entirely suspended when the application running in the foreground is active.

In the multiuser environment, remote terminals share the available memory with the server, which is partitioned among them. The trouble with this approach is that there may not be enough memory — at least 468 Kbytes are needed for major programs — to handle graphics or a good-sized spreadsheet, even with only a few active terminals running off the server. Some systems try to work around this by allowing the active application to use as much memory as it requires, assuming, of course, that the inactivity of other terminals will be enough to allow that to happen.

Error handling is another concern when using multiuser operating systems. You have to be sure, for example, that application errors affect only the crashed terminal and not the serving micro. That way, only the crashed terminal will require rebooting. Ideally, the remote terminal will not crash at all, but will display an error message so you are not confused about what happened. Rather than rebooting, you should be given the opportunity to abort or retry.

Although operating systems may attempt to take advantage of the micro's high-performance hardware to derive multiuser capabilities, the fact remains that multiuser capabilities are added in, rather than designed in. Supermicros are designed from the ground up as servers for the multiuser environment. With the hardware approach to achieving multiuser functionality, much larger configurations may be easily supported because raw CPU power and innovative processing techniques will enhance overall performance.

Some supermicros, like the Universe line of products from Charles River Data Systems Inc. (Framingham, Mass.), use multiple Motorola 68030 processors to deliver performance as high as 12 MIPS for such applications as factory process control, transaction processing, data entry, and teleprocessing. The Universe line offers multiprocessing in two modes: peer multiprocessing, in which up to seven processors execute a program together, and auxiliary multiprocessing, in which one processor can be dedicated to a single task separate from the others.

Intel's 25-MHz 80386 processor makes small multiuser configurations more practical than software-based products alone, but more economical than high-end supermicros, which can support many more users. Supermicrocomputers, like Prime Computer Inc.'s (Natick, Mass.) EXL 325, are capable of supporting 48 users with no noticeable loss in speed. The EXL relies on an Intel 25-MHz 80386 chip to deliver 4-MIPS performance. Moreover, it can run in the UNIX, DOS, or Pick operating environments.

Prime Computer and Charles River Data Systems are among the companies that are in the forefront of pushing the performance of standardized microprocessors, while keeping hardware prices at the personal-computer level. Prime's EXL 325, for example, is priced at $45,900 which equates to under $1,000 per user. The four-processor version of Charles Rivers' Universe supermicro has a price tag of $50,000.

When evaluating supermicrocomputers it is quite easy to get sidetracked by claims of raw performance. Since supermicrocomputers are intended for the multiuser environment, however, you should check into a product's connectivity potential. On the host side (mini or mainframe), the multiuser system should offer a choice of communications links.

If you are in an IBM environment, for example, SNA protocols must be supported for 3270 applications. The system must also be capable of links with other computer systems *via* Ethernet, using such protocols as TCP-IP or DECnet, and if the supermicro comes from a minicomputer or mainframe manufacturer, it should be able to link with the higher-level product using the company's existing communications tools. If not, this could mean that their development paths are unrelated, possibly because the company is undecided about how the supermicro fits into the existing line of products. Without commitment, as expressed in R&D dollars, the supermicro could very well become a dead-end product.

In the multiuser environment, such things as system expansion, administration, and security become issues to consider. You also have to protect the supermicrocomputer against power loss, much as you would any other host in a star configuration. You must also consider the cost of cabling, since each microcomputer or dumb terminal requires a dedicated connection to the supermicro.

Since the performance claims of many supermicros border on the incredible, you should verify this information with existing customers. Find out how many terminals are being served by the supermicro, what kind of applications it is supporting, and how expansion is handled. Also check into how long installation took, and whether the technicians had to come back to fine-tune the system after its initial configuration. Make sure the references have had the system in operation for at least three months, so you can get better feedback on such things as response times, memory partitioning, and terminal crashes. Delve into the weak points of the system and ask the references whether these matters have been discussed with the vendor and what the vendor's response was.

Since most supermicros are sold through value-added resellers (VARs), you will have to perform a background check of the company, as well as explore its relationship with the manufacturer. You will also have to check into service response times and the technical-staffing level of the organization. If third-party maintenance providers are involved, you will have to check into their service record and their relationship with the VAR.

6.3 MINIS, MINISUPERCOMPUTERS, MAINFRAMES

Minicomputers fill the gap between desktop microcomputers and the big mainframes in the corporate data center. Not only can minis be used for stand-alone processing, they are equally suited for departmental-level processing, or as networking hubs in the distributed-processing environment. Either way, minis are capable of supporting large groups of users and providing ties into the centralized data center. The main attraction of the minicomputer is, of course, its low cost. While users gain more immediate access to corporate information and some autonomy in managing their own system, the mini also provides managers with a focal point for centralized operations. Also, rather than let expenses for microcomputers get out of control, managers can use the mini to keep hardware costs within limits and, at the same time, exercise better control over information assets.

Minicomputers have been around for more than 15 years and have amply demonstrated their value as information-processing workhorses many times over. Although today's minicomputers come with richer instruction sets, feature-rich operating systems, higher MIPS performance, a variety of networking methodologies, and are capable of using high-level languages, the technology behind the minicomputer is for the most part highly stable. Of course, some vendors, like Hewlett-Packard, have introduced seemingly radical departures from conventional processing technology with the introduction of the reduced instruction set computer (RISC) architecture for its HP3000 family. Even this, however, has proven to be a reliable method of speeding up data access and processing — so much so that RISC enjoys a wide following among many other companies which are using variations of RISC in their own products, especially workstations.

IBM made a big deal about its 9370 minicomputer two years ago, but it was firmly rooted in its System 370 architecture. This was good for IBM users, because the 9370 gave them much more interconnectivity potential between their IBM PCs and S/370s. The 9370, however, was never a big hit. IBM focused too much on the hardware and not enough on the software, which is still trickling to market.

Another factor in the 9370's poor showing is that IBM announced its "Silverlake" project too soon after introducing the 9370, which stalled sales as potential buyers adopted a "wait-and-see" attitude.

Innovation, if it can be called that, centers around improvements in connectivity, as well as modest improvements in functionality. IBM's AS/400 (code-named "Silverlake" during development), for example, is the follow-on product to the long successful System/36 and System/38. The AS/400 integrates the functionality of both products under a single architecture supported by a single operating system, thus taking a page from DEC's development philosophy. The AS/400 supports more communications lines and higher communications speeds than the S/36 or S/38 and, unlike those two, can support direct Token-Ring connections. Conspicuously absent in the AS/400, however, are connections to other vendors' networks and equipment. Instead, IBM will let third-party vendors provide the AS/400 with Ethernet and TCP-IP support, although DEC has been doing this for a long time.

The companies that dominate the minicomputer market have proven to be very stable over the years. Although the minicomputer market is the battleground for many start-up companies which claim to know how to do things better, the major players continue to be DEC, Wang, Hewlett-Packard, and IBM. The minicomputer segment of the computer industry is perhaps the most conservative, and least affected by the hyper-paced innovation that seems to dominate the workstation and superminicomputer segments of the industry. When it comes to buying a minicomputer, most companies will stay with the vendor already supplying equipment for their data centers. After all, there is little if any risk in staying with proven products. It is when you go outside the mainstream that you have to be careful. When you start falling for the marketing hype of small aggressive companies which claim spectacular performance at a tenth of the cost of conventional machines, you had better be on guard.

Davin Computer Corporation (Irvine, Calif.) tried to jolt the industry with its 64-bit single-processor minicomputer, which claimed 100 MIPS performance. Although the market for plug-compatible devices is growing stronger, not too many corporate managers appear ready to stick their necks out for a minicomputer that relies on one proprietary CPU, cannot link with anything else already on corporate desktops and data centers, and requires massive changes to the buyer's entire investment in software. As if that were not enough, Davin Computer claims that for users willing to buy several hundred units in a given year, the economics will justify using the machine and making the necessary software changes.

Minisupercomputers are a step below the exotic supercomputers that are used for scientific research, but still a cut above the general-purpose mainframe in terms of processing power and main memory. Until recently, minisupercomputers were locked out of most corporate MIS shops, because they were too expensive and lacked applications software, but some companies are discovering that minisupercomputers are well suited to running complex, numerically intensive programs that draw upon mountains of corporate data for such applications as financial forecasting and schedule planning. Minisupercomputers are also being used to simplify and speed up real-time searches of information stored in massive data bases. Of course, there are many technical and research-oriented applications that can be run faster on minisupers than on general-purpose mainframes, such as structural analysis, process simulation, and environmental modeling, which require more processing power than today's mainframes are capable of delivering.

The linear programming algorithms necessary to solve highly complex, numerically intensive tasks require more performance than mainframes like IBM's 3090 running MVS can handle within a reasonable time frame. Enhancing the 3090 with a vector facility (3090 VF) permits multiple processors working in parallel to increase performance fivefold. Such an upgrade costs $300,000. Buying a 3090/200E with one vector costs almost $5 million. The high cost of new mainframes and upgrades, then, makes buying a minisupercomputer look very attractive. For example, Active Memory Technology Inc.'s (Irvine, Calif.) DAP-150 can perform at 60 MFLOPS and is priced at around $120,000. The minisuper is designed to work as the front-end to an existing mainframe to offload numerically intensive applications from the host. It may be accessed with Sun or VAX workstations. AMT uses an architecture called "massively parallel computing," which involves the distribution of a number of processing elements embedded within the computer's memory to insure that the data are positioned close to where the computational power resides. The data path that connects memory to the processing elements is 1,024 bits wide, resulting in an internal data-transfer rate of 1,280 Mb/s. This level of performance makes such machines especially suited for scientific applications, although they are equally adept at handling corporate number crunching, and they are small enough to fit under a desk.

Although AMT and other young, innovative companies are not as well known as IBM, DEC, or Sun, the price-performance of their products is making corporate MIS departments sit up and take notice, especially as their general purpose mainframes reach capacity. Also, at such low prices, MIS people are more willing to explore new ways of information handling, especially if their existing software can be recompiled for use on the new machines, and the product can interface with existing mainframes and workstations. Although buying into products with unconventional architectures is risky, the alternatives are few. Some companies see themselves as being stuck between a rock and a hard place: on one hand, their mainframes are buckling under increased demand and upgrades or changeouts are expensive; on the other hand, the price-performance of some minisupers is such that the risks may be worth taking.

The minisupercomputer market, however, is young, and many of the manufacturers have only recently arrived on the scene with glitzy new technologies and outlandish performance claims that are more theoretical than real. Many of these companies are mired in red ink, or are only marginally profitable, a condition that may be attributed in part to the skepticism of corporate MIS people, who will likely be wedded to their mainframes for many years to come. Venture-capital companies are even hesitant about financing start-up minisupercomputer companies, because they believe that there are already too many players in the game, and that even those with strong products are not assured of survival past the five-year mark. That is the point at which the typical venture-capital firm likes to cash out its original investment with huge profits.

Because minisupercomputer manufacturers employ so many different architectures, chip technologies, and vectorizing capabilities, it is almost impossible to get an accurate reading on which product offers the best price-performance. You almost have to run one of your application programs on each machine before you have enough information on which to make a risk-free decision. Owing to the competitive situation, you may be able to do this more easily than you would think.

Minisupercomputer companies are not only competing with each other, but are also feeling competitive pressures from above and from below. At the high end, supercomputing companies are developing scaled-down versions of their more powerful products in an effort to make them more appealing to corporate users. Moreover, they are trying to reel them in with lower prices and promises of an identical architecture across the product line, from the low end to the high end.

From below, workstation companies like Sun Microsystems and Apollo Computer are touting single-user workstations with minisupercomputing capabilities — and at very competitive prices.

Competitive pressures and the relatively small market for minisupercomputers forces manufacturers to take sales when they can get them, and if this means letting you visit company headquarters to try out your applications, many will accommodate you, especially if they know that they have a fair shot at the sale. It would help, however, if you did your homework and narrowed down your list of vendors to a manageable number before approaching them on the idea. Even if you really do not intend to visit potential vendors to try out your application programs, finding out the vendors' attitudes on this idea may tell you a lot about the confidence level it has in its product. It is worth noting that Multiflow Computer Inc. (Branford, Conn.) routinely allows potential customers to run their own tests on its machines.

When it comes to evaluating specific products and vendors, you should bear in mind the following considerations:

- Some minisupers, like the 64/60 from Floating Point Systems Inc. (Beaverton, Ore.), require the use of front-ends. Unless you already have a front-end to spare, this requirement not only increases the overall cost of the system, but limits configuration flexibility. Many other vendors do not require front-ends, but their products may accommodate those of DEC or IBM, and even those of Apollo or Unisys.

- Some minisupers are totally Cray compatible — source and object code — while others offer only partial compatibility or no compatibility at all. Cray compatibility is important for two reasons: Cray owns the supercomputer market, and more applications are becoming available for Cray supercomputers. This means you will have software portability to the Cray-compatible minisupercomputers of other manufacturers, as well as to Cray's supercomputers when prices fall to the point where they become entirely realistic for the corporate environment. Total compatibility will also allow you to develop applications for the Cray well before the purchase, enabling you to take full advantage of your investment upon its installation and cutover.

- Find out the maximum main memory that the system can accommodate. Memory limits range from 256 to 1,000 Mbytes. The amount of memory is an important consideration, since it will determine the size of the program and data that can be resident at any given time, and whether the machine can operate as a stand-alone system or requires a front-end. The amount of main memory, however, may not correspond to the base price of the system. Floating Point Systems' FPS-64/60 is limited to 360 Mbytes of main memory, while Alliant's FX/8 is limited to 256 Mbytes. Yet the base price for the FX/8 is more than double that for the FPS-64/60.

- Find out the number of processors that are included with the minisuper. Typically, you will have a choice, which differs according to the model you select. The number of processors does correlate with performance and price. Alliant's FX/80, for example, has eight processors and a claimed peak performance of 158.8 MFLOPS. The price is about $653,000. Floating Point Systems' M64/145 Model G has 31 processors with a claimed peak performance of 341 MFLOPS. The cost is roughly $1.75 million.

- Understand that "peak" performance is not the same as the performance you will get running a specific application. Floating Point Systems' M64/145 Model G, for instance, scores only 101 MFLOPS in the Linpack benchmark test, whereas the company claims 341 MFLOPS as the possible performance its product can deliver under ideal conditions.

- If the minisuper will be used by designers and engineers, you do not want the response time at their workstations bogging them down as more users become active on the system. This situation will diminish your investment in both workstations and the minisuper. This possibility can be averted by looking for systems that allow you to add memory incrementally and which offer enough I/O speed and overall bandwidth to support your specific applications, hardware configuration, and number of potential users. Although you may have to pay more, when you have reached the point of needing products of this caliber, it is no time to start penny-pinching.

- Evaluate minisupercomputers on the basis of what applications you can run immediately to increase productivity, but if the vendor only promises the application software you want, keep in mind that you may have to go through a lengthy debugging process before you can reap the benefit of increased productivity. Also find out if the application programs you already have can be recompiled to work on the machine you are considering for purchase. Before committing to a purchase, you should try the vendor's compiler to be sure that your existing applications can be run on its machine. Keep in mind that there are differences in the efficiency of compilers according to application. Some may be very slow for development work, for example, but very efficient for production workloads. You have to decide what level of efficiency you can live with.

- You may want to stay away from vendors who offer only a single product. Having a broad product line that can comply with a range of present and future user needs means that the company will be more likely to survive a prolonged competitive onslaught.

One last thing: before allowing yourself to be enticed into a purchase by a minisupercomputer's MIPS and MFLOPS performance ratings, make sure you obtain the latest copies of Argonne National Laboratory's report, *Performance of Various Computers Using Standard Linear Equations in a Fortran Environment,* as well as Los Alamos National Laboratory's report, *The Performance of Mini-supercomputers: Alliant, Convex and SCS.* Although both were last issued in late 1987, they still contain useful information that will help you in the product-evaluation process.

Chapter 7

Telecom Information-Management Systems

Imagine for a moment that you are suspended in time and space. Your company's financial officer drops a 100-page telephone bill on your desk. You detect no humor in his instructions to sort through more than a thousand telephone numbers for the purpose of cost allocation and recovery. Your only resource is a card file for matching customers and vendors with phone numbers and, of course, an organizational chart — which was last pressed into service as a blotter for spilled coffee . . .

Sound like a screenplay for "Twilight Zone"? Hardly. To a greater or lesser degree, this is pretty much how thousands of businesses nationwide still try to reconcile telecommunications costs with usage, but new or established companies, big or small, do not have to remain stuck in the Twilight Zone forever.

With a little knowledge about the major approaches to telecommunications management, any organization can acquire the tools necessary to identify usage, allocate costs, and generate the data that can be used to configure a more efficient network — all of which can contribute to a healthier bottom line. Beyond mere cost containment, a telecom management system can play a key role in providing organizations with an unprecedented degree of network control and configuration flexibility that can be used to competitive advantage, but an understanding of how each approach fits specific organizational needs is necessary before attempting to evaluate products and vendors.

7.1 FIRST, THE BASICS . . .

With every outgoing call, your organization's PBX or key system generates a call record. If you subscribe to the telephone company's CENTREX service, that, too, generates call records, which are stored on magnetic tape. Whatever the

---

means of generating call records, each manufacturer's equipment uses a different format for the call record. The same manufacturer may even use different formats with each model of PBX or subsequent software release. Although there are about 150 different formats in common use today, they all contain some pretty basic information about each call like the date and time the call was made, the called number, the extension used for the call, the time the call ended, and the type of facility used to transport the call, such as WATS, FX, Tie Line, or interexchange carrier.

If you were to look at these data in their raw form at the end of a typical business day, you probably would not know what to make of them. For one thing, the call-processing equipment generates this information in coded form — sort of a shorthand notation. Also, even if you could read the code, the information would not be in a useful format. Because the PBX generates this information in "laundry-list" fashion, you would not be able to determine such useful things as calling patterns, trunk usage, and call costs by department or employee, and you certainly would not be able to match long-distance calls with project codes for billing back costs to clients.

In its raw form, then, the call data generated by the PBX are totally useless. They must first be collected by a Station Message Detail Recorder (SMDR). This device may be a nine-track magnetic-tape unit, a solid-state store-and-forward device, or a microcomputer equipped with a hard disk. For CENTREX users, on-premises SMDR units are used temporarily to store call data passed to them from the central office. Whatever the source of the call data or the storage media used, the collected data are eventually fed into a computer where they are processed using a proprietary program. These programs make comparisons with various data bases and print the required reports in tabular format, with calls arranged by date and time — for each department, subdepartment, and employee (Figure 7.1). These reports also include the duration and cost for each call. Other reports provide a breakdown of usage by type of facility, or summarize traffic by hour of day and day of week.

With reports like these, telecommunications managers can configure a more efficient network, as in determining the need for additional tie lines to high-traffic areas, documenting the need for higher band WATS lines to extend coverage, or justifying the need for high-bandwidth, bulk-billed services like AT&T's Megacom. These reports can also be used to identify unneeded trunks, or spot trunk outages on remote PBXs. If you have five trunks, for example, with trunks 1, 2, 3, and 4 showing traffic, while trunk 5 shows little or no traffic, chances are that you really do not need trunk 5 and can save some money by having the telephone company take it out of service. On the other hand, if trunks 1, 2, 3, and 5 show traffic, while trunk 4 shows none at all, chances are that trunk 4 was knocked out of service by a component failure. You may have been tipped off that a problem existed by user complaints of blockage during peak traffic hours. With the appropriate report, you

```
Date:    03/07/88
Time:    11:56:20

Report Period:    2/24 - 2/29

Name:    Dan Jones          Division:   Telecommunications
Ext:     1551               Department: Engineering
```

Date	Time	Duration	Charge	Number Called	Facil	City	ST	(1) Acct
2/24	08:01	00:12:15	0.06	616-429-2998	Local	St Joseph	MI	
2/24	11:35	00:25:00	5.86	703-620-0880	WATS	Roanoke	VA	
2/25	08:46	00:00:30	0.06	616-429-4151	Local	St Joseph	MI	
2/25	08:52	01:12:30	25.90	212-829-4272	DDD	New York	NY	
2/25	10:57	00:07:30	0.10		Incmg			
2/25	12:57	00:10:56	4.10	714-525-5252	MCI	Anaheim	CA	
2/25	14:00	00:16:01	6.27	312-577-7901	FX	Chicago	IL	
2/25	14:07	00:01:30	0.10		Incmg			
2/27	09:43	01:05:03	35.12	714-525-5252	DDD	Anaheim	CA	
2/27	12:55	00:01:00	.42	703-620-0880	WATS	Roanoke	VA	
2/27	13:14	00:10:00	0.06	616-429-6241	Local	St Joseph	MI	

```
Totals:   03:42:15   77.15
Fixed:                 5.00
                      82.15

Calls:    11          Cost/Min:   .33
```

Figure 7.1 Extension detail report. (Courtesy Infortext Systems, Inc., Schaumburg, Illinois.)

could determine that the problem is not merely the lack of trunks, but that one trunk is out of service. Until you notify the telephone company, it keeps charging you for the trunk that is going unused.

Other types of report may be used to track down call abuse, as in identifying frequent calling during nonbusiness hours or spotting frequent calls to nonbusiness locations. To help control call costs, a reporting feature flags calls of excessive cost or duration, according to management-defined thresholds. There are reports that will even group calls by project code for client billing, or by account code to identify specific users. Some companies assign separate account codes to employees for their personal calls. This cuts down abuse, while allowing both the company and the employees to take advantage of higher discount rates based on call volume. There is virtually no limit to the practical uses for these kinds of report, and vendors are continually expanding their report offerings to keep pace with customer needs and the changing operating environment. A few vendors even offer the capability of allowing you to customize reports to meet your needs.

Some vendors offer software packages that are designed only to identify usage. Others offer sophisticated software packages that coordinate many aspects of telecommunications management, such as service-order tracking, inventory control (including wire and cable by location and availability), and network optimization. Some integrated packages permit automation of the telephone directory and trouble desk. With a special interface, you can transfer telecommunications cost information directly to your company's general ledger. Organizational requirements and priorities will determine the level of integration. Of course, the more sophisticated the product, the more you can expect to pay for it — and the more diligent you must be with respect to evaluating vendors, which currently number about 200. The sheer number of products available and the variety of pricing schemes used make "apples-to-apples" comparisons very difficult.

7.2 THE MAINFRAME APPROACH

Many large corporations with far-flung, multi-node networks already have a substantial investment in mainframe hardware and MIS-dp personnel. As a result, they already have the basic infrastructure in place to process call-detail information and generate the reports necessary for managing telecommunications resources.

Here is how the mainframe approach typically works. Call records are generated by the PBX (or the telephone company's CENTREX switch) and sent to a temporary storage device where they are held until they can be uploaded to a magnetic-tape unit and written onto tape. The tapes from each network node are brought to a central processing location where they are batch processed into meaningful management reports. There are two variations of mainframe processing: service bureau and license agreement.

A service bureau is a vendor who processes data delivered to it on magnetic tape and provides the customer with finished products like paychecks, insurance claim reports, or mailing labels. As applied to telecommunications management, such vendors process call-detail records delivered to them *via* magnetic tape or floppy disk, or through electronic transmission over phone lines under a polling arrangement (Figure 7.2). The vendor uses its mainframe computer to process the data and generate the reports you want. A reasonable turnaround time for this kind of report processing is five working days.

Under a license arrangement, the user typically signs an agreement with the vendor, which authorizes customer use of the software at one or more PBX locations. This "site license" is common practice among vendors, because it allows them to exercise more control over the proprietary nature of their software than copyright laws usually afford with an outright purchase. Under a license agreement, the user is obligated to maintain confidentiality and adhere to provisions in the agreement governing the use and disclosure of the software product.

The software package usually includes tariff tables required for accurate call rating and a data base of vertical and horizontal (V&H) coordinates to assign city-state locations to long-distance calls made over bulk-rated facilities. (When you subscribe to services like WATS, this level of detail is missing from your bill; the invoice is simply a statement of total minutes per band, and the cost.) The software package is designed to work with a data base defining the characteristics of your network, and includes programs that will do the necessary data-base comparisons to organize this information so that the desired reports can be generated.

7.3 LICENSE *VERSUS* SERVICE BUREAU

Choosing between the license or service bureau may hinge on the capabilities of your organization's MIS-dp shop, specifically the amount of time and effort it can afford to devote to becoming expert at telecom management. Installing new tariff tables, updating the data base to reflect continual changes in your telephone system at each node of your network, and verifying the accuracy of input data and report runs are all time-consuming tasks that require a degree of expertise and staff continuity. Ultimately, it may be easier and more economical to use a service bureau. A good service bureau will not only perform quality control, but also provide assistance in interpreting reports and running special reports when you request them. The drawback to using a service bureau is that special reports take time and will cost extra. With a service bureau, you will typically have call records processed into standard reports on a monthly basis, which gives you less flexibility than running reports in-house with a microcomputer or mainframe.

With multinode networks that utilize PBXs from different manufacturers, remember that each PBX type formats call records differently. Unless you have

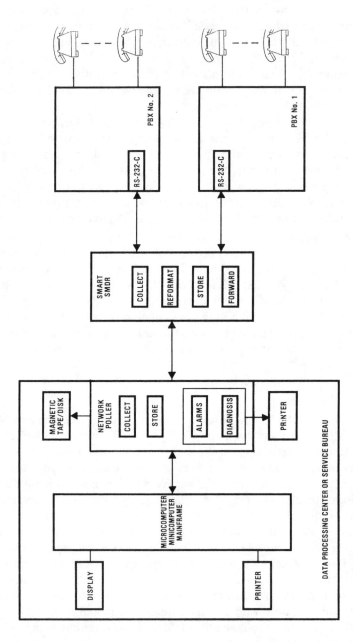

Figure 7.2 Teleprocessing summary.

a front-end processing capability, using the mainframe to preprocess call records into a common format will hog mainframe resources and consume valuable man-hours. Using a service bureau for this purpose will offload your own mainframe.

If your network has a large number of extensions or account codes, and constant equipment moves and changes, you should find out if the vendor offers a micro-to-mainframe communications package that will allow an operator to batch feed this data from floppy disk directly to your mainframe on a daily basis. A menu-driven PC data-entry system permits users to record, review, and edit the data base easily — a more convenient and accurate process than filling out forms for subsequent keyboard entry by the service bureau's clerical staff. Uploading this information to your own mainframe also results in more accurate reports, because they reflect the latest changes to the data base. The amount of daily update activity will determine whether you use your own mainframe under a license agreement or use a service bureau.

Also, if your data-processing center relies heavily on reel-to-reel magnetic tape, be aware that such systems may impose considerable limitations when used in telecom-management applications. There are a number of reasons why.

Magnetic-tape systems are not pollable. With a number of PBX types on your network, preprocessing must be done to put all the data into a common format before mainframe report processing. This could turn into a time-consuming procedure that wastes mainframe resources and increases the chance for human error during tape handling.

Magnetic-tape systems are mechanical devices that require scheduled preventive maintenance. Calls are not recorded during maintenance downtime, unless you have a redundant magnetic-tape system available. If not, acquiring one can add from $25,000 to $50,000 to the start-up cost of in-house processing.

Malfunctioning magnetic-tape drives are not always easy to detect; a misaligned head, for example, can cause call-recording problems. Many times such malfunctions go undetected until discovered by a technician during a preventive-maintenance visit — or when the MIS-dp manager notifies the telecom administrator that a "blank" tape was sent to the data-processing center.

Even more basic, it is easy for inexperienced operators to install a mag tape onto the drive unit improperly. This can result in lost call records, wasted computer time, and delayed report processing. Beyond this, magnetic-tape reels can be lost in transit, damaged from mishandling, or destroyed in an accident — sooner or later it happens!

Mag tapes require careful administration and security to keep track of spares, backups, tapes in transit, tapes awaiting processing, blanks ready to use, *et cetera*. Not only does this burden you with unnecessary overhead costs, but one foul-up can wreak havoc with tightly scheduled data-processing operations. The alternative — submitting finished reports late — may result in stale information that is of limited value for network planning.

The mainframe approach, service bureau or in-house processing, is required if you collect two million or more call records a month. Even if you daisy chain several powerful micros together, you will not have the computing power you need to process that many call records.

7.4 POLLABLE SMDR DEVICES

One way to overcome the problems inherent in using reel-to-reel mag-tape systems is to use pollable solid-state (random access memory — RAM) call-record collection devices. In addition to having no moving parts, which substantially improves the mean time between failures (MTBF), these devices are relatively inexpensive, as in Account-A-Call Corporation's (Burbank, Calif.) store-and-forward Tadpoll, which sells for about $3,000. These units install easily and quickly without PBX modifications or service interruptions, and, because they can be polled, they also eliminate the risk, delay, and high costs associated with shipping reels of magnetic tape.

Some SMDR units provide the additional advantage of recognizing call-record types. Using various data-manipulation techniques, they can compress call data to squeeze more records into the available memory. Compression techniques are relatively simple, consisting of removing unneeded characters and blanks between call records to achieve 50% greater storage efficiency. Retrieving data in a compressed format also reduces the time required for polling, which translates into lower transmission costs. Some vendors compress the call records as they are collected from remote SMDR devices by the microcomputer-based poller. By adding a coprocessor to their multifunctional IBM PC-AT products, for example, call records may be preprocessed into binary format as they are received from remote locations during polling. In storing call records more efficiently, additional workspace is made available for other processing chores.

Pollable SMDR units are ideal for network nodes with less than 60,000 call records per month. They feature integral-battery backup and user-selectable buffer threshold displays, as well as data-checking protocols. In addition to self-test and remote diagnostics, some models monitor critical alarm data from the PBX and send an alert to the polling equipment if the PBX malfunctions. Also, they are designed for continuous, virtually unattended operation.

If, however, you decide to process your own reports on an in-house mainframe and your network is composed of different PBX types, store-and-forward devices do nothing to simplify the task of format translation. Although they operate unattended and may store as many as 60,000 call records, they do not have the intelligence necessary to convert these data into a common call-record format, a chore that will tax your mainframe resources with preprocessing. For smaller volume recording needs on a homogeneous network, however, these low-cost call-data collectors perform this simple task quite reliably.

Alternatively, look for an intelligent Station Message Detail Recorder (SMDR) that will process call records from virtually any PBX into a common format. Although they cost more — as in ComDev Corporation's (Sarasota, Fla.) STU-3B which is available only through distributors for about $6,000 — these "smart" devices make data collection and report processing much easier, and provide flexibility in choosing the PBXs of different manufacturers during network expansion. An added benefit of some intelligent SMDR devices is that they may be used to collect call records from multiple locations simultaneously. ComDev's STU-3B collects data from up to six PBXs simultaneously, while Telco Research Corporation's (Nashville, Tenn.) TRU Recorder collects call records from up to 15 PBXs simultaneously at a rate of about 50,000 calls per hour — without regard to PBX type. These devices may be used to consolidate call records from multiple locations within a fairly narrow geographical area for economical collection by the central-site poller.

If, however, you want to save on the cost for collection equipment, you can do so by making arrangements with a service bureau to poll the store-and-foward devices at various PBX–key system locations. The service bureau will use its own mainframe to assemble the diverse call-record data into a common format for call-costing and report processing. Of course, you will have to compare the monthly cost for this continuing service with the cost of paying for the intelligent SMDR equipment over the life of the contract. You will also have to weigh any cost savings against the delay inherent in using a service bureau for this purpose.

In mainframe processing environments, high-end pollers like ComDev's NP9000 collect call records from the various network nodes. The polling process insures data integrity by auditing the data from the point of origin to the point of delivery. Blocks of data are sequentially numbered at the point of origin and are checkpointed as they are teleprocessed to the destination polling equipment. In this Cyclic Redundancy Check (CRC), the number of data blocks are verified by both the sending and receiving units. Some systems also employ the ACK-NAK (positive acknowledgement–negative acknowledgement) process for confirming the receipt of data. In the event that incomplete or damaged data arrive at the network poller, repolls are initiated automatically until all data blocks are accounted for and arrive at the destination error-free. Polling may be initiated as demand warrants, or according to a routine late-night schedule that takes advantage of lower transmission costs.

7.5 CHOOSING A NETWORK POLLER

The selection of polling equipment requires careful evaluation. Be sure to ask vendors these questions:

- In the event of a power outage, will the poller turn off and stay off, or will it restart automatically and initiate a repoll upon the resumption of power?

- Does the poller have enough intelligence to switch lines automatically or hang up and dial again when it encounters poor line quality?
- Does the poller have the ability to skip SMDR locations when it encounters a busy signal, and return to that location after polling multiple sites in sequential order?
- Does the poller allow remote activation for polling on demand? Can the poller activate powered-off equipment at the distant end to extract call records?
- With different SMDR types on a multinode network, will the poller adjust automatically to the error-checking protocol of the SMDR device being polled?
- Does the poller furnish a complete report of its activities (Figure 7.3)?
- Does the poller have the capability to accept critical alarms from the remote SMDR device, or perform remote diagnostics on PBXs as well as on SMDR devices?
- Can the poller be password protected to permit only authorized access to remote locations?
- If you have a high call volume and plan to use your microcomputer in a multitasking mode, do you have enough disk space or peripheral memory available for polling call records from other locations?

Answers to these questions are critical to the long-term reliability of data collection and, ultimately, to the viability of in-house report processing in the multinode environment. If you have a poller of your own to collect call records from various network nodes, it is a good idea to retain a service bureau as a backup, just in case your own equipment gets knocked out of commission for an undetermined period.

7.6 THE MICRO SOLUTION

Despite persuasive sales pitches from mainframe vendors, the ubiquitous microcomputer is indeed capable of telecommunications-management applications. In fact, their increasing power, memory capacity, and functionality make microcomputers quite useful for performing a wide range of related activities like collecting, polling, and processing call records to provide reports on demand. Infortext Systems, Inc. (Schaumburg, Ill.) provides an idea of the kind of microcomputer-based products that are now available for managing telecommunications. Its CAPS X System polls call records from the SMDR units of up to ten PBX or key-system locations utilizing an IBM PC-AT. This microcomputer is also used to process the

Printed On:
Date: 01-27-1988
Time: 14:10

Comdev
Polling History Report

Report Dates
From: 02-27-1987
To : 01-27-88

Poll Date	Poll Time	Site	Poll Duration	Good Records	Retries	Poll Type	Poll Status
02-27-1987	10:42	1	00:00:36	62	0	Manual_Data	Terminated
02-27-1987	10:46	2	00:01:19	279	0	Manual_Data	Complete
02-27-1987	10:49	3	00:01:16	279	0	Manual_Data	Complete
02-27-1987	10:50	4	00:01:15	279	0	Manual_Data	Complete
02-27-1987	10:51	5	00:01:17	279	0	Manual_Data	Complete
02-27-1987	10:54	1	00:17:38		0	Diagnostic	Complete
02-27-1987	11:12	1	00:01:14	279	0	Manual_Data	Complete
02-27-1987	11:15	1	00:01:49		0	Diagnostic	Incomplete
02-27-1987	11:23	2	00:01:15	279	0	Manual_Data	Complete
02-27-1987	11:24	3	00:01:17	279	0	Manual_Data	Complete
02-27-1987	11:26	4	00:00:38	93	0	Manual_Data	Terminated
02-27-1987	11:27	4	00:00:28		0	Diagnostic	Complete
02-27-1987	11:27	5	00:01:20		0	Diagnostic	Complete
02-27-1987	11:29	5	00:01:18	279	0	Manual_Data	Complete
02-27-1987	11:32	2	00:01:17	279	0	Manual_Data	Complete
02-27-1987	11:35	4	00:01:33	279	0	Manual_Data	Complete
02-27-1987	11:37	1	00:00:14	0	0	Manual_Data	Incomplete
02-27-1987	11:37	1	00:01:32	279	0	Manual_Data	Complete
02-27-1987	11:39	1	00:00:14	0	0	Manual_Data	Incomplete
02-27-1987	11:39	1	00:01:13	279	0	Manual_Data	Complete
11-17-1987	10:46	2	00:00:08		0	Diagnostic	Incomplete
11-17-1987	10:47	2	00:00:08		0	Diagnostic	Incomplete
01-27-1988	11:15	6	00:00:23		0	Diagnostic	Incomplete
Report Totals			0:39:22	3503	0		

Figure 7.3 Polling history report. (Courtesy ComDev Corporation, Sarasota, Florida.)

records into a number of standard reports, providing call information for each extension as well as for entire departments or divisions. One report identifies trunks that are more utilized than others. Another report flags calls of unusual duration and cost. This hardware-software package, which costs just under $20,000, is capable of supporting up to 4,000 extensions, 350 trunks, and 500,000 call records per month. With this many extensions and call records, telecom managers are virtually running their own telephone companies, which underscores the need for selecting the right product for telecommunications management.

Microcomputers have become a viable option for telecommunications management. All that is needed is the appropriate software to do the job, enough storage capacity to handle all of your traffic-analysis requirements, and a modicum of common sense to safeguard the accumulated call records. For example, when equipped with hard disks, microcomputers have the capacity to store hundreds of thousands of call records per month — and they have the speed to access any call record virtually instantaneously for on-demand reporting. This allows reports to be run as required, instead of once-a-month according to the MIS-dp department's schedule. Keep in mind, however, that today's microcomputers are limited to outputting information at a rate that approximates only 30 pages a minute. This may sound like a lot, but printing out 500,000 to one million call records from all of your network nodes may require 20 or more hours, even on a printer capable of outputting 600 lines a minute. Someone has to be around to take action if the paper jams, or if a break occurs in the box of fan-fold paper. In any case, the microcomputer option puts control of telecommunications management where it belongs — with the telecommunications manager.

Inexpensive floppy disks or magnetic-tape cartridge systems can be used to store call records conveniently for later processing, historical inquiry, and traffic analysis. These are much simpler to handle than the magnetic-tape reels commonly used to store data for mainframe processing.

Microcomputers typically accommodate two methods for report retrieval and data-base maintenance. *Via* menu selections, users with little or no experience can be instantly productive, while more experienced operators can implement features *via* direct command without waiting for various levels of prompting. Micros are also multitasking, permiting two or more jobs to be in progress at once.

Vendors accomplish multitasking in a variety of ways. Through the use of MS-DOS interrupt routines, call records may be stored in memory during the brief pauses of an applications program. These pauses are so brief that the user barely notices them. Another multitasking technique involves the use of special multitasking software like Microsoft Corporation's (Redmond, Wash.) "Windows" product.

Windows is a graphics-based operating system that allows the microcomputer to run several applications at once within the windowed areas of the display. The windows "pop up" through simple keyboard commands, allowing the user to move in and out of functions with relative ease. This allows higher levels of productivity with no interruption of tasks already in progress. For example, the operator can open a window to update the on-line telephone directory, or check the status of a service order, or even execute a spreadsheet on the telecom budget — all of this while call records are being collected from the PBX. Such multitasking preempts management concerns about dedicating a microcomputer to call-record collection. The same microcomputer can be used for other telecommunications-management tasks like traffic analysis and network optimization, and if the microcomputer has a UNIX operating system, many users can share it to perform a variety of tasks, providing even more flexibility in managing the telecommunications network.

Finally, hardware support for major brands of microcomputers is widely available and, for the most part, timely and reliable. Sound impressive? It is, but you will not find all these features packaged in one nice, neat little box. The difference is in the vendors. In most cases, the power, convenience, and user-friendliness of today's microcomputers remain untapped because of the limitations of the vendors.

Despite the enormous collection and storage capacities of some microcomputers — as much as one million call records a month — what really counts is the number of call records that can be processed at one time and how long it takes to format and run the final reports. A microcomputer that can store up to 250,000 call records from 1,000 extensions, for example, may only produce summary reports, or retrieve and print call data one station at a time, which may not be adequate for running detailed and consolidated usage reports on a timely basis.

Also, depending on the number and complexity of reports required and the rating scheme used to cost each call, the system may require the supervision of a full-time operator. If you have a high call volume and plan to use your microcomputer in a multitasking mode, polling must be done more frequently, because there will be less memory available for call-record storage.

Many of these problems, however, are the result of poorly structured programs and the failure of some vendors to keep pace with advances in hardware. Many vendor representatives tip-toe around this issue by equating "processing" with "call-record collection" only. Processing actually refers to the systematic execution of operations upon a call record to arrive at a desired result — appropriately costed calls arranged in a meaningful report format. Call collection, then, is only the first step in a much more complicated process of report generation. When a vendor claims that its software package can "process" or "handle" 50,000 calls an hour, what it may really be trying to say is that when the microcomputer functions as a call-record collector, it can collect and store that many raw call records.

Many "bare-bones" systems cannot rate calls according to appropriate tariffs for intra-interstate or intra-interLATA calls. They cannot rate calls to Canada, Mexico, Hawaii, Alaska, and international (011) locations. In trying to generate the cash flow necessary to finance the further development of their stand-alone products, they sacrifice call-rating accuracy in the hope of underpricing the competition.

Some vendors attempt to get away with this by telling prospective customers that other rating schemes are just as good, such as costing calls at a flat rate per minute. Other times, they must approximate rather than accurately rate calls using V&H (vertical and horizontal) tables. These tables provide, among other things, exchange coordinates that are used for calculating the distance of calls. This method can be up to 95% as accurate as rating calls with current tariffs, which may be close enough for all but high-volume users.

If your call costing and reporting needs are relatively simple, the microcomputer-processing option may be your best choice. Using the V&H costing method, Xtend Communications Corporation (New York, N.Y.), for example, can collect and price up to 30,000 calls per hour in real-time and have some reports ready in only 30 seconds. Keep in mind, however, that this kind of performance is more the exception than the rule.

You may want to look into an on-line inquiry capability that allows you to isolate individual calls by the extension used or the number dialed. This capability is extremely useful for tracking down unauthorized computer access, telephone abuse, or trunk problems.

If you attempt to do too many diverse tasks with the microcomputer, however, keep in mind that you may have to prioritize your activities in accordance with disk-storage limitations. For example, your ability to perform historical analysis may be limited by the disk capacity of the microcomputer. Historical analysis involves collecting many months of call records for the purpose of reporting trends in usage and calling patterns. With this information, you can make changes to your network that will result in greater efficiencies and cost savings, but if your microcomputer performs multiple tasks simultaneously, there may not be enough working space to perform a memory-intensive task like historical analysis.

Using the microcomputer for telecommunications management provides many benefits, and, as your needs change and you gain insight and expertise in telecom management, you can always add software modules and memory capacity for only an incremental cost. The microprocessing option is also an effective way for telecom managers to end their dependence on the MIS-dp shop and retain total control of their corporate mission. From top management's perspective, the microprocessing option should end the finger-pointing between MIS-dp and telecom when things get really screwed up.

The micro approach is virtually required if you presently perform such labor-intensive operations as updating the internal telephone directory, tracking inventory, and processing service orders. With a microcomputer, you can update the telephone directory as changes occur; with a service bureau or in-house mainframe, telephone directories are issued monthly and are quickly rendered obsolete with continued changes during the interim between processing, printing, and distribution. Inventory may be tracked on a daily basis, allowing instant access to check the status of all terminal equipment, as well as the status of cables and wires; service bureaus and in-house mainframes cannot handle these operations in nearly so efficient a manner. Service orders may be processed starting with the trouble report and followed all the way to problem resolution by one or more vendors or carriers. Using a service bureau or in-house mainframe for service orders is self-defeating, because timely access to service-order information is the primary benefit of computerizing it in the first place.

Telecommunications Manager, a product of Communications Sciences, Inc. (Iselin, N.J.), illustrates the modularity and sophistication of some microcomputer-based systems. The menu-driven package is designed for use on IBM XT/AT or compatible microcomputers with at least a 20-Mbyte hard disk and 360 Kbytes of RAM. The six-module package consists of Equipment Manager, Private Line Manager, Trouble Manager, Cable Manager, Directory Manager, and Bill Manager.

The Equipment Manager and Private Line Manager modules enable users to keep track of installed communications equipment, including spare parts inventory, and leased lines. Equipment Manager can generate order forms, perform order tracking, and help the user monitor installation progress, and, when the order is filled, the record becomes part of the inventory data base.

The Trouble Manager module keeps a log of all equipment trouble tickets to assist users in identifying vendors with overdue repairs. The module can calculate the rebates owed by the vendor, due to lost equipment or service uptime. It even keeps performance statistics on all equipment, so the user can compare the equipment-failure rates of various vendors.

With the Cable Manager, users have a complete record of cabling by type, end locations, route, and connecting device. The Directory Manager lists all employees, their locations, extensions, departments, and job titles. The telephone operator can look up an extension by performing a data-base search against any of these parameters. The Bill Manager module tracks invoices for equipment purchases and services. It will even compare bills for services with prior bills, so the user can identify vendor overcharges or billing errors. The price of CSI's Telecommunications Manager starts at $9,500.

Microcomputer systems require that the user take responsibility for quality control. It is often a full-time job to monitor equipment and keep it running smoothly. Although microcomputer software costs much less than mainframe software, its limitations should be carefully weighed against the promise of short-term savings. If you really like the idea of using a microcomputer for telecommunications management, however, many of the limitations are easily overcome with the addition of inexpensive add-ons or some vendor customization. Here is how.

Tape-cartridge drives can now hold several hundred megabytes of information, making them an excellent hard-disk back-up medium. That is roughly equivalent to storing about 3,000 pages of *The Wall Street Journal!* On-line back-up speeds have reached five megabytes per minute. Moreover, access to any day's call records takes only a few seconds. Worried about a hard-disk failure? Today's cartridge systems back up files automatically and continue collecting call records after the crash. That sure beats the performance of reel-to-reel mag-tape systems that mainframers are still clinging to.

As with mainframes, call-accounting micros should have a backup power source. Uninterruptible Power Supply (UPS) is available to sit right alongside the micro. Switching to internal battery power not only when line voltage fails, but also when the voltage gets either too high or too low, it provides transition times as fast as one millisecond. This allows micros to operate continuously, if necessary.

If you are concerned about polling call records from powered-off micros at other network nodes, you can buy devices that pull double duty as surge protectors and phone-activated booters. Do not worry about security; these devices require valid codes for access.

If high-capacity storage is a concern, there are 310-Mbyte hard-disk units available that provide access times of as little as 17 milliseconds. Special software allows the entire drive to be partitioned as one volume and permits disks to be joined into a single virtual disk of up to 1,000 Mbytes. Alternatively, outboard hard-disk units are commonly available that provide hundreds of megabytes of call-record storage in high-traffic environments. They connect easily to a controller card that inserts into any of the micro's vacant expansion slots.

If you are worried about having enough hard-disk space to store call records *and* do polling in the background mode, check into the possibility of using a data-encoding technique that packs data tighter and moves data faster. For less than $180 you can simply replace your XT-AT hard-disk controller with a run length limited (RLL) controller, which encodes data on 25 sectors per track instead of 17 to increase storage capacity by 33%. The controller's circuitry also changes the hard disk's 6-to-1 interleave ratio to 3-to-1, resulting in a 50% increase in the data-transfer rate. Other types of RLL controllers double disk capacity and double the data-transfer rate. When using these advanced data-coding schemes, solicit integration advice from your telecom management–system vendor and make sure you have a hard-disk drive that is certified to use RLL.

Over time your microcomputer will slow down because the hard disk scatters call records and other files wherever it can find vacant space. A single call record or file that started out as one seamless piece may become fragmented all over the disk, which means that call-record collection will be slower and report processing will take longer. Hard-disk performance degrades slowly, but you can determine when the problem becomes serious. Watch the hard drive's LED on the front panel of your microcomputer during read or write operations. Each time the LED flickers means that the drive head is moving to another section of the disk, either to pick up or lay down data. If the LED seems to stay on almost continuously, it means the drive head is working harder than it should. This causes extra wear and tear that can shorten the life of the drive. A number of "optimizers" are available, which will quickly reorganize the hard disk by finding fragmented information and consolidating it to permit faster call-record collection, backup, sorting, and retrieval. The cost: less than $60.

If you really want to eliminate worry about the speed of call-record collection and processing, you can add up to one megabyte of RAM and supercharge your microcomputer with cache (pronounced "cash") software. In this way, you can direct some of the vendor's most frequently used programs or addressable subroutines to RAM, permitting their immediate use without accessing the disk, which can be a real time-waster. This frees up the disk for other support operations, resulting in much faster report processing. The cost for caching software is less than $50, but the vendor will probably want an extra charge for "customization" if the source code is involved — sort of "caching in" on the deal.

Off-the-shelf PC software is now available that will turn any table of numbers into a graph, regardless of whether the numbers reside in a word-processing file, a spreadsheet, a data base, or an application program. This means that tables of usage data can be quickly reformatted to provide instant insight and analysis to augment any reports you may already be getting from the call-accounting vendor. The cost: less than $100.

Some vendors will customize their application programs to provide a remote-access feature. This allows the central site's microcomputer to dial up a microcomputer at a remote location and take over its keyboard to update software, load new tariffs, or diagnose problems. All the screen displays are echoed back to the remote terminal.

Many microcomputer products accumulate call records for batch processing once a month. The software of some vendors will process call records as they are collected, allowing the microcomputer to work more efficiently and print management reports more frequently.

High-speed desktop laser printers make true on-demand reporting a reality, if summary reports are all you need. Laser printers can print eight to twelve pages a minute, in easy-to-read tabular or graphic formats. The good news is that prices for laser printers are plummeting fast. Many sophisticated models now cost less than $2,000.

Remember, too, it is not necessary to buy a complete stand-alone microcomputer system to do call accounting. All you need is a specialized plug-in board for your existing IBM XT/AT or compatible, plus the vendor's proprietary software package. Tel Electronics, Inc. (American Fork, Utah), for example, offers an add-in product that provides the RS-232C serial port which connects to the PBX. Call records are processed in background mode. The board comes with 168 Kbytes of RAM, as well as ROM-resident routines that receive incoming call records collected by the PBX. One routine prices the calls, while the other stores the complete call record in RAM. When the RAM comes within 500 records of reaching its limit of 150,000 call records, another software routine interrupts the application program running in the foreground to notify the operator that it is time to store the RAM contents to disk. Unfortunately, if no operator is at the terminal to see the message, the system will stop collecting call records, which points to the need for full-time operator attention with the microcomputer call-accounting option.

7.7 CALL RATING

The comprehensiveness of tariff tables used by the vendor ultimately affects the accuracy of cost allocation. With the implementation of Local Access and Transport Areas (LATAs) in 1984 — which came about as a result of AT&T's divestiture of the Bell Operating Companies (BOCs) and the subsequent need to define the serving areas of local exchange carriers — there are now many different rate structures for interstate and intrastate calls, and still others for inter-LATA or intra-LATA calls. With 164 LATAs defining the service areas of the major telephone companies and another 25 for smaller, independent telephone companies plus a handful of special LATAs that overlap state boundaries (Extended Areas of Service — EAS), the task of rating calls has become quite complex.

If you are dealing with a vendor who was in business before divestiture, look into how they developed and implemented their new rating package when the LATA system went into effect. Find out if their customers experienced any problems related to the changeover. Answers to these questions may determine the reliability of vendor service and indicate future responses to changes in the operating environment.

Whether you choose the microcomputer or mainframe approach to telecom management, consider the advantages in being able to rate calls differently from department to department. In addition to costing calls by tariff, costing may be user-defined as a flat cost-per-call (for local calls) or cost-per-minute, or cost-per-minute with differential rates for the first and each additional minute. Calls placed over bulk facilities may be costed at the equivalent DDD rate, or calculated as a percentage of the total corporate telecommunications cost.

Time-of-day costing allows the application of appropriate evening, weekend, and holiday discounts to direct distance dialed (DDD) calls and calls made over alternative interexchange facilities. When used in conjunction with account codes, any of these costing schemes can be used for client billing. Calls to clients may even be marked up by a fixed percentage to recoup related administrative expenses. Call costs may also be marked up by a fixed percentage to permit departments to pay their fair share for telecommunications resources *via* the company's internal charge-back system. With internal charge-back systems, you may also want to consider the value of being able to rate intra-PBX, directory assistance, local, intercom, and internal-paging calls. Some software vendors are quite adept at providing all of these rating options and will help you define the exact requirements of your organization.

Consider the importance of accurately rating calls originating and terminating in the same state. Calls originated and terminated in Connecticut are rated differently from calls originated and terminated in Alabama. Consequently, if your company has multiple locations, extra tariff files have to be maintained to rate intrastate calls. Although this information is readily available from such reputable organizations as CCMI-McGraw-Hill (New York, N.Y.), some vendors are not properly equipped for such an undertaking and will try to convince you that other, less accurate call-rating schemes may serve your needs just as well.

You must also take care to choose the software-hardware package that can handle your call volume and diverse management tasks. To accommodate future organizational growth, choose a product that will address your short-term and long-term needs — something you will not outgrow in three to five years. Many products are modular, allowing you to add software and hardware incrementally, so that you pay only for what you need now, and add on as your future requirements justify. When choosing modular products, check into their security features. Each module should have password protection, so that the telephone operator, for example, will be able to update the internal directory, but will not be able to access the inventory data base or the call-record files. Insure that there are various levels of authorization available, so you can designate personnel who will have the responsibility for updating various data bases and designate others whose job only requires making inquiries.

7.8 NETWORK OPTIMIZATION

Post-divestiture competition among interexchange carriers has resulted in more frequent and complex tariff changes, making it almost impossible for today's telecommunications managers to evaluate everything available to ensure an optimally configured network. If your network relies extensively on alternative carriers and bulk-billed services, you should perform an optimization study at least annually, or more frequently as changes in available services, equipment, and your network configuration dictate. Here is why:

- Ensure maximum savings with alternative carriers by finding out how they serve specific calling patterns.
- Justify FX, Tie-Line, or WATS services with accurate and comprehensive decision-making information.
- Determine, through "what-if" analysis, the potential effect of proposed tariff changes on the long-distance telephone bill.

- Validate configuration proposals for top management review or approval.
- Sharpen the company's competitive edge with an efficient and economical telecommunications system.

Even if network optimization is only a remote consideration, you may want to choose a service bureau which has compiled a successful track record with such services. The charges for network optimization can be as much as 20% less, because data base set-up has already been done to provide you with call accounting. A reasonable turnaround time for a service-bureau network-optimization study is from four to six weeks, depending on the number of network nodes and the nature and scope of the transmission facilities that will be considered.

Beware of vendors or consultants who promise a certain percentage of savings, or who base their fees on a percentage of savings, because typically they will not allow much input about the type of long-distance and bulk-billed services to be included in the study. These vendors and consultants want to make those decisions for you, because doing so enables them to deliver the greatest amount of savings.

If you like having total control over the network-optimization process and are willing to put in some extra time and effort to become proficient, there are a number of dial-up services and microcomputer-based traffic analysis and network-optimization packages available. They include tariff and availability information for all major interexchange carriers, as well as special access tariffs of local-exchange carriers. The latter would be used to price the local segments of interstate circuits when dealing directly with the local-exchange carrier. These tariffs are used in place of AT&T's Tariff 11, which governs pricing for Accunet T1.5, and Tariff 9, which governs pricing for DDS and subrate services.

These services and packages include a data base of the types of facility available on an exchange-by-exchange basis, including voice-grade lines, digital-grade lines, DDS, and T1 for both point-to-point and multipoint applications. This information is especially useful for comparing the cost of facilities between particular locations. When choosing alternative interexchange carriers, this information can assist you in determining which locations are on and off their networks for the purpose of comparing cost savings. Typically, a discount interexchange carrier will complete calls to "off-net" locations through AT&T, which adds to their cost. Also included in these packages is a program for analyzing real-time calling patterns, which can be used to select the most appropriate long-distance services.

There are two subscription alternatives available to obtain the tariff information necessary to optimize your network: dial-up service and quarterly updates *via* microcomputer disk. The advantage of dial-up services is that you can obtain access to the most up-to-date tariff information available, simply by making a phone call. With large networks, the timeliness of this kind of information is every bit as important as its accuracy and comprehensiveness. Another advantage to dial-up services is that you can have access to the mainframe processing power and storage capacity of the service provider, but these advantages must be weighed against the inherent disadvantages of dial-up services. Over a year's time — after you add up connect-time charges, transaction fees, storage charges, and long-distance call costs — you can probably buy a dedicated microcomputer to run the software and maintain the data base, using it as often and as long as you wish, without regard for usage-based costs.

With microcomputer-based packages, you can have the added advantages of being able to integrate pricing information into other application programs like spreadsheets, data base–management systems, and word processing. Although such packages are effective for optimizing relatively small networks, using them for optimizing complex multinode networks may prove to be quite cumbersome and time-consuming, requiring dedicated staff.

7.9 ASSESSING VENDOR RELIABILITY

Each niche of the computer and communications industries has its own peculiarities, and vendors of telecommunications information-management systems are no exception. For example, you should treat the service bureaus of PBX manufacturers with extra caution when evaluating the mainframe option. Being hardware-oriented, their software products are usually limited to their own brand of PBX, which could limit your equipment choices when it comes time to expand your network, and, when software problems arise, hardware vendors can be very slow with problem resolution.

With any vendor, find out if telecom management is a major or a minor part of their business. If it is only a sideline, the pressing needs of customers may be put on the back burner until difficulties are resolved with other aspects of their business, and, if the vendor appears to be giving away too much to get your business, be suspicious. Getting a telecom information-management system at a bargain-basement price will do you no good, if the vendor goes out of business and leaves you without continuing support. Be aware that the telecom-management industry is extremely volatile; some of the best-known companies are only marginally profitable, while others have actually gone under and resurfaced as new entities with scaled-down product offerings and severely degraded levels of customer service.

Before committing yourself to any vendor, it is a good idea to insist on reviewing the product's technical documentation for scope and clarity — and to verify the claims of salespeople. If your company's MIS-dp shop has problems attracting and keeping qualified staff, the quality of technical documentation may prove to be critically important, if you choose to license mainframe software.

Consider the ability of the vendor to customize software to meet unique organizational requirements. Generally, vendors who sell only software are more receptive to customizing than those who sell complete off-the-shelf packages.

Because each network is configured differently and organizations have varying requirements, you can expect a certain degree of customization when buying telecommunications information-management systems. Therefore, it is very important to obtain from the vendor a list of references whose network configuration and management needs most resemble your own. The references should also use the same type of PBX and mainframe (if applicable) as you do. As recommended in Chapter 1, make sure your reference check includes inquiries about the timeliness with which customization was completed, the cooperativeness of the vendor in ironing out bugs in the program, and whether the final product matched the buyer's expectations. When dealing with privately held firms in this industry, an added measure of caution is called for, because it is these companies that are notorious for management, cash-flow, and product-development problems. Checking into the financial position and creditworthiness of such companies may obviate the need to go any further in the risk-avoidance strategy described in Chapter 1.

If you rely on a consultant for advice, show him no mercy — be alert to potential "under-the-table" relationships with vendors and insist on a thorough justification for all decisions. After all, if you have to make the recommendation to top management, the wrong decision will reflect poorly on you, because it is your ability to handle the consultant that gets called into question.

7.10 ADDING IT UP . . .

These are only a few of the many variables that go into the buying decision. When you are ready to sign on the dotted line, be aware that you do not have to accept the standard contract.

All provisions of the contract are negotiable, including price and terms. Some vendors need reminding that, as in any other business transaction, the seller sets the price and the buyer sets the terms. Just because the contract may be typeset does not mean that you cannot change its provisions or spell out terms in as much detail as you require. Do not be intimidated by vendor rationale that hides behind "company policy" — in the real world of buying and selling there is no such thing!

Also, do not forget to add a weasel clause to the purchase agreement. It is your protection against late delivery or product failure in that it entitles you to a full refund if the vendor cannot remedy such performance during a reasonable time frame. In addition, if you can possibly avoid it, do not put up more than ten percent of the purchase price for any customized software product before it is installed and operating. Software is the most troublesome portion of any product to develop. In recent years, many innovative products have been canceled because of the inability of vendors to come through with the software.

To guard against the loss of technical support, make sure that you are entitled to the program's source code in the event the vendor goes out of business or abandons the product. As noted in Chapter 1, the source code should be deliverable automatically from an escrow account or from a third party specializing in such services. These types of arrangements are relatively new, so it is best to seek the advice of an experienced attorney in setting up this arrangement with the vendor. Make certain that whenever the product is updated, the source code in escrow is also updated.

Be alert to provisions in the license or purchase contract that disclaim performance. This provision seems to be standard practice among software companies. It means that if the software does not work as promised on your computer, that is just tough! When negotiating terms and conditions, strike such clauses and insert new ones that give you the right to a full refund if the product does not perform. If the vendor balks at any of these notions, keep in mind that there are about 200 other vendors who offer telecommunications information-management systems. This is truly a buyer's market.

Whether you choose a micro or a mainframe telecom-management package, keep in mind that you are committing yourself to a long-term relationship in which you are depending on the vendor for such things as tariff updates, software enhancements, and customer service. You must use the risk-avoidance strategy to satisfy yourself that the vendor will be around to support you for many years to come.

These are only a few of the many factors to consider when evaluating telecom-management systems. The choices typically involve economic tradeoffs that are not always very apparent. Some mainframe software may cost hundreds of thousands of dollars to license for multinode network usage and entail organizational commitments to specific hardware and MIS-dp staffing levels. It is not easy to start over after discovering that a particular software package does not perform to expectations, or to regain lost credibility resulting from a poor purchasing decision.

With this in mind, you may want to consider the possibility of starting out with the microcomputer option with the idea of migrating to minicomputer or mainframe products as your organization's reporting needs become better defined or more complex. By that time you will have built up the knowledge base from which to make more informed decisions. If product migration is a possibility, you may want to choose a vendor who offers a migration path from micro to mainframe.

Chapter 8
ISDN Products and Services

ISDN has become one of the most talked about and eagerly awaited developments in the history of the communications industry. In fact, ISDN has been hyped so much by the media and industry analysts that it may never fulfill the expectations of potential users. At the same time, there seems to be an undercurrent of hostility toward those who have pursued ISDN development. Some critics claim that ISDN does not offer anything new that cannot be done with existing technology. Potential users claim that they have been left out of the ISDN planning process, and that the telephone companies and hardware vendors are hogging the show in the hope of lining their own pockets. Uncertain of the market demand for ISDN — and seeming not to care — service providers and vendors alike are rushing headlong into implementing the Integrated Services Digital Network.

8.1 ISDN IN PERSPECTIVE

At the risk of oversimplification, the ISDN concept boils down to the capability of handling voice and data simultaneously over the same transmission medium. Underlying the concept of ISDN is the promise of a worldwide public telecommunications network that will serve a variety of user needs. The ISDN architecture promises several advantages over conventional communications architectures:

- End-to-end digital connectivity, which will provide high-quality transmission with fewer errors;
- User-controlled access to multiple services over the same transmission path through out-of-band signaling;
- Flexible bandwidth allocation to improve transmit and response times, as well as to broaden the scope of possible applications;

- Standardized interfaces to facilitate terminal portability and multivendor interconnection.

A variety of services will be delivered through a standard interface, replacing the multitude of plugs, outlets, jacks, and connectors that seem to hide in every nook and cranny of today's offices and homes. With a standard interface, moving terminals between buildings or within a building will be accomplished merely by unplugging the unit at one location and plugging in the unit at the new location. The resulting economic benefits associated with equipment moves and changes are too compelling to ignore. Moving an analog phone from one floor of an office building to another could cost as much as $70, while the cost of moving a key-system attendant set can cost several hundred dollars because of the 25-pair cable it uses. Since companies typically move people around once or twice a year, such costs accumulate rapidly, but this level of device interconnection should be regarded as a long-term objective.

ISDN itself is not a service; it is an architecture which, when fully implemented, will help users access services in the broad categories of voice, data, text, and image. Although ISDN continues to receive the lion's share of attention in the trade media, the concept behind ISDN is not completely new. Bell Labs articulated the principles of what we now call ISDN about twenty-five years ago as a possible means of extracting more revenue from the public telephone network. The foundation for ISDN has been in place since the early 1960s, with the introduction of digital transmission and digital switching. It may even be argued that many of today's T1 networks already offer many of the characteristics typically associated with ISDN, including end-to-end digital transport, flexible bandwidth allocation, multiple services over the same transmission path, and protocol-conversion facilities for transparent access by users. In fact, much of the controversy about ISDN centers around the inability of potential users to distinguish it from private T1 networks. They do not understand why they should invest in costly interfaces, just to implement services under ISDN that they are already enjoying on their T1 networks. This has lead some communications managers to remark that "ISDN is too late."

Other companies are more forward-thinking in their approach to ISDN. McDonald's Corporation (Oak Brook, Ill.), for instance, participated in one of the first ISDN field trials, which occurred in 1987 and involved Illinois Bell. As a result of its participation in that field trial, McDonald's is now fully committed to implementing ISDN among all of its locations nationwide by the end of 1989. Dozens of large companies are buying their own digital switches to implement ISDN on their private networks. These efforts portend a growing trend among the Fortune 500 to develop IDSN for internal applications, while maintaining an eye toward full interconnection with the public network when ISDN becomes generally available. Such firms plan to have the competitive edge already secured by the time their less progressive competitors are even ready to think about evaluating

ISDN. Although large users will be among the first beneficiaries of ISDN, the network architecture will facilitate new and enhanced services for small businesses and, eventually, residential users.

ISDN alone does not represent the ultimate answer to all user communications needs. The extent to which ISDN may address user needs will rest primarily with users themselves, who will develop applications that utilize and leverage ISDN's inherent capabilities in combination with existing communications tools. This helps explain why the major telephone companies and interexchange carriers are rushing headlong into providing ISDN services and why equipment vendors are developing ISDN-ready products without too much concern for the market demand for such services and products. When asked to explain the key selling points of ISDN, vendors and service providers seem capable of providing only cryptic responses that include references to "flexible access to digital pipes," "new network-control mechanisms," and "distributed network intelligence." It is left up to the user to ponder the wisdom of these words and discover the true meaning behind such apparently gratuitous dispensations.

Ultimately, it will be the responsibility of users to see the potential of ISDN and figure out how best to take advantage of it. For the most part, ISDN service and equipment providers will rely on the publicity of Fortune 500 firms that implement ISDN as the means for demonstrating its possible applications and generating further demand.

Universally available ISDN is not imminent; the CCITT projects a 20-year transition from today's telephone networks to universal ISDN. The size of the capital investment required for replacing existing network facilities and plant suggests that ISDN will evolve well into the next century.

Network connectivity under ISDN will be facilitated by standardized transmission interfaces. ISDN is already in the implementation stage, offering both basic-rate and primary-rate services. These services offer different channel configurations. The 2B + D channel configuration is provided through the basic-rate interface, whereas the 23B + D channel configuration is provided through the primary-rate interface. (In Europe, the primary rate will be 30B + D; the T1 rate in Europe is 2.048 Mb/s *versus* 1.544 Mb/s in the United States.) The B channel in both configurations consist of 64-kb/s bearer channels. The D channel of the basic-rate interface provides 16 kb/s for signaling information, whereas the D channel of the primary-rate interface provides 64 kb/s for signaling. The D channel provides the means to tell the network when to connect or disconnect calls, as well as how to allocate the B channels in terms of bandwidth, facilities, and services. When the D channel is not needed, as in private network point-to-point applications, it may be used for other purposes.

Under the basic-rate interface, the 2B + D channel configuration provides a total bandwidth capacity of 144 kb/s, of which 128 kb/s is intended for the transport of user information, while the 16 kb/s available through the D channel is primarily intended for network signaling. When the 16 kb/s is not required for signaling, however, that capacity may be used for other applications such as telemetry, alarm signaling, and remote meter reading. With current technology, even voice may be carried at 16 kb/s. When 2B + D eventually becomes available to consumers, the possible applications for the D channel include home security, medical monitoring, utility meter reading, electronic banking, and videotext.

Under the primary-rate interface, the 23B + D channel configuration provides a total bandwidth of 1.544 Mb/s, which corresponds to the twenty-four 64-kb/s channels that are provided under T1. The 23 bearer channels will provide for the transport of user information, while the D channel will provide signaling information. Among the applications being considered for primary-rate access are calling-number identification, least-cost routing, high-speed digital facsimile, and computer-to-computer transmission over circuit-switched and packet-switched systems. Although microcomputers may house the actual interface module, 23B + D will be implemented in conjunction with digital PBXs with T1 interfaces.

Teleos Communications Inc. (Redondo Beach, Calif.), for example, offers an array of ISDN products that includes a microcomputer plug-in board which provides the basic interface for 2B + D ISDN. The company also offers a menu-driven ISDN basic-access development system, which helps users develop ISDN application programs by supporting 2B + D signaling software. Another product, the T100U, sits on the trunk side of a PBX to provide T1 connectivity, which may be used as a platform for the 23B + D primary-rate interface.

The primary-rate interface is intended for implementation over private networks. It should come as no surprise, then, that interexchange carriers are pushing the primary-rate interface, whereas the major telephone companies are favoring the basic-rate interface. The first PBX to implement the primary-rate interface was introduced in 1988 by AT&T. Because AT&T's System 85 ANS (Release 2, Version 4) supports flexible channel allocation, voice-data channels, and trunks, users can directly access such AT&T services as Megacom, Software Defined Network (SDN), and Accunet packet-switching data service on a per-call basis, eliminating the need for more costly dedicated lines for each service. These services are offered from an AT&T service node (point-of-presence), either by direct access or *via* leased T1 pipe from a local telephone company. Users of AT&T's System 75 and System 85 may upgrade their PBXs for ISDN by integrating a Universal Module (UM) into their existing system, thus retaining 80 to 90 percent of their previous investment in hardware.

The primary-rate interface allows the user — not the carrier — to allocate network resources on a demand basis. This degree of flexibility enhances the service provider's offerings, while giving the user the ability to address specific needs. For example, the telemarketing department may require the use of thirty 64-kb/s voice paths into the network — one each for its thirty telemarketing representatives. The engineering department may require the use of a 384-kb/s line for a CAD application, while the accounting department may need two 56-kb/s DDS lines for data-base transfers between hosts. The next day, only twenty-five 64-kb/s voice lines may be required by the telemarketing department, while engineering may need two 384-kb/s lines, accounting one 56-kb/s DDS line, and the training group one 768-kb/s line for a video transmission. Organizational needs are ever-changing. The ISDN primary rate gives users the means to access the network in appropriate bandwidth increments and to be charged only for what they use.

In addition to providing out-of-band signaling, the 64-kb/s D channel has other possible applications, including providing access to public packet-switched networks like Telenet and Tymnet. It may also be used to implement electronic mail, energy management, FAX, wide-area networking for microcomputers, and electronic data interchange (EDI). In addition to using the D channel in the packet mode, you can use the B channel for dial-up access to a packet network.

For broadband applications, users will eventually be able to select from among several "H" channels, which provide nonswitched circuits operating at 384 kb/s, 1.536 Mb/s, 1.920 Mb/s, and 135 Mb/s. The current trend, however, is toward SONET (Synchronous Optical Network) for fast-packet formats which will support switched services. The possible applications for these channels include point-to-point high-speed FAX, teleconferencing, high-speed data, and high-performance LAN interconnection.

ISDN promises to yield improvements in transmission quality and to enhance the performance of many existing services through end-to-end digital connectivity (including "last mile" local loop connections) and higher transmission speeds. At the same time, ISDN is expected to result in the introduction of new services, utilizing the advanced signaling capabilities of Signaling System No. 7 (SS#7). Cost savings will accrue through greater efficiency and through integration of services over fewer facilities. The standardization of connection interfaces will provide additional cost savings from terminal portability.

All of this might not sound too impressive until you look at some of the possible applications of ISDN. Companies that are highly dependent upon their customer-service operations stand to benefit most by the initial implementation of ISDN. Upon receipt of an incoming call, the host associates the caller's phone number with a specific account number and file in the data base. At about the same time the account representative picks up the phone, the customer's file is appearing on the terminal screen. In a matter of seconds, the account representative is prepared to speak intelligently about the customer's account and to respond immediately to any customer request. There is no need to put the call on hold, shuffle through file drawers, or call the customer back. The resulting efficiencies reduce overhead at no sacrifice in service quality. Thus, fewer account representatives are needed, while more customers may be served. Brokerage houses, financial institutions, utilities, and insurance companies are among the many service organizations that may benefit from this ISDN capability. The viability of this concept assumes, of course, that the caller is using his or her own phone.

ISDN services may also benefit engineering-intensive companies. For example, two engineers working on a new product design may view the same data on their terminals, while discussing possible revisions. With voice and data traveling simultaneously over the same line, the engineers can make changes, file and retrieve information, and transfer bulk quantities of supporting documents to each other in a matter of seconds. Schematics, parts lists, contracts, and spreadsheets may be accessed *via* windows for simultaneous viewing. This arrangement permits more efficient communication, increasing the performance of the engineers to exponential levels. This results in quality work, under budget and under deadline. Manufacturers, government and defense contractors, and large custom-design firms are among the many organizations that can benefit from this ISDN capability.

In the area of emergency services, ISDN may actually save lives. Let us say that a call comes into the fire department from a distraught caller who is barely intelligible to the operator on duty. The calling number is compared to a data base that includes the unique attributes of commercial and residential structures, potentially hazardous conditions (explosives, chemicals, and radioactive materials), as well as specific information about the occupants (disabled, children, the elderly). Along with the address of the calling party, the terminal displays the nearest police station, hospital, emergency medical service, and poison-control center, along with a map that can zero in to any level of detail showing the nearest locations of such things as power lines, gas and water mains, fire hydrants, and buried cables. With the incoming call, the local police and hospital are alerted automatically. This capability allows fire fighters and rescue teams to arrive at the scene mentally prepared and appropriately equipped to deal with any emergency, but even before such services are dispatched, the emergency situation may be verified through the capture of the calling number, thus eliminating the possibility of a false alarm.

Currently, such capabilities are achieved through an expensive hodgepodge configuration of equipment that includes modems, selective routing systems, automatic number identification (ANI) systems, automatic location information (ALI) systems, ALI record generators, and KTS-PBX systems with call-transfer capabilities. Some of these components are located on telephone-company premises, while others are located at a public service answering point (PSAP) as well as at police and fire departments. ISDN eliminates much of this equipment, reducing the risk of system failure, while permitting the transfer of large amounts of information to the PSAP operator and faster call verification. Municipalities, emergency service providers, and entire communities stand to benefit most from this ISDN capability. Of course, ISDN will not solve the jurisdictional squabbling that currently stalls the dispatch of emergency services under current 911 and E911 arrangements; further proof, as if any more were needed, that technology is only as competent as the people who use it.

In nationwide retail chains, ISDN holds out the promise of simplifying vast

networks, making them easier to manage, while cutting costs by eliminating redundancy. It is quite common for large retail networks to have as many as three
distinct network layers, starting with separate data and voice networks at hundreds
of locations that feed into the second layer of wide-area networks, which then
cross over vertically and horizontally to the third layer of networks that tie together
various accounting centers as well as regional and district offices. In being able to
consolidate the local-area and wide-area networks, ISDN provides an information
outlet to every integrated voice-data terminal. In addition to providing each terminal with such services as electronic mail and host access through standardized
ports and interfaces, tremendous cost savings would accrue to the organization
through the elimination of local coaxial cable.

When it comes to residential ISDN service, the scenario is no less compelling
than the one for business applications. Let us say that you are working overtime
at the office. On an unseasonably cold day, you might want to access your remote
home-monitoring service *via* your office terminal to get a reading on the internal
temperature — just to be sure there is no danger of water pipes freezing. Since
you live in the city, you want to verify that security alarms are activated and all
appliances turned off, except for the front entrance light. While you are at it, you
decide to check phone messages that may have been recorded while you were out,
and then change the greeting to let callers know that you can be reached at the
office. Your real reason for calling home, though, is to retrieve some spreadsheets
from your microcomputer, which you will have to incorporate into the report you
are doing at the office.

When you finally get home, you decide to relax a bit by catching up on the
news. You access UPI *via* the videotext service, and do a major headline scan.
Before you can zero in on anything specific, the phone rings. As you are speaking
to your mother, you stop the headline scan to zero in on a story that catches your
eye — "IRS Prepares for Last Minute Filing Blitz." As the story unfolds on your
screen, a cold shiver runs down your spine. Fortunately, you already filled out the
tax form with the software package you purchased last week, but the filing deadline
is just hours away. Without interrupting your mother, who cannot understand why
you do not call or write more often, you exit the videotext program and access
"communication" from the main menu. A few moments later, your tax return is
off to the IRS regional office, where it is date-stamped and printed out in the IRS-
approved format.

You assure your mother that you will stop by this weekend for a visit, perhaps
a barbecue. After the call, you are feeling good about yourself. You marvel at all
you have accomplished in just a few minutes, and over the same pair of wires,
and, as the headline scan continues, another item catches your eye — "Stock
Market Plunges 300 Points." The phone rings. The terminal displays the calling
number as being that of your broker. A cold shiver runs down your spine . . .

The marketability of ISDN will hinge on the ability of hardware and service providers to demonstrate its value in meeting the needs of specific types of businesses. Meanwhile, communications managers cannot afford to dismiss ISDN entirely, just because they happen to have huge investments in hardware and facilities that offer similar functionality. ISDN offers some very tangible benefits that cannot be ignored, including:

- Elimination of delays in cutting over new lines and services;
- Faster call setup and network response times;
- Better network management and more network control;
- The ability to streamline networks through integration, thus decreasing the complexity and the cost of cabling and equipment.

At the same time, there are a few current problems that ISDN will not solve. It will not improve the attitudes of poor account representatives, for instance, who lack an appreciation of customer needs, or who prefer to sell individual products rather than long-term solutions. ISDN will not improve the critical shortage of qualified technicians, or improve the performance of those who have managed to escape the numerous rounds of budget cutting in recent years. Customers will have to continue hand-holding telco field technicians by finding the problems themselves. Finally, ISDN will not end vendor finger-pointing.

8.2 ARCHITECTURAL ELEMENTS

Although the primary-rate and basic-rate ISDN services consist of different configurations of communications channels, both services require the use of distinct functional elements to provide network connectivity (Table 8.1).

One of these elements is the Network Terminator (NT1), which resides on the customer premise and performs the conversion from 4-wire to 2-wire required by the local loop. This intelligent device terminates the transmission line from the central office and provides functions equivalent to layer one (physical) of the Open Systems Interconnection (OSI) reference model, including line termination, line maintenance and performance monitoring, timing, and multiplexing. NT2 includes all of those functions plus those of OSI layer 2 (also known as Link Access Procedure D, or LAPD) and layer 3, which provide protocol handling and multiplexing, switching, and concentration. NT1 and NT2 may be packaged as a single device, similar to the way CSUs and DSUs are now packaged as a single unit. The NT1-NT2 may be incorporated into PBXs, microcomputers, and key-system attendant consoles.

Terminal equipment (TE) includes functions equivalent to OSI layer 1 and higher. TE1 includes such devices as digital telephones, data-terminal equipment, and integrated workstations, all of which comply with the ISDN user-network interface. TE1 functions include protocol handling, maintenance, and interfacing. The large installed base of non-ISDN devices (TE2), like telephones and micro-

Table 8.1 ISDN at a Glance.

Channel Type	Bit Rate (kb/s)	Switched Network	Primary Applications
A	56	Circuit	• Voice communication (in use today)
B	64	Circuit	• Digitized voice • High-speed data • Digitized FAX
C	8/16	Circuit	• Interactive data transfer • Messaging (Used with A channel during transition to ISDN)
D	16	Packet	• Signaling information for ISDN circuit switching • Low-speed user data • Telemetry
D	64	Packet	• Signaling information for ISDN circuit switching (for PBX interface)
E	64	Packet	• Signaling information for ISDN circuit switching (for use with CCITT Signaling System 7)
H0	384	Circuit	• Fast FAX • Teleconferencing • High-speed data
H11	1,536	Circuit	• Same as H0
H12	1,920	Circuit	• Same as HO
H4	135 Mb/s	Circuit; fast packet	• Connect high-performance LANs • Full-motion video • High-resolution image • Very high-speed data transmission

computers, are made ISDN-compatible through the use of a Terminal Adapter (TA), which takes the place of a modem and includes functions equivalent to OSI layer 1 and higher. A maximum of eight TEs-TAs may be connected to the same NT2 in a multidrop configuration. Hayes Microcomputer Products Inc. (Norcross, Ga.) developed the first plug-in terminal adapter for microcomputers, which combines voice and data for transmission over digital lines.

The architectural elements of ISDN include several reference points that define network demarcations between telco property and customer-premises equipment, as well as the interfaces between various types of CPE. These reference points include:

- R — The reference point separating non-ISDN (TE2) equipment and the Terminal Adapter (TA), which provides TE2 with ISDN compatibility;
- S — The reference point separating terminals (TE1 or TA) from the network terminal (NT2);
- T — The reference point separating NT2 from NT1 (not required if NT2 and NT1 functionality is provided by the same device);
- U — The reference point separating the subscriber's portion of the network (NT1) from the carrier's portion of the network (LT).

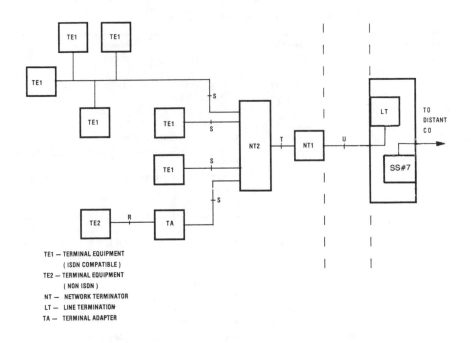

Figure 8.1 Architectural elements of ISDN.

The two B channels and the D channel are multiplexed for transmission using a special bipolar coding scheme. Forty-eight bits are inserted to the multiplexed B and D channels to support multiplexing and synchronization. These signaling bits are arranged in a frame, which is transmitted between the NTs and TEs. A frame contains 32 bits of B-channel data, 4 bits of D-channel data, and 12 bits of signaling data. Some of these signaling bits deliberately introduce bipolar violations, which are used for synchronization. When added to the 144 kb/s already available to users, the 48 signaling bits bring the total data rate to 192 kb/s.

The NTs provide network-control and management functions, while the TEs implement user functions. NTs may interface with multiple TEs. This arrangement is asymmetrical in that the structure of the frames used to enable communication from an NT and a TE is different from the frame structure that enables communication from a TE to an NT. This is due to the different responsibilities of the NT and TE, especially when a communication takes place over a passive bus structure.

In today's telephone network, lines and equipment are in use as soon as a call is initiated. This means that network resources are being tied up, whether or not the call is completed. If the call cannot be completed, the service provider's equipment has been used, but no revenue generated, since uncompleted calls are not billed. In the ISDN environment, calls that cannot be completed are terminated at the source. SS#7 and signal transfer points (STPs) along the network combine to make this possible. The STP can reserve switch resources, which will not actually be used until SS#7 declares end-to-end circuit readiness. Since network resources will not be allocated until the call is connected, service provider costs are reduced and network congestion alleviated. This highlights the importance of signaling. ISDN is totally dependent on signaling to provide digital end-to-end connectivity. In the all-digital network of integrated services, TEs will not be able to process receive messages without the associated information provided by signaling. Since NTs and TEs may differ from network to network, signaling information will be the key to decoding the transmission correctly. It is the signaling protocols that will inform the TE whether an incoming call is digitized voice, binary data, or image, thus allowing it to set itself up to process the information properly.

Signaling is also the key to accessing the multitude of network services. Out-of-band signaling will allow users to choose the appropriate service on a call-by-call basis, and even to access multiple services simultaneously. Signaling services on the D channel may be used in combination with voice, data, and image services. With the basic-rate service, for example, if the two B channels are in use, the D channel may be used to notify the user of an incoming call. Receiving an incoming call number on the D channel and displaying it will inform the user of the call's origin and, if programmed, the identity of the caller. This kind of signaling will also improve the efficiency of 800 services and credit-card calls, and even extend the geographical range of call-back services. In fact, ISDN will bring the advantages of speed, economy, and user control to most communications services.

Today, when a user wants to access the most economical long-distance service, the links to each carrier are either directly connected to the subscriber's equipment or require special access codes for dial-up connections. Under ISDN, the user will be able to access the service of any carrier from the local central office.

The protocol running on the D channel controls the usage of the B channels and has been defined to allow a number of terminals, connected in a multipoint configuration, to gain access to the D channel in an orderly manner. The protocol insures that even in cases where two or more TEs attempt to access the D channel simultaneously, only one TE will succeed in completing the transmission.

The NT transmits the most recently received D-channel bit on the echo bit (E bit). Each TE monitors the echo bit and stops transmitting if the bit received on the echo bit is different from the sent bit. Once transmission is stopped, the TE waits for its turn to transmit again.

To accommodate sophisticated applications, TEs may be assigned priorities: Class A and Class B, for example. Associated with each class is a priority variable. For a TE, D-channel access is determined by this variable, which is indicated by a string of ones. The default priority value of 8 may indicate Class A, while the default priority value of 10 may indicate Class B.

A TE can only transmit after it detects as many continuous ones as its priority variable. When the ones count on the echo bit is equal to the priority variable, the TE begins transmission, at the same time checking for collisions with transmissions from other TEs by checking to see if the received E bit is different from the last bit sent. If there is a collision, the TE stops transmitting and waits for another chance. Once a TE successfully transmits a frame, it increments its priority variable by one. For example, the default value of 8 is incremented to 9 for Class A, which effectively reduces its priority within the same class. When there are 11 continuous ones on the E bit from the NT, all TEs restore the default value. This mechanism insures that Class A TEs will always be able to transmit before Class B TEs. A TE that has just sent a frame will be able to send another one only after all the TEs in its class have had a chance to send a frame. This arrangement is much like that of the token-ring LAN, in which each terminal has a chance to transmit a packet of data upon receiving the token. Additional packets are sent only after the other terminals in the same class on the token-ring have had a chance to transmit packets of data.

As more telephone companies upgrade their digital central offices with ISDN capabilities, ISDN islands will dot the landscape. The full potential of ISDN will not be realized until these islands are linked together to form an intelligent backbone network between telephone company switches. The means to do this are already available through signaling system number seven, or SS#7. SS#7 is basically a point-to-point, bidirectional, application-independent transport medium. The concept originated in the late 1970s, apart from the developmental work leading up to ISDN. Its original design focused on the efficient transport of circuit-switching control messages between central offices. Its primary application would have been to permit telephone companies to take on a higher level of control, particularly with In-WATS (800) and credit-card validation services. Another application of SS#7 is in digital cellular networks. Tracking the movements of subscribers from cell to cell to provide full roaming capabilities requires the kind of high-speed data communications that SS#7 provides. It so happens that SS#7 can also be used to support the end-to-end digital connectivity promised by ISDN and provide such important capabilities as incoming-call identification. Such features will not be limited to ISDN subscribers, because SS#7 will allow delivery of calling-number information from analog voice subscribers as well.

AT&T and many of the major local telephone companies are implementing SS#7, either by adding adjuncts to existing switches, upgrading them with appropriate hardware and software, or through the wholesale replacement of old switches with new ones that incorporate SS#7. MCI and U.S. Sprint are among the alternative carriers that have already started deploying SS#7 on their networks. The signaling scheme immediately halves call setup time, thereby improving the overall efficiency of the network. SS#7 also makes toll fraud extremely difficult, which directly benefits the carrier and indirectly benefits users, because it is they who ultimately pay for misuse of the carrier's network.

At this writing, it is anticipated that AT&T may decide to restrict SS#7 to backbone use only, depriving even its own primary-rate users of these sophisticated capabilities. This strategy is intended to thwart the proliferation of hybrid blends of public and private networks built on the primary-rate interface. This means that a user who wants to allocate the unused bandwidth capacity of the D channel for low-speed packet switching will be prevented from doing so. The user can, however, allocate bandwidth on a B channel for packet-switched data using the DACS facilities at an AT&T service node. It remains to be seen whether AT&T will be allowed to get away with this approach. If user pressure does not force AT&T to open SS#7 for user access, market pressure certainly may. MCI and U.S. Sprint may view this as an opportunity to differentiate themselves from AT&T. From a hardware point of view, Northern Telecom's "SuperNode" architecture may afford users the opportunity to build hybrid networks using the SS#7 as the foundation.

8.3 ISDN IMPLEMENTATIONS

Until ISDN islands can be linked together under SS#7 to fulfill the promise of intelligent networking on a nationwide scale, most primary-rate users will have to be content with implementing ISDN over their private networks. Interconnectivity with the public switched network is not a prerequisite for realizing the many benefits of ISDN. Several PBXs and key systems are now available that provide integral primary rate, while many others can be upgraded with the appropriate hardware and software. While some companies have already implemented basic-rate ISDN as a pilot technology in conjunction with local telephone company offerings, other companies with extensive T1 backbone networks are finding themselves well positioned to benefit from ISDN without waiting for the public network to be upgraded. The first private network applications of primary-rate ISDN were cutover in mid-1988.

Northeast Utilities, for example, linked a 2,000-line Northern Telecom SL-1XN PBX in Rocky Hill, Connecticut, to a 1,000-line SL-1N located ten miles away in Meriden, Connecticut. In implementing primary-rate ISDN, the company reported call setup times of 300 to 400 milliseconds, compared with five seconds for conventional touch-tone dialing and seven seconds for rotary-pulse dialing. Calling-number identification is available for in-house calls originating from the two PBXs, but not for calls originating from customers. That capability is contingent upon the local telephone company's support of primary-rate ISDN. When that day comes, the utility plans to use calling-number identification to have account records retrieved from its data base so that its representatives can serve customers more efficiently. The company also plans to tie in voice mail with calling-number identification to handle routine calls more efficiently. In the event of a local power outage, for example, callers experiencing problems will be greeted with a voice-mail message informing them that work crews have already been dispatched to the scene and that power will be restored shortly.

Houston-based Tenneco Inc. started cutting over its 3,500 lines to ISDN in June 1988. It had previously used the CENTREX service of Southwestern Bell Telephone for voice, as well as for data transfers among several hundred micro-computers, connections to outside data bases, and asynchronous communications with mainframes. Tenneco plans to use ISDN as the backbone of its new corporate information system, rather than install LANs. In doing so, the company will minimize its need for front-end processors for terminal-to-host connectivity. The company claimed that using ISDN at its six Houston locations was less expensive than staying with the CENTREX system, although the company acknowledged that voice-only applications would have made ISDN prohibitively expensive. This appears to contradict the assertions of industry analysts, who earlier in the year predicted that price would not be a motivating factor in moving to ISDN, since it would cost from 1.5 to 2 times the cost of CENTREX. Tenneco, however, acknowledged in press reports that it did negotiate a special deal with Southwestern Bell for a discount off the tariffed ISDN rates, proving once again that everything is negotiable.

Southern Bell initiated the nation's first multiple-customer rollout of ISDN in April 1988. Its first customers were Hayes Microcomputer Products, Inc., SunTrust Service Corporation, Digital Equipment Corporation, Prime Computer Inc., Contel Corporation, and AT&T Network Systems. Some of the applications for the 2B + D service include voice communications, message services, file transfers among microcomputers, laser printer and data-base sharing, packet-data switching, and calling-number identification. Contel is using some of its 35 basic-rate lines to support local-area networking among its three Atlanta offices, while Prime Computer is using 25 basic-rate lines to support software development, office automation, and sales operations.

By mid-1988, ISDN had clearly cast off its image as an experimental technology to emerge as a potentially viable networking strategy. By that time, several major hurdles to ISDN had been cleared, or were well on their way to being cleared. One of the hurdles that had been cleared was the development of ISDN terminal interfaces in the form of new PBXs as well as add-in ISDN upgrades. AT&T has gone a step further in introducing a line of basic-rate interface terminals under what it calls the 7500 series. The intelligent terminals are capable of initiating such calling features as conference call, call transfer, call hold, and call drop. The features are then processed by system software resident in the System 75 or System 85. Although all ISDN terminal standards have not yet been formalized at this writing, AT&T claims that any changes may be taken care of by software. A number of other equipment vendors have introduced ISDN-ready devices that can be software-modified later, if terminal standards change.

Another hurdle that had been cleared was the development of applications for ISDN. For a long time, ISDN had been plagued by a lack of applications to justify its implementation. Early users, however, saw no end to the possible applications of ISDN. Many new applications will emerge as more carriers cutover ISDN and SS#7 becomes more widely available.

Pricing had been another hurdle inhibiting full acceptance of ISDN. Users wanted to know whether ISDN could be obtained at a reasonable cost to support current applications and, more importantly, to see how ISDN stacked up against current service offerings. By mid-1988, AT&T and most of the regional telephone companies had filed ISDN tariffs. Illinois Bell had even packaged a low-cost ISDN "starter kit" to assist its customers who wanted to experiment with the service. The last hurdle, deployment of SS#7, is well on its way to being cleared. The major interexchange carriers and telephone companies have already either begun to upgrade their switching equipment to provide SS#7 on their networks or have announced their intention to do so.

8.4 PLANNING CONSIDERATIONS

Despite the promises associated with ISDN and its implementation progress to date, there is still quite a lot of risk to contend with, particularly when choosing equipment vendors. In addition to following the vendor-evaluation strategy described in Chapter 1, you must exercise extra care in selecting vendors of ISDN equipment such as integrated workstations, terminal adapters, network terminators, and add-in boards that provide microcomputers with ISDN functions. In addition, you must learn to evaluate PBXs, key systems, CENTREX, carrier services, multiplexers, and other network components in a new light — that of providing a smooth migration path to ISDN. Your attention to such matters will help you position your organization to take future advantage of ISDN at the lowest cost.

The mark of a successful ISDN-equipment supplier is its expertise in ISDN, knowledge of standards, commitment to hardware and software development, and a marketing strategy that parallels current and planned carrier developments in ISDN connectivity. The problem for potential equipment buyers is how to establish all this with reasonable accuracy. After all, many of the products that are required for ISDN implementation did not exist a year ago, and because ISDN is only in its earliest stages of implementation, you have the added burden of trying to distinguish between those vendors who are fully committed to ISDN over the long-haul and those who are entering the market just to cash in on the latest trend.

The nature of ISDN requires that equipment vendors be equally adept at meeting the requirements of voice and data. This can be ascertained by reviewing the types of product the company has already developed. If the company is weighted heavily toward either telephony or data communications, you should consider looking elsewhere for your equipment. If the company has developed products for both voice and data applications, you can determine which side the company favors by comparing the sales revenues for each product type.

Look into the networking expertise of the company's R&D people, specifically their experience with a variety of architectures and protocols. You can determine whether key staff members have firsthand knowledge of ISDN by their participation in standards organizations, ISDN field trials, and industry seminars. You can also determine the competence of key staff members by reviewing any books, technical papers, or articles they may have published — by seeing if any of these demonstrate leadership and innovative thinking.

It is not enough to evaluate the hardware that will be used for ISDN applications. After all, PBXs and microcomputers might have excellent track records of reliability in the non-ISDN environment. What really counts is the software. Such products require extensive software development and continual upgrades to make new and existing hardware ISDN-compatible. Rather than focus all of your attention on hardware "futures," you should give more attention to matters of software. Find out what programming language the vendor uses and why that language was selected. Also, find out how easy it is for the vendor to add features and functionality to the software. Find out if upgrades will coincide with changes in standards, how the upgrade will be implemented, and at what cost. Compare the vendor's R&D budgets for hardware and for software; many vendor claims may be validated, if it can be demonstrated that more money goes into software development than goes into hardware development.

Do not be surprised to find as many ways of implementing ISDN as there are vendors. After all, with the long lead time required for software development, vendors had to make some early assumptions about the future direction of standards. This has resulted in slightly different ways of implementing ISDN. You can minimize your concerns by asking vendors for their rationale in doing what they did. In your quest for validation, bounce one vendor's opinions off another vendor. In comparing the answers of several vendors, you can usually arrive at an approach that makes more sense than the others, and dismiss from consideration a vendor who appears to be too far afield. You may even discover two or more approaches that mirror each other, in which case you have narrowed your choices considerably. Since standards committees are dominated by vendors, those who participate on such committees will be more apt to guess right about final standards than those who have not participated. Be aware, however, that if the vendor guesses wrong and you have purchased its products before the standard is set, you have just locked yourself into a proprietary solution and premature product obsolescence.

If ISDN looks like it may be a possibility for your organization sometime in the future, you should determine the vendor's migration plans for products you buy today. This is especially important with high-end purchases of multiplexers, PBXs, and front-end processors that have a useful life of from five to seven years. You should do some detailed probing of the vendor's migration plans, even if you have been dealing with that vendor for many years. If the vendor was the recent target of an acquisition, or recently merged with another firm, make sure that its plans to upgrade newly purchased products have not been sidetracked. The frequency of mergers and acquisitions in the computer and communications industries, and their uncertain outcomes, provides ample justification for seeking written assurances of future compatibility with ISDN before you buy. This is important, because many larger firms have signaled that they will not become too much involved with developing ISDN products until the market for ISDN-derived services begins to assert itself more strongly.

On the other hand, if ISDN is only a very remote possibility for your organization and a new T1 multiplexer will pay for itself within 12 to 18 months, giving too much consideration to future ISDN compatibility may not be very practical. Every situation will be different, of course, but such considerations must be factored into your decision-making to insure risk-free purchases of network products.

If your organization already subscribes to digital CENTREX, you may already have low-cost entrée to ISDN, in which case you might want to consider the advantages of experimenting now with 2B + D service. Telephone companies are very flexible right now in terms of pricing, but that may change as the demand increases for ISDN-provided services. If ISDN is available in your area, now is the time to get some hands-on experience. This will give you the opportunity to compile a knowledge base that can smooth the way toward corporate-wide implementation of ISDN. It will give you the chance to find out what applications are best suited for ISDN, and give company managers and technicians a chance to play "systems engineer" in the development of new applications.

If ISDN is not yet available in your area, you can still stay on top of developments by visiting other companies that are using ISDN. Beyond that, some regional telephone companies have laboratories that simulate business applications using ISDN. Although such labs allow hands-on work, they may not be able to answer all of your system-design questions, however. If none of these avenues for education are open to you, at least attend some of the national conferences that discuss ISDN issues. You can also keep up with the progress of ISDN through the trade press. In reading about prototype ISDN offerings, try to analyze them in terms of your organization's current and projected needs. Start a clipping file of all relevant material, categorized for easy reference. When the time comes to propose ISDN formally to top management, you will have all the basic information at hand with which to develop a background paper and a rudimentary action plan. Accumulating a knowledge base about ISDN is important, because it allows consideration of ISDN when your organization is about to make key decisions, such as where to set up or relocate corporate offices. In such cases, site selection should be determined on its proximity to a digital central office, or to a central office that will soon be upgraded to digital. Of course, if you have been experimenting with ISDN or a transitional offering, your hands-on experience will add credibility to any proposal you make.

Through the use of T1 networks that are already in place, or are being put into place, large users may put together combinations of products and services that provide ISDN-like benefits, such as those that may be derived from DACS-CCR. Even small organizations can experience the look and feel of ISDN through the use of newer services like "Fractional T1" (FT1).

Under the FT1 scheme, existing T1 bandwidth is merely repacked so that customers pay only for what they use. If you require only 192 kb/s of bandwidth to support ten 19.2-kb/s links, for example, you can order only one-eighth of the bandwidth available with T1. If you need only 384 kb/s to provide a backup link for critical facilities, you can order only one-quarter of the bandwidth available with T1, and if you need only 768 kb/s to implement video conferencing, you can order only one-half of the bandwidth available with T1. The carrier can then parcel out unassigned bandwidth to other customers in increments that best meet their needs.

Using AT&T's Accunet T1.5 Service and Dataphone Digital Service (DDS) as the baselines for comparison, the cost savings that can accrue with FT1 become readily apparent. Currently, a 712-mile Accunet T1.5 circuit from New York to Chicago costs about $16,000 a month, including local-loop charges at both ends. Although many businesses have enough traffic between those cities to cost-justify a full T1 pipe, there are many more businesses with such requirements that they may never be able to cost-justify that much bandwidth. Nevertheless, they appreciate the transmission quality that such circuits provide. In the absence of FT1 services, the options available to both small users and the low-density nodes of larger users are not very appealing.

Let us say that a brokerage house has the requirement for forty 9.6-kb/s circuits between its Chicago and New York offices. For the sake of simplicity, let us also assume that the multiplexers at each end are operating at near 100% efficiency. One-quarter FT1 would fit this need more economically than either Accunet T1.5 or multiple DDS circuits, as Table 8.2 illustrates.

As illustrated in Table 8.3, substantial economic benefits accrue to both the customer and the carrier, even when the circuits span great distances — New York to San Francisco, for example. For purposes of discussion, the price for one-eighth FT1 is calculated at 15% of the Accunet T1.5 circuit, whereas one-quarter FT1 is calculated at 30% of the Accunet T1.5 circuit, including local-loop charges at each end. For one-half FT1, the carrier might charge customers only 60% of the current T1 price. Actual pricing is dependent upon regulatory acceptance of cost-based tariffs, and the discounts you can negotiate with individual carriers.

Table 8.2 Circuit Cost Comparison: New York to Chicago.

One circuit: Accunet T1.5 (1.544 Mb/s)	Eight circuits: DDS (56 kb/s)	Forty circuits: Voice-grade private line	One-quarter: FT1 (384 kb/s)
$16,000 per month	$29,864 per month	$39,950 per month	$6,925 per month

Table 8.3 Circuit Cost Comparison: New York to San Francisco.

One circuit: Accunet T1.5 (1.544 Mb/s)	Eight circuits: DDS (56 kb/s)	Forty circuits: Voice-grade private line	One-quarter: FT1 (384 kb/s)
$38,700 per month	$52,604 per month	$61,618 per month	$13,678 per month

Such economies and efficiencies are not limited to data; they apply to voice as well. Let us say that the same customer has the requirement for voice transmission between New York and Chicago. He can order an Accunet T1.5 circuit between the two cities, which will provide enough channel capacity to handle 24 simultaneous conversations under PCM. Instead of paying $16,000 for an Accunet T1.5 circuit, the customer can order one-quarter FT1 service and implement Adaptive Differential Pulse Code Modulation (ADPCM) to obtain 10 toll-quality voice channels, or implement 16-kb/s ADPCM to obtain 20 voice channels of acceptable quality. This arrangement results in a $9,075-per-month savings to the customer on line charges. Assuming that an ADPCM-equipped multiplexer cost about $18,000, the payback period for the necessary hardware would be only two months.

Several vendors now offer both 16-kb/s and 32-kb/s ADPCM with their T1 multiplexers *via* add-in boards. Depending on the vendor, you may even achieve optimal control over bandwidth reconfiguration through software-selectable optioning.

The combination of Fractional T1 and voice compression would yield many other benefits to users. For example, the time frame between planning and implementing network expansion would be reduced considerably, since FT1 service would cost-justify itself a lot sooner than a full T1 pipe. Fractional T1 services would also allow telecom and MIS managers to implement the services they need now, without either one having to consider the needs of the other to achieve optimal utilization of bandwidth for greater cost savings. Fractional T1 can give you the flexibility required to deal effectively with unanticipated growth brought about by mergers, acquisitions, and joint-venture activities that have become so much a part of doing business today. Once access lines are installed, adding incrementally to the existing bandwidth is simply a routine administrative matter.

Fractional T1 at 384 kb/s would provide a logical migration path to ISDN, since this is the bandwidth requirement for the H0 channel, which will provide nonswitched circuits. With the FT1 circuit for 384 kb/s already in place, all that would be needed is the digital interface to derive the full functionality of the H0 channel when connected to the central-office switch offering ISDN services. At this writing, however, interface standards for the H0 channel have not been finalized.

If 384 kb/s is too much to experiment with, try buying a DS0. A few innovative carriers like Cable & Wireless Communications Inc. (Vienna, Vir.) offer Fractional T1 service. In addition to offering individual DS0 channels, C&W offers multiple DS0s which may be bundled to provide 128, 256, and 384 kb/s as part of its Intelli-Flex service. Williams Telecommunications Corporation (Tulsa, Okla.) also provides Fractional T1. Both companies route DS0 bundles over their own DACS networks. Interestingly, in a 1987 survey of its customers, AT&T found that 64-kb/s and 384-kb/s switched services are among those that users want most. AT&T has already announced its intention to offer 64-kb/s switched service, and may offer 384-kb/s service by early 1989.

The concept of FT1 is entirely workable within the framework of today's network architecture. After all, repacking or "grooming" T1 bandwidth is not an entirely new concept; AT&T's Digital Access and Cross-Connect System (DACS), for example, has been available since 1981 with the capability of subdividing bandwidth and routing the traffic to the appropriate destinations. Unlike ISDN, no major investments in the network are required to support Fractional T1 service. In fact, the present architecture facilitates the offering of FT1.

Fractional T1 has the "look and feel" of ISDN in the following ways:

- It allows flexible bandwidth allocation in that DS0s may be bundled according to application, with the user paying only for what is ordered through the Customer Controlled Reconfiguration (CCR) capability of the DACS.
- Voice and data may be carried over the same digital facility.
- Supervisory signaling required for network management may be routed through the public network to remote locations.

An added advantage of FT1 is that cost savings can be documented more readily than with ISDN, which relies on the analysis of efficiency and productivity gains to arrive at cost savings. These criteria are much more difficult to quantify, however.

Despite having the look and feel of ISDN, Fractional T1 differs from ISDN in that there is no secondary channel available for the transmission of signaling information. Under FT1, multiple channels, including the supervisory channel, would be submultiplexed into the standard DS0 (64 kb/s) time slots. The D channel under ISDN enables the user to program the network to dynamically reconfigure itself by adding or subtracting channels and specific services on a call-by-call basis. The DACS provides the look and feel of ISDN *via* the CCR terminal, which may be used to add or substract DS0s as needed and provide access to a variety of public network services. Unlike the D channel of ISDN, the DACS-CCR does not reconfigure bandwidth on a dynamic basis. DACS reconfigurations take from 5 to 15 minutes to put into effect.

Fractional T1 over DACS networks, as a stepping stone to ISDN, offers a more appropriate bandwidth packing scheme for the vast majority of businesses, because it fits a broader range of present and emerging applications than those offered by T1 or its subrates.

Still another way to gain some hands-on experience with ISDN is to participate in CPE trials. Hardware vendors typically beta test planned offerings before starting a full product rollout. This gives them the opportunity to see how their products perform under actual operating conditions, and then modify the product, if necessary. Establishing relationships with vendors during product testing also gives you the opportunity to discuss their plans for future ISDN offerings and to stay abreast of progress in establishing standards. Beyond that, equipment vendors may be a good source of information on some of the major problems still facing ISDN, such as wiring standards and equipment-compatibility testing. In getting involved with such prototype testing, you can get a feel for the applicability of ISDN to your own network architecture. The insights you gain at this stage may help you in evaluating potential products and vendors later.

Participation in user groups is another avenue through which you can stay informed of ISDN developments. The user group sponsored by Southern Bell Telephone Co., for example, helps both vendors and users understand each other's perspectives on ISDN. Such forums provide ample opportunities to ask questions, make known your organization's needs, and learn about the product development plans of vendors.

8.5 ISDN DO'S AND DON'TS

Making intelligent decisions about ISDN is difficult enough, due to the huge capital investment that must be made, and the difficulty in justifying that investment against the realization that the full potential of ISDN will not materialize for another ten to twenty years. That difficulty is compounded when you have to sort through all the vendor hype to arrive at solutions that are important and relevant to your organization. Therefore, when evaluating ISDN, keep these "Do's" in mind:

- Do make every effort to understand the nature of your organization's business and to understand ISDN technology completely. Only then can you make effective recommendations about the use of ISDN to support organizational objectives.

- Do inform and educate senior management on ISDN, framing the concept as a developing architecture and set of standards. Outline some of the possible benefits that can accrue to your organization, relate these benefits to competitive advantages, and discuss the local telephone company's projected introduction of ISDN.

- Do stay in touch with equipment vendors, local carriers, and ISDN users, not only to stay abreast of current developments, but also to determine future opportunities for participation in ISDN. Even if ISDN is not an immediate concern at your company, senior management is probably assuming that you are keeping yourself informed, should they require such information, and since they may be reading the same trade periodicals you do, they are not going to be interested in a mere recital of news; they will want your insights as derived from primary sources.

- Do hone your analytical skills. This will come in handy when evaluating vendors, products, carriers, and services. Since ISDN is such a recent offering, there is little precedent to go by. Much of what vendors, carriers, and industry gurus say about ISDN is only speculative. Moreover, you will have to be alert to the prejudices, contradictions, and hidden agendas of all the players to avoid potential problems later.

- Do solicit the involvement of other department managers in the ISDN planning process. This will foster a cooperative atmosphere that facilitates planning and will give them a stake in the outcome of the project.

- Do include vendors and carriers in the earliest stages of ISDN planning. Developments in ISDN are unfolding so rapidly that you cannot risk incorporating misinformation or stale information in your implementation plan without incurring additional costs.

- Do experiment first with ISDN before making any sizable investments in hardware or services; once satisfied that ISDN works well with selected applications, implement it gradually throughout the rest of your organization.

Now for some of the "Don'ts" about implementing ISDN:

- Don't promote ISDN in your organization as a total communications solution; after all, there are other communications technologies that may be more appropriate for certain applications.

- Don't assume that different vendors' products will work together over the same ISDN network; standards may have changed since they developed their products. Some ISDN-like products will not work when connected to ISDN interfaces.

- Don't anticipate windfall cost savings with the implementation of ISDN; if there are any cost savings at this point, they will come in the form of improvements in the efficiency and productivity of users.

- Don't expect vendors and carriers to hand-hold you forever; although ISDN is relatively new, they cannot afford to devote too much attention to any single customer. You must take responsibility for educating yourself well before the implementation of ISDN and take advantage of any experience-building opportunities that come your way.
- Don't assume others in your organization know as much as you do about ISDN. It is your responsibility to stay up to date and orient others to the concept, and, when you finally obtain access to ISDN services, do not think you can walk away and let users fend for themselves; you must also accept responsibility for training users and providing follow-on assistance.
- Don't ignore current problems with your network in the belief that ISDN is a panacea; ISDN will evolve over time, and perhaps never fulfill all the promises made about it today.

With regard to ISDN, your best strategy for risk-avoidance is educating yourself now, so that you can make informed recommendations and implementation decisions later.

Chapter 9
What to Look for in Maintenance Services

Over the years, computer and communications systems have evolved into expansive, complex networks that require continuous management and control. Whether problems are revealed through diagnostic activity or through users experiencing trouble with their equipment, the need for timely and qualified maintenance services is of critical importance, particularly for organizations in which survival itself hinges on the proper functioning of their networks. Recognizing this concern, many regional and national third-party maintenance companies have surfaced in recent years, some offering economical, seamless service in mixed-vendor installations. These companies offer a viable alternative to vendor-provided services. In fact, managers who know how to shop for maintenance services can get a much better deal than those who merely accept what their equipment vendors offer. At the least, such competition provides users with negotiating leverage that may prove useful in a variety of situations.

There are three distinct methods of handling maintenance services. You can do it yourself with in-house staff, accept vendor-provided services, or contract with a third-party maintenance provider. You may even combine these methods to suit your unique circumstances. Nevertheless, each method entails its own set of benefits and risks, which can affect your ability to address organizational needs, and, considering that annual service charges may amount to from 4 to 15% of the equipment's list price, making the wrong decision can prove to be quite costly.

9.1 DO-IT-YOURSELF MAINTENANCE

Many large companies have chosen to provide their own maintenance services, particularly for microcomputers and peripherals. Scheduled preventive maintenance every hundred hours or so is enough to spot such common problems as misaligned heads on floppy disk drives before extensive damage is done to the disks, rendering stored data irretrievable.

By performing their own troubleshooting and maintenance, many firms are discovering that they can reap substantial savings in both time and money. Off-the-shelf diagnostic software, the availability of technical training and comprehensive troubleshooting guides, as well as relatively inexpensive test equipment, make faulty microcomputer subsystems, boards, and chips fairly easy to isolate and fix. Whatever cannot be fixed, can easily be ordered from mail-order catalogs. In most cases, replacement parts can be same-day shipped for delivery the next day. In extreme cases, in-house technical staff can ship the unit to a third-party maintenance vendor, and provide the user with a loaner until the original equipment is repaired.

With in-house technical staff and a help desk, most problems can be diagnosed and fixed in a short time. Users can be up and running again in a matter of minutes, instead of waiting hours for outside help to arrive, but in deciding whether to handle maintenance in-house, you must determine if you have the resources, time, and the stomach to handle the job. Management must become "diagnostically literate" and demonstrate commitment to the program with a realistic budget.

Many companies perceive no risk at all in providing in-house maintenance services. Contributing to this belief is the fact that advances in technology and production processes have combined to increase greatly the reliability of today's computer and communications products. In addition, most products are modular in design, permitting fast isolation and easy replacement of faulty components from inventory. Some products even come with redundant power supplies, backplanes, and control logic as standard equipment. When subsystem A fails, subsystem B takes over with minimal disruption in performance. While these factors contribute to the timeliness and quality of in-house maintenance, this is not to say that an in-house maintenance program is entirely risk free.

When you assume maintenance for your own equipment, you must invest time and money to train technical staff. In some cases, equipment warranties will not be honored unless your technicians have attended the vendor's classes and become certified to perform routine diagnostics and preventive maintenance. The length of such schooling varies according to the type of product. A central-office digital switch, for example, may require from eight to ten weeks of school at vendor facilities. For a PBX, the training period may last up to four weeks, and for a microcomputer, a full week. Not only must you pay for the technician's travel and living expenses, you must also be prepared to go without the benefit of his services while he is away, and training is not a one-shot deal. With every upgrade, additional training may be necessary to keep your technician current on engineering changes. If the technician decides to leave your employ, which happens quite frequently because these specialized individuals usually do not have a career path, you must be prepared to reinvest time and money in hiring another technician, and having that person undergo vendor training. While the vendor may have thrown in free training to one or two of your technicians as part of the original purchase, most vendors charge hundreds of dollars per week to train additional staff. If you want to reap the benefits of staff continuity, you must offer maintenance technicians a career path — a worthwhile pursuit, but one that does not come cheaply.

There are other costs that must be factored into your decision to provide in-house maintenance services. Consider the high start-up costs associated with stocking and administering an inventory of spare parts, for example. To provide timely service, you cannot rely on next-day deliveries of spare parts from mail-order houses; you must stock the basics. Keeping inventories up-to-date is another challenge. Five years ago, product life cycles ranged from five to seven years. Today's products have a life cycle of only three years. No sooner do you get up to speed in your inventory, than you have to think about restocking to keep pace with technology, and if you use two or three brands of a particular type of equipment, you must keep quite an assortment of replacement parts on hand to meet any contingency.

With regard to spare parts, the types of product your organization uses will determine to what extent you can handle in-house maintenance. The spare parts of some manufacturers, like IBM, are widely available. Parts for DEC computers are difficult to obtain, whereas those for Hewlett-Packard are virtually impossible to obtain even by third-party maintenance vendors.

Be aware that ongoing investments in technical training and inventory tend to be quite high, so much so that the organization may feel locked into certain brands of product to get the most out of these investments, rather than buying what it really needs from another vendor. You can increase the value of your in-house maintenance service by devising a cross-training program whereby a senior technician teaches other technicians to become proficient in servicing other brands of equipment. This will provide the flexibility the organization needs to make future product selections. You can even implement an in-house certification program to motivate higher levels of performance among technical staff, completion of which may weigh heavily in future raises and promotions. All this assumes, of course, that you can find a technician who is both technically competent and reasonably skilled at teaching others. The time spent away from maintenance activities must be weighed against the benefits of the training effort itself.

Consider the opposite side of the spectrum; as equipment ages, the cost of maintenance increases. Instead of trying to keep pace with new technology, you may want to consider in-house maintenance for your installed base of older products. The savings can be enormous. Not only is the need for continuous training eliminated, but you can shop the used-equipment market for spare parts at bargain-basement prices.

Finally, consider whether you even have the stomach or temperament for assuming responsibility for maintenance. After all, dealing with user complaints day in and day out can take its toll on in-house staff. You must also guard against the onset of complacency in your maintenance staff. Servicing fellow employees does not seem to have the same level of urgency that goes into servicing external constituents. This can cause alienation among users, who may take matters into their own hands, or voice complaints to top management. The resulting backlash can become a source of considerable frustration among the members of your staff, causing them to adopt a defensive posture. You cannot allow the situation to deteriorate to the point of instituting rigid procedures. For example, do not insist on the use of multipart forms which must go through several levels of approval before you are willing to act on user problems. This only negates the benefits of in-house maintenance — timely response and quality service. The solution to such problems is to hire technical staff with highly developed interpersonal communications skills, an appreciation for teamwork, and a "can do" attitude, and be prepared to pay for it.

One last thought: do not start an in-house maintenance service operation with the idea of saving money; do it with the idea of improving service quality and hastening the recovery process. As a natural by-product of these efforts, substantial cost savings will accrue over time.

9.2 WHAT IS BEHIND VENDOR-PROVIDED SERVICES?

The easiest way to take care of your maintenance needs is to let the product vendor do it for you. Like in-house maintenance, vendor-provided service can be quite expensive. Nevertheless, there are some key advantages in letting the vendor handle maintenance for you, which may override cost concerns.

The vendor can bring more resources to bear on a problem with its specialized staff of hardware and software engineers, who are experienced in solving a broad range of problems for an entire installed base of customers. The largest equipment vendors, and some third-party maintenance vendors, are able to expedite problem-solving with dedicated links from a central service center to remote customer sites. Using advanced predictive maintenance tools, customer systems can be monitored and analyzed to identify potential problems before they affect system performance. When something goes wrong with a system that is being monitored, the vendor's technicians are dispatched to the scene immediately with the right spare parts and information to correct the problem. Thus, a problem can be corrected before the customer is even aware that a problem exists. Such an arrangement is roughly equivalent to having the vendor on call twenty-four hours a day, seven days a week — only cheaper. This capability may be especially important if you have a large, geographically dispersed network.

Expert systems are just beginning to be used for diagnostic and restoral applications. Basically, expert systems are massive data bases that house the accumulated solutions to a multitude of problems. The technician can input a symptom, and the expert system will propose a course of action based on the stored knowledge gleaned from the past experiences of the entire customer base. Eventually, expert systems will be capable of accepting plain-language input from non-technical people at customer sites, who will use the plain-language output to restore malfunctioning equipment.

Until that time comes, if you rely on a single vendor for most of your equipment needs and have a steady stream of maintenance activities, the vendor can even supply you with a dedicated field engineer who will be available on-site at all times — for a negotiable fee, of course.

Most vendors offer a range of basic service agreements from which to select, according to your requirements. You can get standard 40-hour-a-week service, for example, or pay a premium price for 24-hour, seven-day-a-week service. The price differential between the two plans may be quite high, as much as 30%. Choosing between these plans is no easy matter, especially if the vendor is trying to strong-arm you into making the selection at the time you purchase its products. Users are caught in a dilemma. On the one hand, they feel compelled to bargain for the best maintenance package at the time the equipment is being purchased, and while the salesperson is eager to please. On the other hand, without prior experience with the product, they do not have a clue about how it will perform under their applications.

There are two possible solutions to this problem. Before committing to the purchase, you can visit a customer site that runs applications similar to yours to get a handle on the product's performance. If the product is so new that you are among the first to buy, your best move might be to defer making a decision on the type of plan you want for as long as you can — at least three months, if possible. Meanwhile, your purchase is being serviced under the warranty. During that time, keep records of service calls and translate the free service into dollar amounts. After three months, you can use the accumulated data to extrapolate your service needs over a year's time and do a cost comparison of service plans. If your installation has not been plagued with problems on a bubonic scale, you may want to consider starting out with the less expensive 40-hour-a-week plan, and eat the costs for occasional after-hours service. If the overtime cost approaches the cost for 24-hour, seven-day-a-week service, you might be better off with the more expensive plan; the few extra dollars are a cheap price to pay for peace of mind. You should, however, also try to negotiate the price of service downward, if only because competitive pressures are forcing many vendors to defend themselves against third-party maintenance suppliers, which only enhances your chances for success.

If your situation is such that you do not have the information or the experience with which to determine your maintenance requirements, an effective needs-assessment technique would be to have competing vendors perform this analysis for you as a condition for accepting their bids on equipment. Before the sale, vendors are motivated to design the best possible plan for your needs. After the sale, unscrupulous vendors tend to overstate your maintenance requirements.

There is another way to get the level of service you need from the product vendor, and at a considerably lower cost, but this method usually involves taking more responsibility for diagnosing problems before calling in the vendor. Such plans operate in much the same way as IBM's Corporate Service Amendment (CSA) plan, which was introduced in October 1986. Under the IBM plan, users pay an initial fee of $3,500 for systems-level coverage or $8,600 for network-level coverage, and submit to a 30-question self-evaluation. Part of this process entails the customers' listing potential problems and identifying people who are appropriately qualified to handle them. With systems-level coverage, customers must follow IBM's problem-management and change-control procedures. Under change control, qualified in-house staff must supervise any changes to the system, whether it be something as simple as changing a disk, or something more complicated like making an engineering change to the mainframe.

Network-level coverage requires that the customer set up a "help desk" over which pass the problem-management and change-control procedures required for systems-level coverage. The help desk would be staffed by an in-house technician who can determine the cause of routine problems and offer solutions that will get the user up and running quickly. This minimizes unnecessary equipment downtime, and saves the vendor an unnecessary trip to the customer site. The savings are passed on to the customer in the form of lower maintenance charges.

Further, IBM and other vendors require the customer to undergo a "certification" process, which essentially consists of answering a questionnaire (mentioned earlier) to insure that the installation meets the specifications set by the vendor and that the customer is knowledgeable about proper reporting procedures. With certification, the customer usually must agree to let the vendor review procedures periodically and even to allow the vendor to visit the installation without notice. Certification may be withdrawn by the vendor, if the installation falls out of specification, or if the customer fails to adhere to established procedures. Certification may be reestablished for an additional fee. Incidentally, if your installation is based around IBM's System/36 or System/38 minicomputers, you may qualify for IBM's Midrange Service Agreement (MRSA), which does not require an initialization fee and entails only a ten-question self-evaluation in preparation for certification.

In standardizing reporting procedures and structuring maintenance charges uniformly over its entire customer base, the vendor can reduce the cost of providing service under such plans, and pass the resulting cost savings on to its customers. For this reason, the vendor may exhibit little willingness to entertain customer proposals to lower fees even more, but do not be fooled by vendor posturing.

The fact is, you can negotiate the price of such plans. Even if "Big Blue" steadfastly refuses to lower the price of its CSA, you can negotiate the terms of the contract to get concessions that achieve the same result. For example, if the vendor refuses to remove its start-up fee, you may want to point out that your installation contains only its product line, and that you do not see the need for the customary initialization fee when you already meet the specifications that qualify you for the plan. If the vendor will not compromise on that issue, it is probably because it must protect its fee structure. After all, once word gets out that you succeeded in negotiating away the front-end fee, other potential customers will be motivated to make the same demand. The next step in your negotiations, then, is to offer an alternative that allows the vendor to save face. You point out again that your installation already falls within specifications and, therefore, you deserve special consideration over customers who do not. Propose that the vendor give you free maintenance for a month to compensate you for your efforts, and another month's free maintenance for saving the vendor the time and trouble of extensively evaluating your installation for compliance. Even if the vendor agrees to only one month of free maintenance, that may be enough to negate the initialization fee you found objectionable. From the vendor's point of view, it can publicly claim that nobody has ever succeeded in having the start-up fee removed, which will discourage others from even trying. This is only one example of how you can get what you want by applying some creativity to negotiations. Creativity is limited only by your imagination — channeled, of course, by the contract terms the vendor shoves under your nose.

In addition to monthly cost savings, the CSA-type plans afford other advantages that can provide you with operational flexibility which, in turn, can save you even more money. IBM, for example, in late 1987 eliminated the penalties for withdrawing equipment under contract, provided that customers give sufficient notice. A five-year IBM contract now requires six months' notice, while the three-year and one-year contracts require only three months' notice. Without sufficient notice, however, you may leave yourself open to termination charges, which run as high as a month's maintenance charge. You can come up with some pretty good justification for eliminating penalties altogether, and offer the vendor an incentive not to impose them. You can bait them with promises of future purchases through a "letter of intent," which, incidentally, does not bind you to a purchase. Failing that, you can request equipment upgrades and options ahead of schedule. As long as you give the vendor something, no matter how trivial it may seem at the time, you can pretty much have your way, especially if your first point of contact in the negotiation process is a salesman. Since salespeople are under pressure to meet sales quotas, they may constitute your best allies in any dealings with the vendor.

Attempting to reduce termination penalties, as well as testing your ability to eliminate them entirely, is important, because it allows you to shop the third-party maintenance market for better service deals on certain types of equipment like printers, terminals, and microcomputers where competition is fiercest. IBM even makes this easy to do, because the CSA does not prevent you from using third-party maintenance services; you are still eligible for the CSA discounts.

The original intent of the CSA was to shut out competitors from IBM shops, but other maintenance vendors like Control Data Engineering Services (Minneapolis, Minn.) and Bell Atlantic subsidiary Sorbus (Frazer, Penn.) responded with clone CSA packages of their own, which offer discounts of up to 20% over IBM's CSA, and entail no charge for certification. Unlike IBM, Sorbus and Control Data do not reserve the right to drop products from eligibility.

Despite the apparent economies of CSA-type services, if you are not willing to increase your involvement in maintenance, for whatever reason, bargains abound among third-party maintenance providers, and many do not base discounts on user participation. The TRW Customer Service Division (Fairfield, N.J.), for example, does not require that users implement a help desk. Instead, it provides whatever services you need when you call. The cost of TRW's services averages 40% below IBM's non-CSA schedule.

9.3 THIRD-PARTY MAINTENANCE SERVICES

Price and breadth of equipment coverage are the principal reasons users opt for third-party maintenance vendors. It is oftentimes easier to negotiate with third-party vendors on such matters as the amount of coverage, price, and the types of equipment to be serviced. This option, however, entails the greatest amount of risk to uninformed users.

While cost savings and equipment coverage are important, you must guard against sacrificing service quality. Anything less, is not only shortsighted, but will also end up costing you more in the long run in terms of equipment downtime. In choosing this type of vendor, inquire about response times, particularly for your remote locations. Check into how many customers it supports in your immediate area. Find out about staffing levels and the locations of the nearest spare parts inventories. All of this information will confirm vendor statements about response times.

The more types of equipment the vendor supports, the better. This will provide you with more flexibility in future equipment selections, since it eliminates the need to contract with other vendors. In a mixed-vendor environment, a third-party maintenance provider who can handle everything eliminates finger-pointing among competing vendors, which can waste your time and test your patience. In your check of references, inquire about the ability of the vendor to work with prime vendors, who may be servicing equipment still under warranty.

If you have a widely dispersed network, check the vendor's capability for performing remote diagnostics; this will limit unnecessary downtime and save you money over the long term. Also inquire as to whether the vendor has a computerized spare parts–management system. After all, qualified technicians are of little value unless extensive spare-parts inventories and a parts-distribution network are in place to insure fast delivery of the right parts. Find out if the third-party maintenance vendor can provide computer-generated reports on equipment service histories and reliability factors by product line and customer location. This capability can assist you with future equipment selections, and if the vendor's computerized inventory and dispatch system is also used for billing, so much the better, because it means that there will be less chance of errors on your invoices.

You should also screen potential third-party maintenance vendors for organizational depth to be sure that they have the resources with which to provide comprehensive support. This means checking into the qualifications of the technical people who service your type of equipment. Find out if the technicians have the required expertise to assist you with upgrades, reconfigurations, and relocations.

If the third-party maintenance vendor has a repair and refurbishment facility, you can save quite a lot of money by having equipment and printed circuit boards fixed or rebuilt, instead of replaced. Sorbus, for example, operates two such facilities totaling 170,000 square feet — one in Tustin, California, and the other just outside of Philadelphia, Pennsylvania. That is in addition to its 18 strategically located computer-repair centers. The company even offers a limited 90-day warranty on items sent in for repair.

Some third-party maintenance vendors even stage new equipment and perform thorough systems-level testing before installing it on customer premises. If any problems are revealed, they can interface directly with the equipment vendor to resolve them. Meanwhile, the customer is saved from protracted dealings with the equipment vendor, and his network is spared the burden of malfunctioning equipment.

If your organization depends on its electronic data-processing assets for its continued survival, verify the capability of the third-party maintenance vendor to implement a Disaster Recovery Plan (DRP), which protects entire nodes of your network against loss. Many large primary vendors offer such services, under a separate annual fee, as a supplement to the *force majeure* clause within the standard maintenance contract. The type of disasters covered under such programs include loss by fire, water, vandalism, theft, power surge, or air-conditioning malfunction.

A key feature of a disaster recovery plan is the replacement of nonrepairable equipment with like equipment within a specified time frame, which may vary with your choice of vendor from 48 to 72 hours. This is particularly significant if your installation relies heavily on products that are discontinued or in long-term production cycles. Replacement parts for repairable equipment should be shipped by the vendor within 24 hours. A technician should be dispatched to the customer location within four hours.

Although insurance covers the replacement costs of equipment, you may be responsible for such things as on-site installation fees and rush delivery charges for parts or system equipment, but many third-party maintenance vendors are ill prepared to address critical problems on short notice. Others are slow in responding to trouble calls after normal business hours. Neither qualifies as offering "disaster recovery" services.

If you have doubts about the claims of a third-party vendor, but cannot resist the promise of big cost savings, use the vendor on a trial basis at selected sites to determine its response times and quality of service. Having satisfied yourself that you can work with that vendor, gradually turn over more sites, rather than give out all your business at once. The rationale of this strategy is to keep rewarding the vendor for good performance; in giving the vendor all your business at once, you take away the incentive to perform. If this strategy is not workable in your situation, stick to short-term contracts with third-party maintenance vendors to keep your options open and to motivate the vendor to perform top quality work on a consistent basis.

Another strategy for insuring that you get the best possible service at all times is to match vendors with equipment. For example, you may elect to have IBM maintain the out-of-warranty mainframe and front-end processors, while you partition the rest of your equipment types among the third-party vendors best qualified to handle them. This arrangement keeps everybody on their toes, because the hidden agendas of most vendors include stealing your business from competitors. After they get their foot in your door, there is only one way they can steal business from each other, and that is to outshine their nearest rival.

Be careful about choosing a third-party maintenance vendor whose parent is, or was, the subsidiary of an equipment manufacturer. While the list of products such companies claim to service may look impressive, the quality of service is going to be skewed toward a very narrow brand of products. If you need proof, just look at their source of revenues. In many cases, you will find that at least 80% or more comes from customers of the parent firm. If the maintenance provider is trying to broaden its customer base, it will do so by offering huge discounts, as much as 50% off IBM's non-CSA schedule. If such price cuts look attractive to you, just be aware that you may be getting what you pay for in terms of service quality. Setting yourself up as a guinea pig invites vendors to use you as one.

Because such vendors are overly eager to get your business, however, you may want to use this opportunity to test your negotiation skills with contract terms and conditions. If you succeed in obtaining outrageously favorable terms, however, do not let your swollen head obscure the need to fend off unnecessary risk. The risk is that the vendor really will not have the resources or the expertise to support you during times of critical need. If the resources and expertise are there when you need them, the danger of a one-sided agreement is that the vendor will provide what you need, but only grudgingly. Before committing to the vendor on a broad scale, you may want to test its capabilities by assigning it one or two of your sites. Make the vendor earn any further business.

Finally, do not overlook the benefits of a cooperative approach to maintenance, whereby in-house staff work right alongside primary vendors and third-party maintenance firms. Oftentimes such arrangements produce a synergistic effect among the various players that manifests itself in performance of the highest quality, resulting in maximum equipment uptime, which, in turn, insures a high level of satisfaction among the numerous users and work-groups of your organization.

9.4 CONTRACTUAL CONSIDERATIONS

Whether you opt for the maintenance services of the primary equipment supplier or a third-party vendor, be aware of the natural adversarial relationship that exists between you and the service provider. You must remember that the vendor's objective is to maximize opportunities for profit, while minimizing exposure to contractual risk and unnecessary obligations to customers. Not surprisingly, then, the service contract is going to favor the vendor over the customer. This means that you must carefully scrutinize the service contract, preferably with assistance from legal counsel.

Never sign a contract at the time the vendor presents it to you, even if the vendor assures you that it is the standard agreement that everyone signs. Review the contract with the vendor from top to bottom. Check your understanding of the various clauses by having the vendor explain them to you. Even if the language strikes you as simple and clear-cut, compare your interpretations with those of the vendor. For example, find out what the vendor means when it says that it will use "best efforts" to correct any problem reported by the customer within 24 hours of its occurrence. Does the vendor mean that it will have a technician at your site in an hour? Four hours? The next day? Or sometime soon?

You will be surprised how many times both of you differ on such matters. If the disparity between the interpretations is too great, the language in the contract should be changed to something both parties can live with. Even with such changes, it is a good idea to hold onto the contract for a week or two, just in case other questions occur to you.

Have a secretary keep minutes of any meetings at which terms and conditions are discussed with the vendor. Send a copy to the vendor to aid in drafting an agreement that fulfills your needs. Keep a copy for yourself in case the vendor has to be reminded in the future of the original intent behind certain clauses.

Relations between vendor and customer are never better than at the start of a business relationship. As soon as the customer invokes provisions of the service agreement, however, the facade of cordiality is dropped, as the vendor tries to interpret the agreement anew to achieve its objectives: maximize opportunities for profit, while minimizing exposure to contractual risk and unnecessary obligations to customers. Your job is to obtain an agreement that does not put you in a position of risk while the vendor pursues those objectives.

If you are in doubt about the maintenance offerings of equipment manufacturers or third-party vendors, The Ledgeway Group (Lexington, Mass.) has available a subscription service that includes comprehensive descriptions of their support programs, support plans, customized maintenance plans, pricing, and performance assessments. In fact, many vendors subscribe to Ledgeway's reports to determine how best to position themselves against their competitors in the area of maintenance and support services.

Chapter 10
How to Choose the Right Consultant

Consultants as a group fare poorly in terms of image. They are frequently the targets of criticism and ridicule. "A consultant is someone who knows half as much you, but charges double for it," is a familiar jest, along with: "How do you tell when a consultant is snowballing you? When he moves his lips." Other jokes define consultants as "anyone who is out of work" or "anyone who is fifty miles from home."

A combination of factors contributes to the poor image of consultants. Sometimes there is a poor match between consultant and client, just as there sometimes is between employee and employer. Personalities may clash, political issues may surface, or hidden agendas may conflict. The consultant's status as an "outsider" may fuel speculation, skepticism, and suspicion among client staff, which brings into play a wide range of human emotions that may adversely affect the consultant's performance. The very act of bringing in a consultant may put employees into a defensive posture. They may interpret it as an act of "no confidence" in their own abilities. Employees may go so far as to perceive the consultant's presence as a prelude to a department-wide, or company-wide, shakeup and, consequently, as a threat to job security.

Some employees use the opportunity to spar with consultants on technical issues in an attempt to "set the record straight," put the consultant in his place, or to show who's really the boss — all this to feed their own egos, or enhance their credibility with top management. Other employees are only skilled at pointing out flaws in the consultant's recommendations, and in the most arrogant and deprecating manner imaginable. When asked for constructive alternatives, these people delight in such refrains as, "Hey, I'm not the high-priced consultant. Don't ask me."

When it comes to nonconstructive feedback, I am reminded of a client who continually accused me of "lying." She and I differed on how a certain product would perform. Instead of asking me to explain why I thought as I did, she would review my work and exclaim, "That's a lie . . . That's a lie . . . That's a lie!" Eventually I had to bow out of that assignment because this client's real problem was well beyond my ability to solve.

Although these are extreme cases of things gone awry, it does point to the need for clients to start shouldering their share of the blame for failed client-consultant relationships — and for the low esteem in which consultants are held. This is not to say that consultants are entirely blameless. Some are clearly not qualified to render services in their claimed area of specialization, lacking the prerequisite experience and depth of technical understanding to make the kind of contribution you expect of an expert. Other consultants may not perform up to expectations or within agreed upon time frames because they lack appropriate support staff or technical resources. Others are deficient in interpersonal communication skills, which reflects badly on anything they do, regardless of the quality of their work. A small minority of consultants are just plain incompetent. Regulating consultants or qualifying them through a certification process will not eliminate this problem. We can look at the medical, legal, and teaching professions to see that regulation has not eliminated incompetence. In fact, a strong argument can be made that regulating such professions has done more to exclude promising candidates from admission by setting up barriers to entry than it has in protecting the public from incompetent practitioners.

Of all business relationships, the one between client and consultant is among the most complex, if only because so many human factors come into play. As in product selection, it is your responsibility to qualify the consultant before buying his or her services. Nobody can do that for you — certainly no regulatory agency or professional association. You cannot even rely on the academic credentials of consultants. College degrees only indicate that a person has read the required books and passed the required tests — all of which may bear no relationship to your organizational needs. This chapter offers advice on how to select and manage consultants to achieve your objectives, thereby minimizing the risk of making the wrong decision.

10.1 WHEN SHOULD YOU USE A CONSULTANT?

Consultants are most often used to find a solution to a known problem. For example, the problem might be excessive delay on your four-node regional network. Both vendors say that their newly installed multiplexers and data sets are working properly; therefore, the problem is with the interexchange service provider which must be rerouting your T1 links over longer distances in an effort to balance the load on its network. You check with the service provider who confirms that there has been some rerouting, but claims the distances are not enough to affect your network. Not making any headway, you call in a consultant to track down the source of the delay and to meet with the hardware vendors and the service provider to resolve the problem.

In this example, the value of the consultant is in the objectivity he can bring to bear on the problem. As an independent party in the dispute, he also has more credibility with the vendors and the service provider. After verifying the proper operation of the vendor's equipment and calculating the effects of the service provider's reroute scheme, the consultant takes a look at your network configuration. His findings: the combination of nodal delay and path delay exceeds 25 milliseconds, which can only be remedied with an echo canceller. The problem is attributable to poor network design on your part. In choosing a multiplexer that was responsible for five milliseconds of delay per node, you did not have enough flexibility to withstand even a slight increase in path delay, should the service provider ever decide to change the routes of your T1 links to balance the load on its network, which every service provider has a right to do.

Sometimes, however, consultants are used to help define the problem. Top management may have an uneasy feeling that things are not going well in the organization, but cannot quite get a handle on what it is. A consultant is called in to do some probing. Maybe employees will open up to someone who treats their conversations as confidential and who cannot use the information against them in a future performance evaluation. After extensive discussions with managers and employees in all departments, the consultant begins to notice a trend: despite the presence of sophisticated workstations tied together on a LAN equipped with E-mail, people are not communicating efficiently. In fact, a lot of work is redundant from one department to another. People complain that their tasks require input from others who do not return calls. Others complain that they are not invited to meetings to discuss factors that affect their projects. Managers are frustrated by subordinates who do not seem to value the importance of meeting deadlines.

It becomes apparent to the consultant that people are not using the productivity tools top management has provided. The E-mail capability of the LAN could improve interoffice communication, but the consultant has discovered that many employees have not been trained in how to use it. Others are too busy to learn.

Still others are too intimidated by the LAN's complexity to figure out how to access E-mail, and feel dumb in asking. Those who do use E-mail are frustrated by the fact that the people they want to communicate with do not know how to use it. The problem, then, is that the company dumped all this technology on everyone's desk, but neither took the time to explain its benefits nor offered to provide training for those who needed it. As a result, the company was not reaping an adequate return on its capital investment.

There are other good reasons for turning to the services of an outside consultant. In-house staff may lack the experience, the expertise, or the time to devote to certain projects. MIS staff may do an admirable job in running the company's information systems, for example, but be thoroughly unfamiliar with what is required to implement microcomputer-based LANs. The telecom manager may do a fine job of running the company's PBX, but lack the expertise to integrate voice and data for transmission over the public switched network. Although the MIS and telecom staffs are quite knowledgeable about their particular areas, a consultant may be able to provide the kind of broad-based view necessary to bring the two disparate functions together on certain types of projects.

Provided that the consultant has been properly introduced to company staff, another benefit of consultants is that they are recognized within the client organization as being apolitical and, therefore, may elicit more cooperation than someone on staff who has tried to accomplish the same task. For this reason, the consultant can more easily sidestep turf issues, while bringing a new perspective to the situation that is firmly grounded in his cumulative experiences with other companies. If you structure the consultant's role properly, he will take on the role of a senior executive with planning authority over the various departments, offering top management a discrete source of advice about company operations.

Sometimes companies may have to act quickly in a competitive situation and do not have the time to develop expertise in-house, or to acquire it from another source. A qualified consultant, on the other hand, will already possess detailed, up-to-date information on available equipment and services, as well as hands-on experience with the latest technologies. As such, he will not feel locked in by internal attitudes, prejudices, and policies. The consultant also has a base of information on which to make sound purchasing recommendations. With a qualified consultant in your corner, there is less chance that vendors will take advantage of your situation or lack of knowledge to gouge you on price, or that they will attempt to sell you on products and services you really do not need.

Even when in-house staff have the necessary knowledge and expertise to follow through on a project, it is oftentimes a good idea to bring in a consultant for independent verification of their plan of action. The cost of using a consultant for this purpose may be paltry compared to the cost of making a wrong decision. As in medicine, a second opinion cannot hurt.

10.2 WHERE TO FIND CONSULTANTS

Good consultants are not at all hard to find. You can start by getting recommendations from peers at other companies within your industry. Make sure that the consultants they have used performed work similar to the kind of work you require.

You can also obtain directories of consultants from associations and commercial publishers (Table 10.1) and, if time permits, begin your search for qualified consultants systematically by phone or letter. Some directories, like the one published by the Society for Telecommunications Consultants (New York, N.Y.),

Table 10.1 Sources of Information About Consultants.

American Business Directories, Inc.
Computer Systems Designers & Consultants Directory
P. O. Box 27347
Omaha, NE 68127
402-593-4600
Cost: $160.00

American Business Directories, Inc.
Data Systems Consultants & Designers Directory
P. O. Box 27347
Omaha, NE 68127
402-593-4600
Cost: $100.00

International Computer Consultants Association
933 Gardenview Office Parkway
St. Louis, MO 63141
314-997-4633
[ICCA does not publish a membership directory, but will direct you to the nearest association chapter for consultant referrals.]

Professional and Technical Consultants Association
Membership Directory
1030 South Bascom Avenue, Suite D
San Jose, CA 95128
408-287-8703
Cost: $15.00

Society of Telecommunications Consultants, Inc.
Membership Directory
1841 Broadway, Suite 1203
New York, NY 10023
212-582-3909
Cost: $15.00

provide a brief description of members' experience, qualifications, and affiliations, as well as of how many years they have been in the industry and what kind of experience they have had as consultants.

The computer and communications trade press is another good source of information about consultants. When consultants are quoted regularly, this is a clear indication of their stature in the industry. Also look for relevant technical articles that are authored by consultants. Aside from the accuracy of their specialized knowledge, a carefully crafted article can tell you a lot about the consultant's ability to communicate ideas to those who may not be as technically gifted. As further substantiation of their qualifications, some consultants have regular columns in the trade press.

Do not neglect trade shows as a source of qualified consultants. Most trade shows feature consultants on their programs, either as speakers or as participants in panels. In most cases, consultants must compete for a spot on the speaking program by submitting papers that are judged by a committee. Papers are judged according to specific criteria, not the least of which is the timeliness and relevance of their topics to the association's membership. Not only do you get to to see them in action, but you also have the opportunity to meet them in person for a brief chat, at which time you can ascertain their interest in a future meeting to discuss your requirements.

10.3 HOW TO EVALUATE CONSULTANTS

Do not be so impressed by a consultant's public role, however, that you neglect to perform a thorough evaluation of his or her qualifications and references. You still have to be sure that there is a good fit between the project at hand and what the consultant can bring to the table.

If you are considering the technical consulting services of a Big Eight accounting firm, you can be reasonably assured from the start that it has adequate resources with which to support client projects, but if the consultant is an individual practitioner, or belongs to a small firm, you should inquire about the availability of resources that can be brought to bear on behalf of clients. Many consultants devote as much time to servicing clients as to marketing their services to secure future business. If consultants do not answer your correspondence or return your phone calls, that is a good indication of how you will be treated as a client.

You have to determine how many hours of the week an individual practitioner can devote to meeting your requirements. Depending on the nature and scope of your project, a small consultancy of five to ten practitioners may offer you more security in that if one gets sick, or a new client is picked up, your requirements will not be neglected. Find out if the consultant will be doing the work personally, or if your project will be handed off to another consultant, possibly one less experienced. Check into the consultant's use of researchers, analysts, and outside experts. Aside from manpower, check into the consultant's technical support infrastructure, including the use of computers, facsimile, electronic mail, and appropriate applications software like PERT and GANTT planning tools, and spreadsheets. Your objective with this line of inquiry is to determine whether the consultant has the resources to follow through on your project to its completion. After all, changing boats in midstream can cost you more than you bargained for.

Once you have determined that the consultant has the resources to do the job, find out about his availability for your project. Because consultants do not always know when and where their next clients will turn up, they have a tendency to overextend themselves, but, realistically, a consultant with heavy commitments to existing clients, who contributes articles to the trade press, has a book contract, and speaks at industry trade shows, probably will not have much time left to give your project the attention it deserves. In this case, settling for someone just as qualified, but less busy and well known, may be the wisest decision you can make.

Next, check the consultant's references, even if he or she is considered one of the big names in the industry. Without checking references, you cannot be sure if that reputation is just hype, or well deserved.

Probe references in very general terms; assuming that you do not know them personally, do not expect them to answer specific questions about their operations. Tactfully inquire as to whether the consultant worked well with company staff. If references experienced any problems in this area, but you are still impressed with the consultant's technical ability, you may want to try him out on a smaller project to see how he gets along with others in your company. If things do not work out, you can cut him loose without jeopardizing more important projects.

Was the consultant able to justify product and vendor selections? This is important, because some consultants seem to deal with the same vendors over and over again. As a result, they become closed minded to considering other products and technologies that may be more appropriate for your application. When you are footing the bill, you should not tolerate this kind of stagnation, and, if the consultant is new to the game, you should ask him or her what vendors they have recommended in their last five to ten projects. You want to be sure they are not consistently recommending the products of their former employers.

Did the consultant exhibit a willingness to challenge vendors or service providers? One of the reasons for hiring a consultant is to have someone on your team who knows when vendors and service providers are trying to pull the wool over your eyes. The last thing you need is a shy consultant who is not up to playing the role of task master.

Did any unexpected problems develop during the project? If so, how did the consultant respond? Answers to this line of questioning will tell you about the consultant's project-management skills. Look for any indication that the consultant has a high tolerance for ambiguity — the ability to function effectively, despite the absence of key information. You are also looking for flexibility — whether the consultant can accommodate new developments into his plan without starting from scratch. If things did go wrong for the client, find out if the consultant had a contingency plan ready.

Did the company follow the recommendations of the consultant? If not, why? This line of questioning may reveal whether the consultant's proposed solution was too "creative," "risky," or just plain "off target."

Did the consultant complete the project on time, within budget, and to expectations? How did the consultant structure his fees — by the hour, day, or project? Were these fees reasonable in relation to the work done? If there was a retainer arrangement with the consultant, find out if the company was satisfied with it. Did the consultant build into his final invoice any extra items that were not anticipated at the start of the project?

Do not just check the consultant's references, check with the vendors he or she has recommended to clients. You want to find out if the vendor had to pay the consultant a "finder's fee" or a commission for recommending its products. Believe it or not, some consultants still operate like this — and some vendors still play along.

10.4 HOW TO GET THE MOST OUT OF CONSULTANTS

After the evaluation process, you may be ready to hire the most qualified consultant available, but your work does not stop there. To get off to a good start, you have to delineate the consultant's role, as well as describe the nature and scope of the project, so that you both understand exactly what has to be done. Anything less can mean the difference between a frustrating, costly experience and a job completed on time, within budget.

Start off by outlining your organization's problems. Describe some of the solutions you may have tried and the reasons why they did not work. Explain in as much detail as is required what you are trying to accomplish. Be forthcoming with information; share with the consultant any historical, statistical, or performance data that may have been collected which support your conclusions. Tell the consultant about your organization's future growth plans. With proper authorization, you should provide the consultant with a copy of your company's business plan. Consultants will not hesitate to sign a nondisclosure agreement, if one is required as a condition of release. Also, fill in the consultant about some of the staff members with whom he might have to work closely. This might be necessary to prevent the consultant from inadvertently offending people with quirky personalities, or those who are hypersensitive about issues of intelligence, gender, turf, personal appearance, or any of the myriad trendy issues that may be gripping today's social conscience. At the same time, acquaint the consultant with the latest political battles in your organization to prevent him from stepping unknowingly into no-win situations. The consultant will treat such conversations as confidential, so you need not worry that this information will fall into the wrong hands.

On the other hand, if you fail to update the consultant on these matters, your project may be delayed and will likely go over budget, because too much of the consultant's time is consumed in sorting things out. At the same time, if you are not very specific about your needs, have no project plan, and have not bothered to gather the basic information needed to do the job, the consultant will view the project as risky and inflate his fee accordingly. Remember, the more groundwork you can do yourself, the less money it will cost for a consultant, and the faster the project will reach a satisfactory conclusion.

Chances are, though, that if you could do all this, you would not need the services of a consultant. One way around this dilemma might be to hire a consultant to perform your needs analysis, write up the objectives, and structure the project. Sometimes this kind of research will reveal that a simple upgrade is all that is necessary, instead of new equipment or systems. If not, you will still have all the relevant information in hand, so you can send out bids to several qualified consultants. For this to work, however, you must not allow the consultant who scoped out the job to compete for the final project. Other consultants will not compete, if they sense that the bidding is fixed.

It is usually not a good idea to ask consultants to scope out the project within the proposals they submit for the job. First, you are not likely to get anything more than marketing hype about the capabilities of the consultancy. Second, once you get the successful bidder under contract, there will likely be a lot of deviation from the original plan anyway.

After the consultant is chosen, you must insist that periodic status reports be submitted of work in progress. The purpose of such reports is to insure that acceptable levels of productivity are indeed being maintained and that significant progress is being made toward reaching the various project milestones. If demonstrable progress is not being made, you have the opportunity to take corrective action. Perhaps the consultant is experiencing problems with uncooperative members of your staff. Maybe the consultant has lost perspective, and needs your assistance in getting back on track. Quite possibly, the consultant has delegated to junior members of his firm work which is beyond their level of expertise to handle. In addition to status reports, use frequent meetings with the consultant to allow such problems to surface.

Regardless of the cause for nonperformance, status reports will reveal problems. If the problems are beyond corrective action, you have the option of terminating the relationship to avoid wasting organizational resources and squandering any more valuable time. If the problem is diagnosed early on, you may still be able to complete the project on time and under budget with another consultant.

10.5 HOW TO PAY CONSULTANTS

Consultants expect you to raise the issue of fees before agreeing to hire them. Many kinds of fee structure are possible. Per hour, per day, and per project fee structures are among the most common. Sometimes a combination of all three is possible. Sometimes the consultant's fee is a percentage of the total purchase price of equipment or systems. It is not uncommon to have incentives and penalties built into fee structures. There are also retainer arrangements, in which the consultant accepts a monthly or yearly fee to be on call and to provide a specified minimum number of hours to the client on an as-needed basis. There have even been cases of consultants accepting stock in lieu of cash.

Consultants may prefer one type of fee structure over another. Clients, too, may have their preferences. Clients may prefer a fixed fee per project, because then they know exactly what the project will cost over an entire fiscal year and can work that figure into their budgets. Consultants, on the other hand, usually avoid fixed fees, because they really do not know how long a project will take until they are well into it. This is especially true when clients cannot articulate their objectives. In such cases, hourly rates may be more appropriate. Sometimes the lead consultant will charge for his services by the day, and bill for staff research and support on an hourly basis.

If client needs are precise, the consultant can offer a flat fee based on an estimate of hours and overhead consumed. Sometimes projects are pretty straight-forward, as in the purchase of a PBX or key system. In this case, the consultant's fee may amount to from seven to ten percent of the total purchase price. That is not bad, if the consultant has negotiated a competitive price, as well as guaranteed price protection for additions, enhancements, and maintenance. In this case, the consultant may have saved you ten times his fee over the useful life of the switch.

If your project has a tight deadline for completion, you may want to consider offering an incentive for finishing the job ahead of schedule. If the project comes through under budget, you may consider offering the consultant a percentage of the savings. Such arrangements are common with government contracts and are growing in popularity in the private sector.

10.6 A MODEL FOR INTERPERSONAL COMMUNICATION

No consultant can rely on technical expertise alone to bring a project to successful completion. Interpersonal communication skills are equally important, because they facilitate the problem-solving process, but — perhaps because it is so obvious — both new and experienced consultants tend to overlook the importance of finely honed interpersonal skills, preferring instead to get by on technical know-how, business acumen, and plain old horse sense.

As the client, you must be concerned with a consultant's interpersonal skills to avoid adding to your "problem." A common complaint among those who have used consultants is that the consultant starts to discuss one problem, only to end up with another — such as the frustration of not being heard, or of not being respected. A consultant's poor interpersonal skills, then, can create problems that delay projects, drive up costs, and create ulcers.

Despite the seeming complexity of interpersonal communications, the process can be broken down into identifiable stages, which require specific skills to support. This section offers a model for the interpersonal communication process against which consultants may be evaluated for future projects, or for continuing with projects already in progress. Interestingly, the model may be put into practice very simply during the consultant-evaluation stage, helping you to minimize the risk of a wrong decision.

10.6.1 Structuring the Setting

In your first meeting with the consultant, the first element that requires attention is the setting. Ideally, most information during the initial meeting should come from you. The consultant's responsibility is to listen for the problems before jumping to conclusions, questioning your actions, or proposing solutions.

If you are the type who relates information in a casual, roundabout way, look for the consultant to engage in active listening, notetaking, and requests for clarification. It is the consultant's responsibility to analyze and develop the information you have given. To encourage this process, the consultant should have structured the setting in such a way as to make you feel comfortable about talking, and to make you feel that your problem is his problem.

There are several ways in which good consultants can structure the setting, even though the meeting may be in your own office. Face-to-face meetings quite obviously require that enough *time* be set aside to develop a meaningful dialog. When the consultant gives you all the time you need to explain your problem, this is evidence that he is client-oriented, and that he regards clients as important.

Privacy is another element that helps structure the setting. Since you may be divulging sensitive or proprietary information, you have every right to expect that the consultant will treat the discussion as confidential. Obviously, you will not feel that you are at the center of the consultive process if you think privacy will not be protected. One way consultants destroy their credibility is by dropping the names of present or past clients and discussing their problems in too much detail. Some consultants have even been known to demean their past clients in an effort to score points for themselves as they go after new business. While this is intended to impress you, what you have to consider is how this consultant will be talking about you, when he feels the need to impress others. A good consultant will volunteer to keep information confidential, when he senses that you are ready to get specific.

Besides time and privacy, the consultant should insure that the *physical setting* supports the client-centered process. Be attentive to what the consultant is doing to reaffirm the personal nature of the meeting. Such things as maintaining appropriate physical distance, eye contact, and body attention encourage the free flow of information. They also reinforce the client's feeling that the meeting is taking place just for him.

10.6.2 Managing the Process

It is not enough for a consultant to structure the setting of the meeting to make you feel at ease. The consultant must also use this opportunity for your benefit, not to fulfill his own needs. When a consultant's actions reflect his own needs, he undermines the whole consultive process. Beware of a consultant who tries to "set the record straight," or who is preoccupied with trying to establish who is technically superior. Be alert to remarks that belittle your circumstances, level of understanding, or lack of foresight. Any of these indicate that the process is being controlled by the consultant's ego needs, rather than by the needs of your organization. If that is the case, terminate the relationship immediately. The consultant has already demonstrated his inability to gain the acceptance and cooperation of your staff; letting the situation run its course is pointless.

Take note of a consultant who knows how to bridge the inherent psychological distance that quite naturally exists between "insiders" and "outsiders." The three qualities that help reduce that distance are comfort, rapport, and trust.

A consultant can first promote *comfort* by making sure that the setting is a good one. He can reinforce it with cordial greetings and appropriate physical contact, such as a handshake, or a friendly slap on the back. A skilled communicator also displays good body attention, such as nodding in agreement, or furrowing the eyebrows while asking a question to convey sincere interest.

Rapport presumes some level of comfort, but goes deeper and tends to encourage more exchange of information. Consultants commonly achieve rapport by sharing common experiences — such as having worked for the same company, knowing the same people, or using the same product or vendor.

A third strategy for bridging psychological distance is the establishment of *trust*. From your point of view, trust is the sense that you can believe in the consultant's competence and integrity. It carries a deeper level of commitment than does mere rapport. Alternatively, if you think you have been manipulated, or that you cannot trust the consultant, you may have yet another problem to compound whatever technical problems you had originally.

10.6.3 A Closer Look at Bridging Skills

A consultant's nonverbal communications are critical to his effectiveness in relating to the client, as well as to the client's staff. Body language is one facet of nonverbal communication that is widely recognized as a critical part of the total communication package.

The consultant who tells a client, "Go on, I'm listening," while checking his watch, presents the client with physical behavior that contradicts verbal behavior. When verbal and nonverbal messages are delivered at the same time, but contradict each other, the listener tends to believe the nonverbal message.

There are definite actions that comprise positive body language: facial expressions and head movements that encourage; eye contact that is natural and continuous; body positioning; open posture; leaning forward; and maintaining appropriate physical distance. In your initial meetings with consultants, look for these communication skills and use them as evaluation criteria. Remember, you must select a consultant who can interface well with your staff. Good communication skills will go a long way toward clearing information bottlenecks.

While body language tells a client that the consultant is listening sincerely, mental attention lets the client know that the consultant is using all his senses to appreciate all parts of the message fully. The message's verbal component is the simplest and most direct: the words. The words themselves, however, may not convey the whole message. The simple statement, "I'm looking for another third-party maintenance vendor," might indicate despair, frustration, anger, or optimism. Here, what is included in the verbal package becomes the message's critical element.

You can think of verbal packaging as the message's emotional content. You can determine that content by observing how fast or how loud a consultant speaks, and by listening for tone of voice. Everything about the consultant, including his competence, integrity, and well-being, is wrapped in these emotional components.

The message's nonverbal components are particularly useful to you in your initial meeting with the consultant, because they provide you with real-time feedback that can be brought to bear on your decision to hire the consultant. Through nonverbal signals, you can tell how quickly a consultant will succeed in bridging the psychological distance between himself and your staff. A consultant with a tense jaw, clenched fists, slumped shoulders, or one who avoids eye contact, is sending subtle messages that will be interpreted by others as "stay away." The proper evaluation of a consultant entails paying attention to his whole message, integrating its verbal, verbal packaging, and nonverbal elements into the "big picture."

Everything the consultant does to show that he is interested in problem-solving demonstrates respect for clients. Respect begins by managing a meeting's setting so that clients instantly know they are at the center of the problem-solving process. Without respect, there is little chance that the client will want to continue the meeting, much less offer helpful information and participate in possible solutions to technical problems.

Look for the consultant to demonstrate his ability to draw out answers, helping people to speak freely. This comes naturally to most consultants. "Where would you like to start?" "How can I assist you?" These are straightforward invitations to talk, but note that there is a danger in those otherwise valuable comments, especially when they are coupled with inane remarks like, "After all, that's what I'm here for," or "That's what I'm paid for."

In such cases, you and your staff will not know if they are talking to a person, or to a role. "Role" comments serve only to widen the psychological gap. Remember, until the psychological gap is bridged, solutions to technical problems will take longer, which will cost you more money.

Look for a consultant who acknowledges what you are saying. This is a clear indication that you are being understood and that you should continue speaking. "Acknowledgement" signals may be as simple as nodding in agreement or uttering a few "uh-huhs," but watch out for their overuse. Continually bobbing the head, for example, eventually loses its effect, and might even be a turnoff to some people.

These are some of the communication skills consultants should demonstrate during the evaluation process. They are important skills for enlisting the confidence and cooperation of your staff, who may or may not forgive affronts to their intelligence, skills, knowledge, status, or abilities. Let us turn to a discussion of how to evaluate a consultant's "integration" skills.

10.6.4 Integrating the Information

A separate step in the consultive process is integration. Here, the client continues to talk about his organization's needs with the consultant's encouragement and shared insights. The consultant must demonstrate not only his ability to bridge psychological distance, but also to encourage and stimulate the disclosure of more and more information until the client gains new insights into the nature of the technical problem at hand. There are at least six skills consultants can use to help their clients arrive at this point and, during the screening process, it should become very apparent which consultants use them.

Reflection is a response that lets clients know they are being heard. Unlike mere "acknowledgement," which is passive, reflection is an active process which helps clients focus on areas that require further exploration. For example, a frustrated MIS manager might talk about personnel-turnover problems in her department, which she believes are responsible for low morale and missed deadlines on even routine equipment move and change requests. A skilled consultant can use reflection to zero in on an unstated problem, as in the following scenario:

MIS Manager: I have so many problems in my department, so much turnover. Now top management gives me responsibility for processing half a million call records per month to account for telecommunications costs.
Consultant: You're struggling to improve your operations, but you feel that taking on the additional burden of call accounting at this time may hamper your efforts?

The consultant should employ this strategy when doing so can help the client examine certain aspects of the problem in more depth. When the client acknowledges the consultant with an appreciative "yes," or a nod, and then continues, that is an unmistakable sign that the strategy has worked effectively.

By using *self-disclosure,* a consultant encourages a client to continue by interjecting a relevant comment which indicates that the consultant has also run into a similar problem. Self-disclosure reminds the client that the consultant is also a human being with a wide range of experience — which is what makes problem-solving possible. Some consultants are reluctant to use self-disclosure, because it deals with here-and-now feelings and issues generated during face-to-face meetings. They are afraid to provoke or embarrass a client, appear intimidating, or they just cannot bring themselves confidently to propose an appropriate solution. That is a critical shortcoming, especially when you are considering a consultant for the purpose of riding roughshod over vendors and service providers — a task that is definitely not for the squeamish.

The skill should be used with discretion, however. Its injudicious use could interrupt the client session entirely by drawing too much attention to the consultant, which may be misinterpreted by listeners as "ego speak." Consultants who engage in this behavior waste your time and your money.

Immediacy deals with behavior that your staff may create or display to the consultant. The way a consultant handles immediacy may very well determine the outcome of your project. For example, a key staff member may appear anxious about the consultant's role. The staff person may think that top management is giving him or her a vote of "no confidence" by inviting the consultant to take such an active posture. If unaddressed, those pent-up feelings may undermine the consultant's efforts by hampering his or her ability to obtain important information and staff participation.

A good consultant will not let such problems fester for too long. At the most opportune time, the consultant must defuse that time bomb, perhaps with the pointed observation, "I have the feeling that it is difficult for you to be as candid as you would like, possibly because you resent the fact that I have been asked by your boss to look into these matters and outline some possible solutions."

If any of your staff have the tendency to ramble on aimlessly, the consultant can use the skill of immediacy to help that person zero in on the problem. An appropriate remedy could come from a statement like: "Tom, I'm feeling a bit confused right now. We seem to be covering a lot of ground, but not getting down to the real problem."

As with self-disclosure, the skill of immediacy must be used judiciously. Any hint of cleverness or of adopting a superior attitude could negate whatever benefits the strategy was intended to achieve in the first place.

Probing refers to the consultant's request that client staff explore further some area or issue that they have stated directly. It can help a consultant develop key points, define problems more completely, or identify patterns in nonconstructive thinking.

With *checking*, the consultant structures responses that confirm his understanding of what the staff member has said. For example, a consultant might begin with the lead phrase, "It seems to me that you have identified at least three network configurations worth looking into . . . ," and then repeat back what he thinks he has heard from the staff member.

Unlike probing, *confrontation* deals with information that has been only hinted at and not directly stated. Using confrontation assumes that a satisfactory relationship already has been established. A consultant should start with the least difficult or threatening point and move to the more difficult or threatening subject. Confrontation must be tentative, but specific, to be effective. Phrases that convey tentativeness include: "I wonder if . . ." or "Could we talk about . . . ?"

The staff member should know that confrontation has taken place, but should not feel boxed in or under attack. A sign that confrontation has not registered is when the staff member ignores the statement, or becomes tight-lipped or defensive.

Confrontation can serve many purposes. It can help you and your staff look at problems from different points of view, or see possible consequences you have overlooked. It also aids in uncovering a fundamental need or an underlying problem. It may even help a staff member "own" a statement or feeling. For example, a network manager who feels threatened by the presence of the consultant may say something like, "Some of the staff are really going to be turned off by your vendor recommendation." The consultant should be skilled in the use of confrontation, perhaps handling the problem in this manner: "Bill, it would help me a lot if I knew you were describing your own feelings. Do you know something about this vendor that should be taken into account?"

10.6.5 Client Support

In the support phase of a client relationship, consultants are supposed to help define problems, deal with problem ownership, and develop action plans.

During *problem definition*, consultants work with clients to develop statements that accurately describe the problem or problems they intend to address. After appropriate discussions and research, the consultant should test the best course or courses of action with the client — through statements and restatements — until problems are specific and realistic. The consultant should also make sure that the client can "own the problem," at least temporarily, and that the client has enough resources to solve the problem.

It will not be possible to develop action plans until there are specific problem statements. This can be a difficult skill for some consultants to master, particularly for those who are used to acting on imprecise client statements like "This is a lousy way to increase T1 bandwidth capacity . . . ," "I'm hearing bad things about that resale carrier . . . ," "I can't see myself committing to that network-management system . . . ." All such client statements beg for clarification. By themselves, they can never be translated into specific action plans that clients will accept. In evaluating consultants, it is pretty easy to determine which ones will fall prey to emotional statements like these; just throw out some half-baked ideas to see how they react. Weed out those who do not request clarification. In the end, their "solutions" will be off target and cost you a bundle.

In helping you define the problem, a sharp consultant will deftly steer you from the general to the specific, as in the following progression of statements made by the MIS manager in one of the previous illustrations:

I don't like the idea of having responsibility for corporate telecommunications. I don't have time for that sort of thing.
My staff are specialized and dedicated to other priorities.
I don't know anything about telecommunications; this assignment is an invitation to fail so they can get rid of me.

Until the consultant can arrive at the crux of the problem — which may boil down to simple and unfounded staff insecurities — any product-based solutions offered by the consultant without this vital information may prove to be far off base. By moving toward specific problem statements, the consultant can dramatically change his understanding of the problem's fundamental nature and scope.

Problem ownership is a prerequisite to developing action plans. It is achieved through a common search for solutions. Consultants can encourage clients to own problems by presenting solutions with "we" comments.

Clients will assume ownership as they work with consultants to develop problem statements. Once those statements have been worked out and the client indicates acceptance, both parties can turn the statements into goals, and then agree on a strategy for reaching them.

There are a variety of methods for *developing action plans*. For highly technical problems, consultant and client should take these minimum steps:

- Develop alternative solutions;
- Select the alternatives which are most acceptable to the client's organizational needs, budget, and personnel resources;
- Develop milestones with mutually agreed upon time frames for completion;
- Plan follow-up meetings to monitor satisfaction with progress and to make appropriate refinements in the action plan as new knowledge becomes available or as new developments in technology may warrant.

In the consultant-evaluation process, ask candidates how they typically go about developing action plans. Make sure that their explanation roughly corresponds to the process outlined above. Your goal with this question is twofold: to find out if the consultant even has such a process, and to determine his ability to articulate that process in terms that you can understand.

10.6.6 Consultant Withdrawal

Having implemented the solution, the consultant will initiate a withdrawal process that weans the client from ongoing dependence on him for support. Two skills are involved in that phase, centering and appreciation.

When a consultant uses *centering,* he identifies and makes positive comments about client strengths, especially those displayed in productive meetings. This is not public-relations gimmickry. Instead, it is an attempt to help client staff feel more secure and confident about implementing the consultant's recommendations. Client strengths that may deserve mention by the consultant include candor, analytical skills, commitment to problem-solving, and progress at arriving at the most appropriate solutions.

Client staff may have played an important role in the consultive process, by choice or by chance. A final skill in the withdrawal phase is for the consultant to express *appreciation* to each staff member for the time and effort he or she put into the sessions.

Why should you care about these skills, or even about the withdrawal phase in general? Quite simply, the consultant's behavior at the conclusion of an assignment will determine what level of staff support you are likely to get in the future. If, for example, the consultant does not acknowledge the contributions of others in arriving at a solution, they are less likely to lend their cooperation in the future. After all, it is only human nature to want credit for one's efforts. Nothing is more demoralizing to staff members than to watch an outsider repackage their ideas for top management's consumption, and then charge a hefty fee for it. Allowing such a charade to go unchallenged will reflect badly on your credibility with your staff, peers, and superiors.

When choosing a consultant, be aware that he must have the skills necessary to deal with "people problems" before the people can effectively address technical issues. Recognizing the skills that are crucial to interpersonal communication will lessen the risk of making the wrong decision, which can cause you to miss project deadlines, increase staff turmoil, and eat into your budget.

Chapter 11
Mail-Order Buying

As recently as two years ago, corporate buyers looked with disdain upon mail-order firms, preferring instead to buy computers and peripherals from rock-solid, blue-chip suppliers like IBM, which could be counted on to support their products for many years to come and provide a relatively smooth migration path from one generation of product to another. Never mind that IBM and others may have prodded customers into stepping up to the next level of computing before they were ready; in retrospect that may have been a blessing in disguise, because it started many companies thinking of their information resources as strategic assets that could be used for competitive advantage, but a sequence of events in 1987 pushed corporations headlong into mail order.

That spring, IBM introduced the Personal System/2 (PS/2), a microcomputer that would greatly surpass the performance of the PC-XT class of microcomputer, but the architecture of the PS/2 was radically different from the XT, right down to the storage media. The PS/2 used 3.5-inch disks instead of the industry standard 5.25-inch disks, and to demonstrate its commitment to the PS/2, IBM dropped the XT from its product line. This caught corporate buyers by surprise. They had huge installed bases of XTs, much of which had not been fully depreciated.

By the summer of 1987, the XT-clone industry was in full flower. After a few disruptive shakeouts, the remaining clone manufacturers, like Leading Edge Hardware Products, Inc. (Canton, Mass.), were turning out some pretty inexpensive, yet reliable, clones. After some successful experimentation with clones bought through the mail, corporations were no longer treating mail order with contempt, but with cautious interest. The turning point came in October of 1987.

After a stock-market boom that brought the Dow Industrial Averages well beyond the magic 2000 mark, a series of events, including news of a record trade deficit, sent the Dow plummeting to losses that rivaled the crash of 1929. Faced with an uncertain business climate, corporate America had to get lean and mean in a hurry, and that meant toeing the line on capital expenditures. Suddenly, buying computer and communications equipment — and related consumables — through the mail was looked upon as a viable way to meet productivity goals, while living within a tight budget. At the same time, mail-order firms continued to clean up their act in an effort to attract corporate customers in what has become a thriving $500 million a year business. Some companies are so pleased with mail order that their experiences are routinely written about in the computer trade press — to enlighten colleagues who remain skeptical.

11.1 THE PROBLEM WITH MAIL ORDER

Many people still remember the highly publicized cases of a few decades ago in which average Americans were swindled out of their life savings by sending money through the mail for plots in planned retirement communities. Many of them did not know they were snookered until they drove out to see their "investments" months, or years, later, only to find out that what they had purchased was desert, swampland, or government-owned property. A variation of the scheme was that the property was, indeed, everything the advertisements promised, except that it was sold to hundreds of unsuspecting buyers, each of whom received a counterfeit deed. There are still many cases of mail fraud today. When one scheme is exposed, two new ones seem to take its place. For this reason, the mail-order business still has a sleazy reputation in the minds of corporate buyers.

Now there are hundreds of laws on the books that are designed to protect consumers against mail-order fraud. Many computer and communications trade publications routinely screen ads placed by mail-order firms to preserve their credibility with readers. Despite these added precautions, flimflam artists have been known to sneak in through the back door, causing just enough damage to give the mail-order industry a black eye. Such was the case in 1987, when several leading computer magazines ran ads from an outfit calling itself Compusystems Co.

The bogus firm circumvented the close scrutiny ordinarily given to mail-order advertisers by providing the publications with phony bank records that established its credit history. The advertisements lured customers with low prices, offering substantial discounts on major brands of computers, software, and peripherals. The catch was that all orders had to be prepaid by check, and many people innocently complied.

Fraud aside, there are other risks in dealing with mail-order firms. It is not easy, for example, to get a handle on the financial stability of such firms, even when it appears that they are doing well. At one time, PC Network (Chicago, Ill.) was among the largest mail-order houses in the industry. Then it declared bankruptcy by filing for Chapter 11 with the U.S. Bankruptcy Court in Chicago. According to Steve Dukkar, the firm's president, PC Network resorted to bankruptcy to discontinue servicing older loans, thereby freeing enough cash to meet its order backlog. [1] While the company reorganized its operations, customers with prepaid orders had to wait, wait, and wait . . . The company has since resurfaced under the name "The New PC Network" — without Steve Dukkar. Although it offers a wide variety of computer and communications hardware and software, its "no refunds" policy does nothing to bolster confidence in the new operation.

Another aspect of mail order is that you do not really know with whom you are dealing, except when talking with sales representatives over the phone. You do not know for sure what you are buying, until it arrives at your doorstep, and if the item is exactly what you expected but does not work, what do you do next? Over the years, mail-order houses have committed just about every conceivable offense against consumers, including: misleading advertising; promoting items not in stock; pushing faulty, untested, discontinued, or counterfeit products; delaying shipments; failing to return customers' calls; delivering the wrong items; and providing inadequate repair services. All this has understandably eroded consumers' confidence in purchasing products through the mail. In comparing mail-order purchasing with purchasing through dealers, however, similar complaints surface.

In going to a dealer, you will typically find that sales staff push what they have in stock, rather than what you really need. There is no large selection of brands, because it is expensive to stock too many products. As a result, dealers only stock items that move fast. Besides, with the increasing rate of technological change, stores cannot afford to weigh themselves down with big inventories. They must restock frequently so that they can offer products that employ the latest technology.

There was a time when dealers distinguished themselves from mail-order houses by the level of support they could provide to customers, but with dealers being pressured by manufacturers to provide more support for more sophisticated machines on rapidly shrinking margins, there is no longer any advantage to buying from your neighborhood dealer. Dealers, trying to match mail-order prices, will charge you extra for support, particularly for setup and installation services.

If the dealer does not stock what you need, its sales representatives will usually offer to have it for you in two weeks, but you can do that yourself, and save a bundle, and, if you need technical advice, dealer staff members typically lack the right combination of education and experience to help you. For the most part, they are salespeople, not technicians. Stores are not immune from going out of business; it is a very competitive industry that affects every distribution channel.

Although incidents of mail-order fraud have conditioned many people not to buy anything through the mail, believing that mail-order houses are only in business to rip them off, buying computer and communications equipment through the mail can be a pretty safe transaction, once you understand the fundamentals of the mail-order business, know your rights, ask relevant questions, and observe some simple guidelines regarding payment.

The mail-order business is quite simple, conceptually. The mail-order firm buys products in bulk directly from the manufacturers and sells them in any quantity they can, well below the retail price. The more they buy, the bigger discount they can get. Since the company transacts business over the phone and through the mail, it does not have to keep up appearances by maintaining a high-rent storefront, displaying merchandise, and hiring articulate, well-groomed salespeople. They can get by with a run-down warehouse, an 800 number, and a few order fulfillment people. In qualifying for bigger discounts and eliminating a great deal of overhead, there is enough margin flexibility that they can undercut retail stores by thirty percent or more. These margins give you bargaining leverage, in that you can quite easily negotiate discounts from advertised prices, especially when ordering bulk quantities. Many firms even invite such inquiries. An added benefit of mail order is that out-of-state purchases are exempt from sales taxes, which can run as high as eight percent, as in Illinois and Texas. Some corporate buyers like the idea of dealing with mail-order firms that are only a few blocks away, because service is timely, and it makes them feel more secure. Be aware that if you do business with a mail-order firm in the same state, you must pay the sales tax, which negates a key benefit of mail-order buying.

As you can imagine by now, buying through the mail is still not entirely risk-free, but knowing how to purchase through the mail is the first step in minimizing risk. Actually, you should not expect it any other way. The thrust of this book has been that you alone bear responsibility for the purchasing decision, and making the right decision means asking the right questions, as well as scrutinizing the answers. Let us start with a discussion of what the Federal Trade Commission (FTC) has to say about your mail-order rights.

11.2 KNOW YOUR RIGHTS

Mail-order sales are governed by the Federal Trade Commission (FTC), as well as by the laws of each state. According to the FTC rules, the mail-order firm must ship your order within 30 days of receiving it, unless the advertisement clearly states that it takes longer. It sometimes happens, however, that a mail-order firm expects delivery of the product from the manufacturer on a certain date. In anticipation of that date, the product is advertised with no disclaimer that it will take longer than 30 days to ship. All right, these things happen, and there is no intent on the part of the mail-order firm to deceive consumers. Meanwhile, if you prepaid *via* credit card, you start accruing interest charges immediately and must fork over to the credit-card company the first installment payment before you even have the item in your possession.

You can avoid this potential scenario quite easily. Call first and ask if the item is in stock now. If so, ask when it will be shipped. Most reputable mail-order firms will ship within 24 hours after receiving the order, provided the item is in stock.

For an added measure of safety, have the item shipped COD (cash on delivery). If it is not in stock, find out when the firm expects to have it in stock, so you can call in your order at that time. Never prepay for items that are not in stock, unless you are willing to have your money tied up for an indeterminate period. Regardless of what the FTC says about shipping within 30 days, high-demand items that are on back order can take a lot longer to deliver.

The FTC rules also say that if it appears that your order will not be shipped when promised, the mail-order firm must notify you of that fact in writing, in advance of the promised date. Further, the firm must offer a definite new date, if known, and give you the opportunity to cancel the order with a full refund. On the other hand, you may consent to the delayed shipment date, or to an indefinite delay. Before accepting any of these alternatives, you should check around with other mail-order houses to see if they are experiencing delivery problems with the same brand of product. If not, ask if they can match the price of their competitor. If they cannot match the price, you have to decide if the item is worth waiting for, or worth the extra cost of obtaining it sooner.

If you are notified of a shipping delay after ordering, it may be the case that the item you want to buy is in such demand that the manufacturer cannot fill the distribution pipeline fast enough. In this case, you may want to withdraw your original order, wait for the pipeline to be filled, and get a better discount when the product overflows warehouse shelves. Generally, it is not a good idea to agree to an indefinite delay if you have prepaid on the order, if only because you are tying up your money. It is also a bad idea for another reason: once you have agreed to an indefinite ship date, the mail-order firm has no real incentive to fulfill the order as soon as it can. Since it already has your money, there is nothing to stop it from fulfilling new orders, when the item finally becomes available. Another reason not to agree to an indefinite ship date is that the price can fall dramatically between the time you placed the order and the date the item is finally shipped. In prepaying, you limit your flexibility in taking advantage of better deals. Of course, you retain the right to a full refund, even if you previously consented to a delay, but by then you will probably find that the mail-order firm is suddenly ready to ship your order — at the price you already paid. They do not have to offer you the current discount. The entire transaction has complied with all the FTC rules. Even when the mail-order firm cancels your order and agrees to a prompt refund, the delay in processing your request and posting a credit to your credit-card account, or cutting a refund check, could cause you to miss out on better deals. All of this assumes, of course, that the firm has not gone out of business during the interim.

In notifying you of a change in the ship date, or informing you of an indefinite delay, the mail-order house must include a return card on which to indicate your preference. If you do not respond to this notice, under FTC rules the firm is allowed to assume that you agree to the delay, but the firm must either ship your order within 30 days after the original shipping date, or cancel the order. If the order is canceled, you are due a prompt refund.

The mail-order firm must comply with FTC rules regarding the timeliness of refunds. If your order was prepaid by check or money order, the refund must be mailed to you within seven business days. Orders prepaid by credit card must be credited to your account within one billing cycle following receipt of your request. The mail-order firm is not allowed to send you an alternative brand of product, or a voucher in lieu of a refund, without your consent.

When the FTC does get involved in resolving consumer complaints, it thoroughly investigates the charges made against mail-order firms. Depending on its findings, and the level of cooperation it gets from the offending firm, it may decide to resolve the issues in federal court. Alternatively, the FTC may give the firm an opportunity to sign a consent decree in which it agrees to pay a civil penalty, while promising not to violate FTC rules in the future. A consent decree is not an admission of guilt, however. Whatever the outcome, it could have a devastating effect on the firm's future sales, if the situation generates enough publicity.

11.3 MAIL-ORDER BUYING TIPS

There are a number of ways to minimize the risk of buying computer and communications products through the mail. Start by knowing exactly what products you want. There are numerous sources of information available to help you determine your requirements. Many trade publications, for example, routinely test and offer benchmark comparisons on a broad range of products. Make sure any such comparisons offer a complete explanation of the benchmark tests, that the benchmark applies to your application, and that the reviews list the names of the people who wrote them. The computer press is also a good source of information about compatibility issues. You can find out what computers are compatible with various software packages, and what problems may surface when using new versions of your favorite application programs. You should also talk with the vendors of your current software and hardware to find out if they are aware of any problems with mail-order products that are used with their systems.

On-line services are available that provide product evaluation information. Byte Information Exchange (Peterborough, N.H.), for example, provides market feedback on computer and communications hardware and software. It also researches technical problems and finds the solutions, and offers a "Microbytes" newswire that provides analysis of new products, merger and acquisition activity, as well as seminar and conference news. Another service, PC Magazine Interactive Reader Service (Northport, N.Y.), allows you to search for product reviews by key word or product name. It then provides 20- to 30-word summaries of the reviews and the issue in which they were published, whether it was *PC Magazine*, or any one of nine other computer publications.

You can also determine which products to buy by test driving them in dealer showrooms, getting recommendations from peers at other companies, seeing demonstrations at trade shows, and checking with employees who may be using such products at home. You can very easily persuade local vendors to leave their products at your office for as long as 30 days, so you can try them under actual workloads and using your own application programs; or you can buy a single unit from the mail-order firm and try it yourself as part of your evaluation process.

Once you have identified the products that fit your requirements, you can state precisely what you want by brand name, specifying such details as model, make, size, component parts, and subsystem or subsystems. You are better off sticking with name brands, or alternative brands that have received favorable reviews in independent benchmark tests. Your next task is to find out which mail-order firms stock the products you want, and then do a background check before placing an order. It is often helpful to devise a checklist to aid in comparing products and vendors.

In your investigation of mail-order firms, it is a good idea to check with the publication which carried the firm's advertisement. Find out if any complaints have been received about the firm's practices or about the products advertised. Have your company's financial people check on the mail-order firm's rating with the Dun & Bradstreet Report or the TRW Business Credit Report. Since many mail-order firms are not listed on either of these reports, you should conclude that they are too small to process orders in the volumes you require or to provide an adequate level of support.

Also check the Better Business Bureau closest to the firm from which you plan to order. While you are at it, check the membership or research office of the local chamber of commerce. The purpose of this background check is to determine if there are any outstanding complaints. If you do find that there are complaints, find out how long they have remained unresolved. If problems are resolved fairly quickly, that is a good sign, but if the same problems persist, or if they are left unresolved for long periods, you have to assume that the firm does not care. All this investigative work may seem like overkill, but it will save you a lot of problems later. Your goal in all this is to identify one or two mail-order firms that you can feel comfortable dealing with over the long term. Besides, it only takes a few minutes when you "let your fingers do the walking."

Find out how long the firm has been in business, how many employees it has, and if it can furnish you with corporate references who have bought the same products that you want to buy. You should also confirm that all the product's components are from the same manufacturer, and that all the components are the latest versions. Ask if the product will be shipped in the manufacturer's original sealed container. This provides ample assurance that brand name components have not been replaced with those of an inferior brand, enabling unscrupulous firms to sell the originals separately.

Buying software through the mail is slightly less risky than buying hardware. Hardware involves complex circuitry and moving parts, which takes a certain level of expertise to fix. Software, on the other hand, is easily replaced. If the mail order firm will not help you, you can usually go straight to the vendor for a replacement package. You can save yourself a lot of grief, however, by making sure that the software will work on your computer and monitor type, that it includes the driver for your type of printer, and that the version you are buying is the latest release offered by the developer. Keep in mind that most mail-order firms are not staffed to help you install, configure, or use the software, and do not expect refunds on software that is not copy protected.

The low prices of many products may strike you as fantastic bargains, especially PC-XT clones. When buying "no-name" brands originating from offshore locations, or "house-brand" hardware that is not available in stores, check with third-party maintenance organizations like TRW Customer Service Division (Fairfield, N.J.), Sorbus (Frazer, Penn.), or Honeywell Bull Inc. (Waltham, Mass.) to determine if they can support them, just in case the mail-order firm goes out of business. Combined, these service companies have hundreds of locations nationwide, and, if they can support the products you want to buy, you can be reasonably assured that there will not be any unusual service problems waiting in the wings. Do not be lulled into a false sense of security with the mail-order firm's one-year parts-and-labor warranty; the firm may go out of business before you can take advantage of it.

Find out to whom you should address correspondence in the event you experience problems. If you have not received your order as promised, notify the mail-order firm immediately by telephone, and then reiterate your discussion in a letter. Include all pertinent details about the order, including the date the order was placed, the name of the person who took your order, the order number, and a complete description of the item you ordered. Do not forget to keep a copy of that letter. In fact, save all correspondence with the mail-order firm, including your canceled check (photocopy the money order before you send it), credit-card receipt, packing slip, and the written confirmation letter sent by the mail-order firm. The more documentation you have, the easier it is to get the firm's cooperation when you have a dispute, or prove a claim before a consumer-complaint agency in the firm's own state. When experiencing a problem on a credit-card order, notify the credit-card company immediately. Depending on the problem, you may have the right to withhold payment under the Fair Credit Billing Act. Certainly, if the firm has declared bankruptcy and you have not received ordered items, the credit-card company should have no qualms about withholding its payment and crediting your account.

In your initial contacts with mail-order firms, ask about the availability of toll-free technical support. Try out the number to make sure it is in operation. Ask the technician about the procedure for returning items for repair. Find out the turnaround time for repairs that involve swapping out key components or subsystems, and to what extent you can do them yourself, without voiding the warranty. Compare these answers to what the sales representatives tell you. If their stories are not consistent, it is safer to eliminate them from further consideration than to get bounced around later, when you encounter problems with your purchase.

If you are looking to buy complete systems with optional boards and drive subsystems, find out if trained technicians can assemble the whole package and test it for you before shipping. If there is an extra charge for this service, you may be able to eliminate it entirely by reminding the national-accounts representatives that you will probably buy again in quantity, if you can be accommodated on this issue. Remember, the name of the game in mail order is to move products; for the most part, such firms do not want to quibble over a few dollars.

Some products, like hard disks, are more prone to trouble than others. Considering that they may cost between $400 and $2,000, depending on capacity, comparing warranty information is vital. Most manufacturers' warranties vary from six months to one year. If you run into problems and return the new drive for repair, you may have to accept a repaired or rebuilt unit, instead of a new one. There are exceptions, however. In addition to a one-year parts-and-labor warranty, Qubie (Camarillo, Calif.) offers a 30-day money-back guarantee, and will pick up the shipping costs on returned drives. For items still under warranty, it also offers a 48-hour turnaround on repairs, and ships back repaired items *via* Federal Express. Not to be outdone, Core International (Boca Raton, Fla.) not only offers a three-year parts-and-labor warranty, but ships out a new drive, if the original drive proves defective.

It is not enough to inquire about the availability of warranties; find out what recourse you have, if the product you order is delivered "dead on arrival" (DOA). One firm, Micro-to-Go (West Babylon, N.Y.), goes so far as to send a free replacement within 48 hours, if a defective laptop computer arrives DOA. It will even pay all shipping and handling charges, including your cost to ship the defective unit back to them.

If the item performs all right, but is still below your expectations, find out the firm's policy on such returns. In such cases, can you get a complete refund, or must you accept a replacement? Find out if delivery costs are refundable. Find out if there are any "restocking" charges on returned items. Some firms offer a 30-day or a 60-day money-back guarantee on their products, no questions asked, but when receiving the goods under such arrangements, run them extensively during that period. If there is any weak componentry, it will be revealed within that time frame, but only if the equipment is run for long periods to allow maximum heat buildup.

At the same time, do not be too hasty about judging an item defective; check the product's instruction manuals. Setup and initialization procedures may differ markedly from one model to another within the same product line. You may think you know what you are doing, but unless you check the manual you may believe the product is defective. If checking the manual does not help, call the mail-order firm's service hotline. Go over the problem with the technician. Many times the problem will turn out to be a switch that was improperly set at the factory. Do not send the product back to the mail-order firm until you have been instructed to do so. You will be given a return authorization number; without it, the firm may not accept the returned item upon delivery, or may slap you with a restocking charge.

If you are buying brand name products, call the manufacturer to confirm that the mail-order firm is authorized to offer those products. If not, you may have stumbled across a "gray-market" situation, in which the mail-order firm has acquired the products through an under-the-table arrangement with a dealer trying to meet a sales quota, which may be against the manufacturer's policy. In retaining control of distribution, the manufacturer may be trying to protect consumers by offering its products only through authorized dealers who have the resources to provide an acceptable level of support, or maybe the manufacturer just wants to preserve the high price of its products to enhance its image of quality. Either way, the manufacturer may not honor the warranty of products purchased under any other distribution arrangement.

The warranty issue may be complicated by manufacturer-distributor arrangements. It is up to you to seek clarification. For example, some manufacturers provide one-year product warranties to their distributors, but the distributors may only offer six-month warranties to their customers. This means that if the mail-order firm goes out of business during, or slightly after, its warranty period, you are stuck. You will have to return the unit to the manufacturer and pay on a fixed-fee or parts-and-labor basis, just like you would if the product were out of warranty.

When it comes to warranties, have your company's legal counsel review them for overly restrictive language. Sometimes mail-order firms fill them with escape hatches that relieve the firms of any responsibility for servicing defective products. Some warranties are considered void if you perform even simple tasks like installing expansion boards or adding extra RAM.

When it comes to buying computers, PC's Limited (Austin, Tex.) offers a warranty that cannot be beat. Under its 12-month After Sales Assurance Program, which comes with your purchase, if you encounter a problem that cannot be resolved over the phone by its technical support staff, that person will have a Honeywell-Bull technician on your site by the next business day. If your location is more than 100 miles from one of the 160 Honeywell-Bull locations nationwide, however, you can return the unit to PC's Limited to be fixed or replaced at no charge. Other firms, like Discount Micro Products Inc. (Farmington Hills, Mich.), offer extended warranties on their products. At a reasonable cost, DMP will warrant the product to be free of defects in material or manufacture for an additional year, parts and labor included. In addition, the extended warranty is transferrable, in case you sell the product before you have a chance to use the warranty.

Do not be taken in by so-called "creative-pricing" schemes; they cost you money, instead of saving you money. The pricing of The New PC Network borders on the ridiculous. After its advertisements get your attention with attractive prices, you are told at the bottom of the page that "members" pay just eight percent more. (Membership costs $8 annually.) Nonmembers must pay an additional two percent "surcharge." If you want to use a credit card, be prepared to pay another three percent. The advertisement says that there is a $4.00 minimum shipping charge per order, which means that it can be quite a bit more, and the firm will not accept COD orders. Although the firm will repair or replace defective items if returned within 15 days, it will not grant refunds. After that period, the manufacturer's warranty applies.

Granted, even mail-order firms must look for ways to differentiate themselves from their competitors, but complicated pricing is not the way to do it. After you have been "nickeled-and-dimed" like this, you really have not saved much, if anything at all.

If you are paying by check or money order, always call the mail-order house before placing an order, even if you have had successful dealings with it before — just to be sure it is still in business. You also want to call first to find out if the advertised price is still in effect. Remember, in the mail-order business prices fluctuate almost daily in keeping with supply and competitive pressures. When you have determined that the item or items you want are in stock at a price you are willing to pay, establish the delivery date, making sure that the information corresponds to what is offered in the firm's advertisements. Do not forget to request the name of the person with whom you are talking. Make a record of your conversation and follow up with a letter that confirms the details of the order. The letter should state whether you will accept a substitute product if the one you ordered is not in stock by the time the order is processed. If you will not accept a substitute product, state that you expect a prompt refund.

One final tip: never send cash. No reputable mail-order firm ever requests cash as a condition of fulfilling the order, or to hold an order during backlog. With cash, you will not have a receipt, which will make it difficult to prove your claim if anything goes wrong. Look for firms which offer a variety of payment methods, especially those which will establish accounts with your company and accept purchase orders. Look for firms that do not charge extra when ordering by credit card and that do not charge your credit-card account until the order is shipped. You cannot let the mail-order firm decide these matters for you; ask the right questions. If you get the wrong answers, negotiate better terms. There are too many mail-order firms to choose from; deal with the one that best meets your needs.

11.4 PRODUCT EVALUATION

Another factor contributing to the increasing acceptance of mail order by corporate buyers is the fact that many companies already have in-house technical support people on their MIS staff. These technicians usually set up new equipment, load software, perform scheduled preventive maintenance, install cabling and wiring, troubleshoot hardware, and perform equipment moves and changes. This means that the company can get along quite well without relying on mail-order firms for technical support, which provides the company with a great deal of flexibility in the kinds of equipment and subsystems it can buy.

As mentioned before, you should never purchase in volume from a mail-order firm without first obtaining a unit for evaluation. Even though you may have in-house technical support, you must be sure that the equipment you purchase will not pose so many problems that staff time is totally consumed in troubleshooting. Furthermore, the evaluation unit will provide a clue as to what componentry is used. You can then check the availability of replacement parts with the manufacturers. These items can be purchased in any quantities required through manufacturers' catalogs, further lessening dependence on the mail-order firm for technical support. The do-it-yourself method also saves time in that it gets equipment back into action in a matter of days, rather than of weeks.

Since companies are purchasing XT clones in record numbers, here is an evaluation procedure that can be used to qualify obscure brands of microcomputers and subsystems for volume purchases. Upon receipt of the evaluation unit, check for FCC certification and UL (Underwriter's Laboratories) approval stickers on the rear panel. If the FCC sticker is missing, chances are you are going to have problems with radio-frequency radiation, which might interfere with the proper operation of other equipment in the vicinity such as radio, television, cordless phones, and wireless modems and LANs. Even if you do not have these items in the workplace, buying non-FCC approved hardware limits your flexibility to add them later. Radio-frequency radiation also affects the performance of other computers that may be in close proximity. You may be so plagued with error messages that office operations are severely hampered, until you rearrange cubicles and desks in an effort to get computers far enough away from each other.

If the unit does not carry UL approval, chances are that you will experience short circuits that wipe out much of the unit's componentry, because the unit is not built to withstand electrical-power problems. You can save time and effort by not even ordering evaluation units, unless the mail-order firm provides assurance over the phone that the unit complies with FCC and UL standards.

Check the quality of the keyboard to be sure it can withstand continuous use. The keyboard layout should match the IBM XT/AT, and include all the function keys, a numeric keypad, and cursor keys. The Caps Lock and Num Lock keys should have indicator lights. Make sure the keys are uniformly spaced across the keyboard, that they do not stick, and that the space bar works when depressed at each end, instead of just the middle. You should also determine if the unit's keyboard is interchangeable with other IBM-standard keyboards, just in case it needs replacement later.

Make sure the unit's case is made of metal, instead of plastic, so it can withstand frequent moves and changes. The outside seams should fit tightly together to prevent cigarette smoke, dust, and other debris from getting inside the unit, causing the premature death of internal components.

The motherboard should contain a slot for a math coprocessor for numeric-intensive applications, as well as readily accessible and fully documented DIP (dual in-line package) switches. Check the motherboard for solder splashes, discoloration, and loose components — signs of little or no quality control. Check for "spider webs," exposed wires that are meant as a quick fix to faulty connections on the board. Check for gray residue around chips, an indication that they are burned out. Also check for any chips that may be sitting astride other chips, which is a slip-shod way of altering the operation of the original chip on the bottom. Make sure there is enough room on the motherboard for at least 640K of RAM, if it is not already included. The RAM chips should be of the same speed, no greater than 150 nanoseconds, and come from the same manufacturer. All board componentry should be labeled to facilitate fault isolation and replacement.

Check that the power supply is of sufficient size to operate the unit when adding specialized boards. A power supply of at least 135 watts is required for the XT, while a power supply of at least 150 watts is required for the AT. Make sure that the power supply has connections to run two floppy disk drives, as well as a hard disk drive, preferably four connections in all, so you can add an internal tape-cartridge drive without upgrading the power supply. The power supply should have its own UL stamp.

Look for XT clones that come in AT boxes. This will improve heat dissipation when adding boards and drives. Make sure that there are eight expansion slots — standard for the XT — and that they are properly sized to accommodate expansion boards.

Check the quality of the unit's disk drives by running disk-intensive operations like software boots and database searches, comparing completion times with those of an IBM XT. Also, run these operations for long periods to exercise the drive thoroughly with the idea of exposing weaknesses in the drive mechanism. Do a long series of reads and writes to check the quality of the disk drive controller. It is not enough to read data from disk; crashes most frequently occur when writing to disk. When it comes to disk drives — floppy or hard disk — users will be much better off in the long run with brand names. Although cheaper offshore models can easily be replaced, the data that is wiped out during a crash is often irreplaceable.

When your technicians have completed this cursory inspection, they can fill the unit with expansion boards and load software to begin compatibility testing. This kind of testing is necessary to reveal weaknesses in the BIOS (Basic Input-Output System). It is not enough to use spreadsheet or word-processing programs for this purpose, however. You will have to use software that exercises the BIOS more rigorously, such as Borland International's (Scott's Valley, Calif.) Sidekick or Quarterdeck Office Systems' (Santa Monica, Calif.) Desqview. Only when this evaluation process is complete to your satisfaction, can you be reasonably assured of a risk-free volume purchase.

Granted, an extensive evaluation process is usually beyond the capability of most corporate buyers to carry out. If your organization lacks the technical staff or the facilities to perform such evaluations, you may find it worthwhile to hire a consultant to do them for you, in which case, this chapter may help you evaluate the consultant.

11.5 CHECK THE FINE PRINT

You can avoid many problems from the start by reading the fine print accompanying the advertisements of mail-order firms. Sometimes the policies are so heavily weighted against the customer that it is hard to believe that these firms really want to do business at all. Here is a sampling of "fine print" provisions that are commonly used by mail-order firms, with appropriate commentary in brackets.

- Due to currency fluctuations and other factors beyond our control, prices are subject to change without notice. [This is usually a tipoff that the products come from offshore. In this case, make sure the warranty comes from the mail-order firm and not from the offshore manufacturer.]
- All product claims and warranties are handled by the manufacturer only. [If the manufacturer is offshore, do not count on ever getting the problem solved.]
- Sorry, no refunds. [This means that the mail-order firm has the option of repairing or replacing defective items, but does not have to return your money. Months may go by before you finally get a product that works properly.]
- All sales final except that defective items will be repaired or replaced with identical merchandise if returned within 15 days. To qualify, the merchandise and all packing materials must be in new, unused, fresh and saleable con-

dition. After 15 days, manufacturer's warranty applies. [This is risky because it may take 30 days of intensive use to reveal weak componentry. This is also a tipoff that the firm is selling repaired items as if they were new!]

- Products subject to availability. [This means that the mail-order firm can take your order and hold your money without having the product on its warehouse shelves. Although they must comply with FTC rules to notify you of the delay, your money can be tied up for at least two months.]

- Compatibility of products not guaranteed. [This may indicate that the mail-order firm is not too choosy in selecting the products it sells, and needs to protect itself with a disclaimer. If no compatibility problems were revealed in discussions with your hardware and software vendors, buy an evaluation unit and test it yourself before buying in quantity.]

- We reserve the right to limit quantities and to substitute manufacturer. [Doing business with a firm that uses this provision in its advertising is dangerous. According to the fine print, it has the right to substitute the name brand products you ordered with inferior brands from offshore. Although FTC rules say that this is illegal, mail-order firms can get away with it by incorporating such language in their advertising, which makes it a condition of the sale.]

- Add 5% for COD orders. [This clause is intended to dissuade you from using cash on delivery as the method of payment. After all, you retain the option of refusing delivery if the item does not arrive by the agreed-upon date.]

- Refused shipments subject to 20% charge. [This clause is aimed at COD customers who refuse their orders upon delivery. One of the advantages to using COD is to insure that your money does not get tied up in case of shipping delays. If the item does not arrive during the time frame promised, you can cancel the order without negotiating and waiting for a refund, but with this provision, the mail-order firm wants to bill you 20% of the purchase price. On an order of $1,000 that's $200!]

- All returned nondefective merchandise subject to 20% restocking charge. [This provision points to the need to investigate compatibility issues before you buy. If you return an item because it is not compatible with your current hardware or software, this does not mean the item is defective. The mail-order firm feels justified in collecting a restocking charge, because it has to recover the cost of repackaging the item for resale.]

- All returns are subject to our approval. [This clause gives the mail-order firm total discretion on whether it will accept returned items. This gives the firm an opportunity to resolve the problem over the phone with technical support, or to sell you on another product, if the one you have appears defective. Remember, the last thing mail-order companies want to do is return money.]

- Prepaid mail orders sent with corporate or personal check are subject to 15-day fund clearance. [This reveals that the mail-order firm is very much into the "float" game to supplement its sales revenues. They count on the fact that most people do not know that it only takes three working days to clear checks electronically through the Federal Reserve System. Meanwhile, they have free use of your money for 12 days before they even get to processing your order. The cumulative interest over a year's time can produce substantial extra income for some mail-order executives. Federal rules that went into effect on September 1, 1988, prohibit banks from holding personal checks for more than four business days. This rule does not apply to businesses, however.]

- Assembly charges for computer systems may apply. [When you specify that certain brands of components be used in the computer you are purchasing, do not overlook the fact that you may have to pay more than just the price differential. Many firms will add a hefty assembly charge to cover the cost of such customization. Although some mail-order firms tell you about this charge in their advertisements, or call you about it after receiving a purchase order, many others prefer to bill first and explain later.]

These are some of the fine points about buying computer and communications products through the mail; they will help you reap substantial cost savings, but eliminate the risk.

REFERENCE

[1]   *PC World,* August 1987, p. 284.

PART THREE

Know what it is you want, and when a vendor responds, know what it is he's responding with. Be sure the two of you are singing from the same page of the same hymnbook. [1]

Sid Smith
Manager of Corporate Telecommunications
Crowly Maritime Corporation
San Francisco, California

Chapter 12
Structuring the RFP for Risk Avoidance

Many companies considering the purchase of complex products like cross-connects, multiplexers, and local area networks do not have dedicated staff to devote to the task of writing the Request for Proposal (RFP), particularly smaller companies with 500 or less employees. Even among the communications staff of major corporations, writing an RFP can be quite a time-consuming and tedious chore — something to be put off until tomorrow.

The RFP must take into account the different needs of many people within the organization. Internal political issues may complicate the development of the RFP, as in the case of writing an RFP for a PBX that carries both voice and data. The MIS-dp group may perceive encroachment into its domain by the telecom group. The telecom group, on the other hand, may not appreciate the needs of MIS-dp, and dismiss their concerns as mere blather.

There may be different operating philosophies within the company that must be reconciled before the RFP can be developed. For example, in considering the need to changeout a mainframe computer for something more powerful, MIS-dp may feel justified in developing the RFP without too much input from other departments, only insuring that the needs of the company as a whole are well served. There may, however, be a growing feeling within the organization that departmental or work-group computers are necessary, which would offload the mainframe enough to negate the need for its immediate replacement.

Once these issues have been flushed out during your assessment of internal requirements, writing an RFP can be a rewarding and intensely educational experience, particularly for those writing an RFP for the first time. With an understanding of your organization's technical requirements in hand, all that is needed is a place to start — some guidance on structure that is consistent with the risk-avoidance strategy discussed throughout this book. A properly structured RFP containing appropriate details about your present network configuration, as well as future user requirements based on projected growth patterns, will insure that vendors interpret the RFP correctly and address the unique needs of your company with precise information that can be integrated into your buying decision.

There is no standard format for developing an RFP; in fact, each tends to take on a life of its own, as input is gathered from a variety of sources, and as organizational needs change from one purchase to another. Keep in mind, as you develop the RFP, that it should be as thorough as possible, with everything expected of the vendor fully spelled out. If your problem has multiple solutions, or if you are not quite sure what products are most appropriate for your application, provide vendors with as much raw data (traffic studies, modeling diagrams, *et cetera*) as you can, and ask them for their recommendations.

You should also specify your organization's responsibilities to the vendor. With the responsibilities of all parties well documented, the contract negotiation process should proceed smoothly. For those who do not know how to start the first draft of an RFP, this section provides the essential structural elements (see Table 12.1). Additional levels of detail may be added to this "boilerplate" as your situation may warrant. But a word of warning: to facilitate understanding, this framework uses a minimum of legal and financial jargon. As such, the specific provisions should not be used verbatim in any RFP or contract without appropriate assistance.

Table 12.1 Essential Elements of the RFP.

Section I General Information
  1. Statement of purpose
  2. Scope
  3. Schedule of events:
     ●RFP issued
     ●Vendor meeting
     ●Proposal deadline
     ●Vendor presentations
     ●Evaluation procedures and criteria
     ●Contract award
     ●Letter of intent

Section II Contract Terms and Conditions
  1. Liabilities
  2. Mechanical clauses
  3. System specifications
  4. Project support
  5. Costs and charges
  6. Reliability and warranty
  7. Maintenance
  8. Product delivery

12.1 GENERAL INFORMATION

Section I of the RFP, entitled General Information, outlines the intent of the RFP and establishes the ground rules for vendor participation.

12.1.1 Statement of Purpose

This section should begin with a statement of purpose which describes your intent in very broad terms, such as:

The XYZ Company requests proposals to provide products and services necessary to install, implement, and maintain a corporate-wide data-communications network which will replace the existing network. This request will provide interested vendors with appropriate information with which to prepare and submit proposals for consideration by XYZ. It is the intent of XYZ to select the best proposal based on an evaluation of responses and other considerations described in this RFP. XYZ reserves the right to reject any and all proposals received as the result of this RFP prior to the execution of a contract.

The second from last sentence in this example concerning "other considerations" is very important from a risk-avoidance standpoint because it puts vendors on notice that factors apart from the proposal will play a part in the selection process. Such considerations might include findings that reflect negatively on the vendor which come to light as the result of implementing the risk-avoidance strategy described in Chapter 1. This statement will also discourage vendors from challenging your purchase decision solely on the basis of point-by-point comparisons with the proposals of their competitors, assuming that they know how their competitors have responded to your RFP, and, if your organization is a government agency, such statements strategically placed throughout the RFP will minimize the risk of legal challenges from unsuccessful vendors.

12.1.2 Scope

The next heading of Section I should be labeled Scope. This will provide the vendor with an idea of how the RFP is organized. Simply list the various sections of the RFP, including appendices, with their respective page numbers.

12.1.3 Schedule of Events

After the scope should come a schedule of events section, which lists your decision-making milestones, along with their dates, starting with the RFP itself:

- RFP issued;
- Vendor meeting;
- Proposal deadline;
- Vendor presentations;
- Evaluation procedures and criteria;
- Contract award;
- Letter of intent.

Each of these milestones deserves a subhead of its own in Section I of the RFP to relate additional information to the vendors. Under "RFP issued," include the name, address, and telephone number of the person to whom all questions, correspondence, and proposals should be directed.

Regarding the vendor meeting section, include the date, time, and location of the meeting. The purpose of this meeting is to provide interested vendors with an opportunity to ask questions arising from their review of the system requirements. Sometimes the RFP may contain ambiguous terms or statements — or not enough information. This meeting may also benefit you, since vendors may bring up points you had not considered, and which merit inclusion into the RFP. It is important to issue addenda, because you will want the entire RFP and the vendor's proposal to be included as parts of the final contract. In this paragraph, explain the meeting's ground rules, including the procedure for amending the RFP and distributing the changes to attendees, should that possibility arise.

Under the proposal deadline heading, state the number of copies of the proposal that must be submitted, as well as the due date. Provide your organization's point of contact, including telephone number. Specify the procedure vendors must use for obtaining extensions, if any. State whether you will accept multiple proposals from the same vendor, and with what stipulations. Describe the procedure vendors should use to update their proposals. Define what constitutes a complete proposal. For example, if proposals will not be considered without pricing information, say so. On the other hand, if you will accept pricing information in a separate package, as long as it arrives by the proposal due date, say that, too. You should take this opportunity to warn vendors that you will not allow price increases after the proposal is submitted.

Under the vendor presentations heading, specify the time frame you plan to allot to having semifinalists make presentations in support of their proposals. You may want to consider going to the vendors' corporate offices for the presentations, which will give you the opportunity to evaluate the vendor according to the twelve-point risk-avoidance strategy discussed in Chapter 1, and, if a vendor's proposal has included subcontractors, you may want to specify that they are to have a representative in attendance at the presentation. After narrowing down the list of vendors, you can always invite them to your offices for the contract signing.

In the evaluation procedures and criteria section, explain the process that will be used to evaluate vendors. Will one vendor be chosen from among the submitted proposals, or will the two or three strongest proposals be selected for follow-up presentations by the vendors? Will proposals be evaluated in-house, or will outside consultants play a role? What evaluation criteria will be used?

Evaluation criteria may include any or all of the following items, which should be listed in the RFP:

- Prior experience of the vendor in successfully completing undertakings similar in nature and in scope;
- Understanding of the technical requirements and the magnitude of the work to be accomplished, as defined in the proposal and in subsequent meetings with the vendor;
- Arranging the demonstration of a similar system currently in use at a customer site;
- The completeness of the proposal with regard to information requested in the RFP, its level of detail, and its conformance to specifications;
- The vendor's ability to respond with a viable alternate solution, if it cannot precisely address the specifications described in this RFP;
- The vendor's willingness to accommodate changes during installation;
- Experience, qualifications, and professionalism of the vendor's staff assigned to the project;
- The vendor's ability to comply with the terms, conditions, and other provisions of the RFP;
- The vendor's workplan for delivery, installation, and acceptance testing;
- The total cost of fulfilling the proposal;
- The willingness of the vendor to provide information relating to organizational structure and departmental or work-group capabilities (as discussed in Chapter 1).
- During the time that submitted proposals are being evaluated, vendors should be prepared to demonstrate at any time that all aspects of the RFP's requirements can be met or exceeded.

Include in this section a statement that tells vendors that the evaluation procedure is intended to screen out nonresponsive and incomplete proposals to allow your evaluation committee to concentrate its efforts on those proposals which are responsive and complete. In keeping with the concept of risk avoidance, carefully document the reasons for rejecting vendor proposals. A written record of such decisions holds more weight in court than vague recollections, should a vendor want to challenge your decision.

Vendors hate clauses known as "reservations," but they are absolutely required in the evaluation-procedures and criteria section of the RFP as part of your risk-avoidance plan. Inform the vendors that you will not necessarily be bound by the evaluation criteria in making the final selection. Among the many possible contingencies, you may reserve the right to:

- Reject any and all proposals received in response to this RFP;
- Enter into a contract with a vendor other than the one whose proposal offered the lowest cost;
- Adjust any vendor's proposed costs based on a determination that selecting a particular vendor will involve incurring additional costs;
- Waive or change any formalities, irregularities, or inconsistencies in proposals received;
- Consider a late modification of a proposal, if the proposal itself was submitted on time, and if the modification makes the terms of the proposal more favorable to XYZ Company;
- Negotiate any aspect of a proposal with any vendor, and negotiate with more than one vendor at a time;
- Accept any proposal submitted, whether or not there are contract negotiations with other vendors already in progress;
- Extend the time for submission of all proposals;
- Select the next most responsive vendor, if negotiations with the vendor of choice fail to result in an agreement within two weeks.

The importance of the last clause cannot be overstated. In case negotiations with the first-choice vendor fail, it is a good idea to designate second and third choices, so that you do not have to issue the RFP again. In naming alternate vendors, you not only keep their interest alive, you keep their proposals — including pricing and scheduling — in force. In any case, add to your list of "reservations" that you may:

- Prepare and release a new RFP, or take any other action XYZ Company may deem appropriate to insure the best contract.

It is quite common for organizations, especially government agencies, to have two committees perform separate evaluations of the same proposal, one financial and one technical. In this case, state that separate preliminary reviews will be conducted of the vendor's pricing and technical packages to insure that all mandatory requirements have been met.

For each package, describe the review process that will be used to qualify the vendors. For example, if proposals will be assigned points for various evaluation factors, list those factors and the maximum number of points that can be scored on each.

The purpose of describing your evaluation procedure is to convey an image of fairness to vendors, so that the most qualified among them will be encouraged to respond with a proposal. This, in turn, will help insure that your organization's network needs will be met with the best products available at the most reasonable cost.

Under the contract award heading, state how all parties will be notified of your decision. For example, will the announcement be public, or will vendors be notified privately *via* letter? In any case, state that you have no obligation to disclose to any vendor the results of the evaluation process, or the reason why particular vendors were or were not successful. Do not forget to notify the second and third place vendors of their status. This is your escape hatch, if contract negotiations go awry with the first choice vendor.

It is a good idea to include a statement about how you plan to handle proprietary information from vendors. Obviously, proposals contain sensitive information that vendors do not want falling into the hands of competitors. Unless your organization is a government agency, there is only one way to handle this issue, and that is to state up front that any specifications, drawings, documentation, or pricing information, and any other information pertaining to the business of the vendor which is submitted as a result of the RFP will be treated as confidential. Vendors will usually have a copyright on their proposals, along with a caveat that obligates the recipient not to disclose the contents to a third party without prior written authorization.

Proposals submitted to government agencies, however, are usually considered to be in the public domain. If your organization is a government agency, you should remind vendors that if they submit sensitive information as part of their proposals and they want it protected from public disclosure, they should appropriately mark the relevant pages at the top and bottom.

To clarify the ground rules for awarding the contract, state the subsequent steps in the contract-award process. If the purchase entails a long delivery cycle, you might mention that a letter of intent will be issued at some point before or during the contract negotiations. As its name implies, this document is used only to establish your intention to purchase products from a specific vendor — it carries with it no obligation on your part to follow through with an actual purchase and, as such, may be canceled at any time.

To guard against the possibility of any vendor using your organization's name in any self-serving publicity campaign, you may also want to stipulate that vendors will not be permitted to issue press releases or to issue public statements of any kind about the project under bid without prior approval.

12.2 CONTRACT TERMS AND CONDITIONS

Section II of the RFP, entitled Contract Terms and Conditions, is typically reviewed first by vendors. Before committing their resources to developing a proposal, they want to check the RFP for anything out of the ordinary, like an unrealistic delivery schedule, hefty penalties for missed deadlines, or unreasonable claims, such as the right to future product enhancements at no charge. With guidance from your organization's financial and legal officers, this section of the RFP should outline the terms and conditions of the contract which the successful vendor will be expected to enter into with your company. The reason for including this information in the RFP is to notify the successful vendor of the kind of contract you expect it to sign. Your goal is to minimize the time spent in over-the-table haggling, which may jeopardize your time frame for project completion, and to avoid being outnegotiated by the vendor into accepting a compromise solution. With contract terms and conditions spelled out in the RFP, subsequent contract negotiations should end up being a mere formality. Appropriate legal and financial language should always be used to prevent misunderstandings down the road, and to discourage vendors from attempting to get contract clauses voided in court when things do not go according to plan.

This section of the RFP should include a clear statement to vendors that for their proposal to qualify for further consideration they must include a specific response to these terms and conditions, either by indicating complete and unconditional acceptance, or by including in their proposal specific language to replace those provisions to which exception is taken.

12.2.1 Liabilities

The contract's terms and conditions should include a set of liability clauses that specify who is responsible for what, and who pays whom for whose failure, and under what circumstances. Perhaps an explanation will help . . .

- Proposal acceptance: The vendor agrees that the submitted proposal, including separately submitted product-pricing and proposal addenda, constitutes a part of the final contract.
- Financial terms: Neither party will assign this agreement or its rights or obligations, or subcontract its performance to any person, firm, or corporation without the prior written consent of the other party. This consent will not be unreasonably withheld.
- Proprietary rights: The vendor warrants that the products furnished under this contract do not infringe upon or violate any patent, copyright, trade

secret, or the proprietary rights of any third party. In the event of any claim by any party against XYZ Company, the vendor will defend the claim in XYZ Company's name, but at the vendor's own expense, and will indemnify XYZ against any loss, cost, expense, or liability arising out of the claim, whether or not the claim is successful. If any product furnished is likely to or does become the subject of a claim of infringement of a patent or copyright, then, without negating or diminishing the vendor's obligation to satisfy the final award, vendor may, at its discretion, obtain for XYZ Company the right to continue using the alleged infringing product or modify the product so that it becomes noninfringing. In the absence of these options, or if the use of the product by XYZ Company is prevented by permanent injunction, the vendor agrees to take back the product and furnish a replacement that closely matches the performance of the infringing product.

- Consent to jurisdiction: The contract will be deemed to be executed in the City of XYZ, State of XYZ, regardless of the location of the vendor, and will be governed by and be interpreted in accordance with the laws of the State of XYZ. With respect to any action between XYZ Company and the vendor in XYZ State Court, the vendor waives any right it might have to move the case to Federal Court or move for a change of venue to an XYZ State Court outside the City of XYZ. With respect to any action between XYZ Company and the vendor in Federal Court located in the City of XYZ, the vendor waives any right it might have to move for a change of venue to a United States Court outside the City of XYZ.

- Hold harmless: The vendor will hold harmless and defend XYZ Company and its agents and assigns from all claims, suits, or actions brought for or on account of any damage, injury, or death, loss, expense, civil rights, or discrimination claims, inconvenience, or delay which may result from the performance of this contract.

- Injury or damage: The vendor will be liable for injury to persons employed by XYZ Company, persons designated by XYZ Company for training, or any other person or persons designated by XYZ Company for any purpose who are not the agents or employees of the vendor. The vendor will be liable for damage to the property of XYZ Company or any of its users prior to or subsequent to the delivery, installation, acceptance, and use of the equipment either at the vendor's site or at XYZ Company or its users' places of business. Liability results when such injury or damage is caused from the fault or negligence of the vendor.

The vendor will not be liable for injury to persons or damage to property arising out of or caused by an equipment modification or an attachment, or for damage to modifications or attachments that may result from the normal operation and maintenance of the vendor's equipment by XYZ Company or its agents.

Nothing in this contract will limit the vendor's direct liability, if any, to third parties and employees of XYZ Company for any remedy which may exist under law in the event a defect in the manufacture of the vendor's equipment causes injury to such persons or damage to such property.

- Forces beyond control: Neither party will be held responsible for delays or failures in performance caused by acts of God, riots, acts of war or terrorism, earthquakes, or other natural disasters.

- Litigation expenses: The parties agree that in the event of litigation to enforce this contract, or its terms, provisions, and covenants; to terminate this contract; to collect damages for breach or default; or to enforce any warranty or representation described in this agreement, the prevailing party will be entitled to all costs and expenses, including reasonable attorney fees, associated with such litigation.

In addition to the liability clauses listed above, government agencies typically include these protective measures:

- Nonappropriation: If the Department of XYZ does not receive adequate funding for the next succeeding fiscal period and is unable to continue lease, rental, or purchase payments covered by this contract, the contract will automatically terminate, without penalty, at the end of the current fiscal period for which funds have been allocated. Such termination will not constitute default under any provision of this contract, but the Department of XYZ will be obligated to pay all charges incurred through the end of such fiscal period, up to and including the formal notice given to the vendor. The Department of XYZ will give the vendor written notice of such non-availability of funds within thirty (30) days after it receives notice of such nonavailability.

- Performance bond: Upon execution of a contract for lease, rental, or purchase, the Department of XYZ will require the vendor to furnish and maintain, until the product or system has been accepted, a performance bond in an amount equivalent to ten (10) percent of the purchase price.

12.2.2 Mechanical Clauses

There are a number of mechanical clauses that should be included in the RFP's contract terms and conditions section. These mechanical clauses clarify the relationship of the contract's format and individual clauses to the whole of the contract, so that neither party can use it out of context to support a claim against the other.

- Headings not controlling: The headings and table of contents used in this contract are for reference purposes only and will not be deemed a part of this contract.

- Severability: If any term or condition of this contract or its application to any person or persons or circumstances is held invalid, this invalidity will not affect other terms, conditions, or applications, which will remain in effect without the invalid term, condition, or application. Only to this extent may the terms and conditions of this contract be declared severable.
- Waiver: Waiver of any breach of any term or condition of this contract will not be deemed a waiver of any prior or subsequent breach. No term or condition of this contract will be held to be waived, modified, or deleted except as mutually agreed in writing.
- Authority: Each party has full power and authority to enter into and perform this contract. The representative(s) signing this contract on behalf of each party has been properly authorized and empowered to enter into this contract. Each party further acknowledges that it has read this agreement, understands it, and agrees to be bound by it.
- Compliance: The vendor agrees, during the performance of work under this contract, to comply with all provisions of the laws and Constitution of the State of XYZ, and that any provision of this contract that conflicts with them is void. The parties also agree that any action or suit involving the terms and conditions of this contract must be brought in the courts of the State of XYZ or the United States District Court for the State of XYZ.

12.2.3 System Specifications

The contract terms and conditions section of the RFP should include provisions that address system specifications. The following clauses are provided as examples only. As such, they are weighted to favor you, the buyer. If vendors would like to negotiate terms more favorable to themselves, they may do so by proposing alternative language in their proposals. Be careful not to word these clauses too restrictively; your object is not to force reputable vendors from issuing a proposal, but only to insure that you are adequately protected under a variety of adverse circumstances which may arise in the future.

- System warranty: The use of equipment and any software will be under XYZ Company's exclusive management and control. XYZ Company agrees that the vendor will not be liable for any damages caused by XYZ Company's failure to fulfill any of XYZ Company's responsibilities.

The vendor warrants that the proposed equipment and any software, when installed, will be in good working order and will conform to the specifications described in XYZ Company's Request for Proposal, the vendor's official published specifications, the contract specifications, and the vendor's proposal.

In lieu of this warranty of fitness, XYZ Company will have the right, to be exercised in writing, to cancel the procurement within ninety (90) days of installation. XYZ Company shall pay a reasonable lease charge for the time the products were used.

- System configuration: The equipment and any software components to be supplied under this contract, for purposes of delivery and performance, will be grouped together in one or more configurations, as defined in this Request for Proposal (cite the appropriate section or appendix of the RFP). Any such configurations will be deemed incomplete and undelivered if any component in that configuration has not been delivered, or, if delivered, is not operable.
- System performance: Vendor will certify in writing to XYZ Company the date the equipment will be installed and ready for use. The performance period will commence on the first day following certification, at which time operational control becomes the responsibility of XYZ Company.

XYZ Company will make no payments until the system has been in satisfactory operation for at least thirty (30) days after installation.

If successful completion of the performance period is not attained within ninety (90) days of the installation date, XYZ Company will have the option of terminating the contract without penalty or continuing the performance tests. XYZ Company's option to terminate the contract will remain in effect until such time as a successful completion of the performance period is attained. The vendor will be liable for all outbound preparation and shipping costs for contracted items returned under this clause.

- Access to diagnostic information: XYZ Company will have, during the life of the equipment, access to diagnostic procedures and the information derived from them.
- Equipment interfacing: XYZ Company will have the right to connect the products contracted for to any equipment manufactured or supplied by others which is compatible with the vendor's system. Such equipment includes, but is not limited to, peripheral equipment, terminal devices, computers, and communications equipment. XYZ Company may require that the vendor supply interface specifications and supervise the connection of equipment.
- Field service: Vendor will warrant that, in any case where equipment is installed or modified on the premises of XYZ Company, which is contracted for under this agreement, vendor will make such installation at charges in effect at the time of the request by XYZ Company.

12.2.4 Project Support

In keeping with the strategy for risk avoidance, these project-support clauses should be included in the contract terms and conditions section of the RFP:

- Staff quality: The vendor will exercise due care to choose and manage its personnel so that only suitably disciplined and responsible representatives will be operating at XYZ Company and user locations.
- Training: The vendor will provide appropriate training to XYZ Company on the operation, maintenance, and management of the vendor's products, as described in the proposal and its attachments and appendices.
- Documentation: The vendor will provide XYZ Company with three (3) sets of each manual required to operate the system effectively as described in the vendor's proposal. Vendor represents that these manuals are the only manuals necessary for the operation of the system. The vendor will include any other manuals and program descriptions it considers helpful to XYZ Company. All documentation and printed materials provided by the vendor may be reproduced by XYZ Company, provided that such reproduction is made solely for the internal use of XYZ Company and that no charge is made to anyone for such reproductions.
- Emergency response: XYZ Company will be provided with access to an answering service or operator for the purpose of requesting vendor assistance during times of emergency. A vendor representative must have a response time of one hour or less during nonwork hours, weekends, and holidays until full acceptance of the installed system by XYZ Company.

12.2.5 Costs and Charges

The contract terms and conditions section of the RFP should include a set of provisions that clarify costs and charges, so that all parties understand their financial obligations under the agreement:

- Term of agreement: The terms, provisions, representations, and warranties contained in this contract will survive the delivery of the equipment; payment of any lease, rental, or purchase price; and transfer of title.
- Payment procedure: All payments otherwise due under this contract will not be payable until thirty (30) days after receipt of invoice from the vendor.
- Transfer of title: Before any payment is made, the vendor will provide a statement warranting that all equipment and materials, including those of its subcontractors, are free of mechanical liens or encumbrances.
- Failure to perform: In the event that the vendor fails to perform any substantial obligation under this agreement and the failure has not been satisfactorily remedied within thirty (30) days after written notice is provided to the vendor, XYZ Company may withhold all amounts due and payable to the vendor, without penalty, until such failure to perform is remedied or finally adjudicated.

- Default: XYZ Company may, with thirty (30) days prior written notice of default to the vendor, terminate the whole or any part of this contract in any one of the following circumstances:
  — If the vendor fails to perform the services within the time specified in the contract or within the time specified under subsequent extensions;
  — If the vendor fails to perform any of the other provisions of this contract, or fails to make satisfactory progress in the performance of this contract in accordance with its terms. Or the vendor does not remedy such failure within the thirty (30) days — or as mutually agreed in writing — after receipt of notice from XYZ Company specifying the failure.
  If this contract is terminated pursuant to the provisions above, XYZ Company's sole obligation will be to:
  — Continue any installment contracted payments due for products previously delivered and accepted; or
  — Purchase for title, as agreed, any products previously delivered and accepted for payment with principal outstanding.
  XYZ Company may, in addition, procure from the vendor goods specifically procured or acquired by the vendor for the performance of such part of this contract as has been terminated.
- Taxes: XYZ Company will not be responsible for any taxes coming due as a result of this agreement, whether federal, state, or local. The contractor will anticipate such taxes and include them in the proposal. In the case of leased products, the lessor will be responsible for any personal property taxes and will adjust prices accordingly.
- New equipment warranty: The vendor warrants that all equipment and software, when installed, will be new and in good working order and will perform to the vendor's official published specifications and the functional specifications described in this contract. Further, the vendor will make all necessary adjustments, repairs, and replacements without charge to maintain the equipment in this condition for a period of not less than one year after the standard of performance has been met and the product accepted by XYZ Company.
- Prices and terms: All prices, terms, warranties, and benefits granted by the vendor in this contract are comparable to or better than the equivalent terms offered by the vendor to any other public or private entity purchasing equipment of the same quality and quantity. If the vendor offers, during the term of this contract, greater benefits or more favorable terms to any other public or private entity, those benefits and terms will be made available to XYZ Company upon their effective date. Failure to do so will constitute a breach under this contract.

12.2.6 Reliability and Warranty

The contract terms and conditions section of the RFP should include appropriate clauses concerning the product's reliability and warranty. The following items are offered as essential requirements:

- Equipment reliability: In all situations involving performance or nonperformance of equipment or software furnished under this contract, the remedy available to XYZ Company will consist of:
  - The adjustment or repair of the system or replacement of parts by the vendor or, at the vendor's option, replacement of the system or correction of programming errors; or
  - If the vendor is unable to install the system or replacement system or otherwise restore it to good working order or make the software operate as required under this contract, XYZ Company will be entitled to recover actual damages as set forth in this contract. For any other claim concerning performance or nonperformance by the vendor pursuant to or in any other way related to provisions of this contract, XYZ Company will be entitled to recover actual damages to the limits set forth in this section.
- Acceptance testing: In addition to operational performance testing by the vendor, XYZ Company reserves the right to perform additional testing, prior to acceptance, to insure compliance with the requirements and functional specifications of this contract. All attachments may be inspected for compliance with the Federal Communications Commission (FCC) Registration Program. All wiring may be inspected for compliance with state and local electrical codes.
- Building modifications: The vendor will perform all work required to make the product or its several parts come together properly to fit the space allocated for its placement and to make provisions for the equipment to be received for work by other vendors. This work will include all cutting of floors, walls, and ceilings which may be necessary to install equipment and cabling, as well as the restoration of such surfaces to an approved condition.
- Building repairs: The vendor will take all the necessary precautions to protect the building areas adjacent to its work. The vendor will be responsible and liable for any building repairs required as a result of its work and caused by the negligence of its employees. Repairs of any kind that may be required will be made and charged to the vendor or, at XYZ Company's option, deducted from its final payment.
- Clean-up: As ordered by XYZ Company, and immediately upon completion of the work, the vendor will, at its own expense, clean up and remove all

refuse and unused materials from the work site. Upon failure to do so within forty-eight (48) hours after written notification, the work may be done by others, the cost of which will be charged to the vendor or, at XYZ Company's option, deducted from its final payment.

- Additional work: Without invalidating this contract, XYZ Company may order extra work or make changes by altering, adding to, or deducting from the work and causing the contract sum to be adjusted accordingly. All such work will be executed under the conditions of the original contract by a change order. Under no circumstances will extra work or any change be made in the contract unless through a written change order to the vendor stating that XYZ Company has authorized the extra work or change. Any change order which involves a ten (10) percent deviation from the total contract amount may require a new agreement.

  In the event the extra work or change involves materials and labor for which unit prices have not been established, such prices will be determined by:
  — The unit price agreed upon subsequent to the execution of the contract; or
  — By estimate and acceptance in a lump sum.
- Use of premises by vendor: The vendor will confine all apparatus, storage of materials, and operation of this work to the limits specified by law, ordinances, or permits, and shall not unreasonably encumber the premises with materials. The vendor will comply with the laws, ordinances, permits, or instructions of the state regarding signs, advertisements, fires, smoking, and vehicular parking. The vendor will not load or permit any part of the structure to be loaded with weight that will endanger its safety.
- Use of premises by owner: XYZ Company and its users reserve the right to enter upon the premises, to use same, and to have work done by other vendors, or to use parts of the work of this vendor before the final completion of the work, it being understood that such use by XYZ Company or its users in no way relieves the vendor from full responsibility for the entire work until final completion of the contract. XYZ Company reserves the right to enter into other contracts in connection with this work.
- Recovery from disaster: In the event the system or any component of the system is rendered permanently inoperative as a result of a natural occurrence or disaster, the vendor will deliver a replacement within thirty (30) days from the date of XYZ Company's request. In such event, the vendor agrees to waive any delivery schedule priorities and to make the replacement system available from the manufacturing facility currently producing such equipment, or from inventory. The price for replacement equipment will be the price payable under this contract. If the inoperability is due to the negligence or fault of the vendor or its subcontractors, replacement equipment will be delivered at no cost to XYZ Company.

12.2.7 Maintenance

The following provisions concerning maintenance should be incorporated into the RFP's contract terms and provisions section:

- Vendor responsibilities: Vendor will provide maintenance, including associated travel, labor, and parts, either under a maintenance contract, or on a time-and-materials basis at the prices listed in the proposal. This provision does not apply to the repair of damage resulting from accident, transportation between XYZ Company sites, neglect, misuse, or causes other than ordinary use.

- Maintenance personnel: Hardware maintenance will be performed by qualified maintenance personnel totally familiar with all of the equipment installed by the vendor at XYZ Company and its user sites. Maintenance personnel will be given access to the equipment when necessary for the purposes of performing maintenance services under the terms of this agreement.

- Term of maintenance services: Maintenance services will be provided at the prices quoted in the vendor's cost proposal, and may be renewed annually for up to two years at the original prices. XYZ Company may elect to terminate maintenance services at any time upon thirty (30) days prior written notice to the vendor.

- Maintenance documentation: For purchased equipment, the vendor will, upon request, provide to XYZ Company such current diagrams, schematics, manuals, and other documents as are necessary for the maintenance of the system by XYZ Company or its subcontractor(s). There will be no additional charge for these maintenance documents, except for reasonable administrative costs involved for reproduction.

- Right to purchase spares: Vendor guarantees the availability of long-term spare parts for all equipment acquired under this contract for a minimum period of six (6) years following the date vendor provides written notification to XYZ Company that the equipment is out of production, but in no case less than ten (10) years from the date of this contract. Such sales will be made at the prices then in effect, except that prices will not be increased per year by more than the National Consumer Price Index, calculated at a simple rate of increase for each year between the date of acceptance of the equipment purchased under this contract and any order for spare parts.

- Replacement parts: The vendor warrants that only new standard parts or parts equal in performance to new parts will be used in effecting repairs.

- Request for maintenance: XYZ Company will be provided with continuous access to an answering service or operator for the purpose of notifying the vendor of the need for immediate maintenance services. The vendor will have a response time of two hours or less, and have the ability to restore service within three hours of notification.

- Remote diagnostics: It is desirable, but not necessary, that remote diagnostics be performed from the vendor's site. If this type of monitoring is not available, the vendor must describe to what degree its local point of contact will provide diagnostic support to XYZ Company.
- Maintenance and repair log: The vendor will keep a maintenance and repair log for recording each incident of equipment malfunction, as well as the date, time, and duration of all maintenance and repair work performed on the equipment. Each unit of equipment worked on will be identified by type, model, and serial number. A description of the malfunction will be provided, as well as the remedial action taken to restore the unit of equipment to proper operation. This report will be signed by vendor's representative and XYZ Company's representative, with one copy sent or retained at XYZ Company. All response time and downtime credits to XYZ Company will be based on this jointly signed document. Failure to provide XYZ Company a properly completed and signed document will render any claims by the vendor invalid.
- Response time credits: If the vendor's maintenance personnel fail to arrive at the site requiring such services within the designated response time, the vendor will grant a credit to XYZ Company. The amount of creditable hours will be accumulated for the month and adjusted to the nearest hour. Each hour in excess of the specified response time will be computed at the rate of 1/30th of the monthly full service maintenance agreement charge.
- Component downtime credits: If the faulty component cannot perform due to a malfunction through no fault or negligence of XYZ Company for a period of eight (8) consecutive hours or more than sixteen (16) nonconsecutive hours during a twenty-four (24) hour period, XYZ Company will be granted a credit toward monthly maintenance (or rental, if leased). For each hour of downtime, credit will accrue in the amount of five (5) percent of the total monthly charges for all components due under the proposed contract. Downtime will commence from the time of initial notification of the vendor that maintenance is required. The credit for component downtime will be computed to the nearest half or whole hour.
- Equipment replacement: If any unit of equipment fails to perform, and the total number of inoperative hours exceeds twenty-seven (27) hours over a period of three (3) consecutive calendar months, the vendor will, at the option of XYZ Company, provide:
  — A back-up unit of equipment at no additional cost; or
  — On-site technical support at no additional cost; or
  — Replacement of the malfunctioning unit of equipment with a functionally equivalent unit of equipment in good operating condition at no additional cost to XYZ Company. In this case, accrued response time credits and downtime credits will be transferred to this unit of equipment.
- Preventive maintenance: Preventive maintenance, if required, will be sched-

uled by XYZ Company and the vendor at a mutually agreeable time. In the event XYZ Company decides that equipment performance warrants an increase or decrease in frequency or hours, vendor will so increase or decrease such maintenance, provided such request is reasonable.

12.2.8 Product Delivery

In the contract terms and conditions section of the RFP, discuss the vendor's responsibilities related to product delivery:

- Installation responsibility: The vendor will be responsible for unpacking, uncrating, and installing the equipment, including making arrangements for all necessary cabling, connection with power, utility and communications services, and in all respects making the equipment ready for operational use. Upon completion, vendor will notify XYZ Company that the equipment is ready for use.
- Risk of loss prior to installation: During the period that the equipment is in transit and until the equipment is installed and ready for use on XYZ Company and its users' premises and acceptance tests are successfully completed, vendor and its insurers, if any, relieve XYZ Company of all risks of loss or damage to the equipment. After the equipment is installed, ready for use, and has been accepted, all risk of loss or damage will be borne by XYZ Company, except where the damage is attributable to vendor's negligence or to defects XYZ Company could not reasonably have discovered.
- Liquidated damages: If the vendor does not install all the equipment specified in the agreement, including the special features and accessories included on the same order with the equipment, the vendor will pay to XYZ Company liquidated damages for each item of equipment, whether or not installed. For each day's delay, beginning with the installation date but not for more than 180 days, the vendor will pay to XYZ Company 1/30th of the basic monthly rental or maintenance charges or 1/1000th of the purchase price of all equipment listed in the order, whichever is greater.

  If XYZ Company operates any units of equipment during the time liquidated damages become applicable, liquidated damages will not accrue against the equipment in use.

  If the delay is more than forty-five (45) days, XYZ Company may terminate the agreement with the vendor and enter into an agreement with another vendor. In this event, the terminated vendor will be liable for liquidated damages until the substitute vendor's equipment is installed, or 180 days from the original installation date, whichever occurs first.

12.2.9 Rights and Options

Various rights and options clauses should be included in the contract terms and conditions section of the RFP to take into account various unknowns which may arise in the future and which may have adverse consequences:

- Equipment upgrades: XYZ Company may at any time, upon demand, require the vendor to substitute upgraded equipment for any component purchased under the provisions of this contract, including spares and replacement components, with XYZ Company paying the base price of the original item as well as the difference between the price of the equipment installed under this contract and the price in effect for the upgraded equipment.
- Equipment changes and attachments: XYZ Company will have the right to make changes and attachments to the equipment and any software, provided that such changes or attachments do not lessen the performance or value of the equipment or prohibit the proper maintenance from being performed.
- Software ownership: The vendor agrees that any software and accompanying literature developed specifically to implement this agreement will be the sole property of XYZ Company. The vendor further agrees that all such material constitutes a trade secret, and must use its best efforts in the selection and assignment of personnel to work on the development of such software to prevent unauthorized dissemination or disclosure of information related to its development.
- Rights to new ideas: The parties acknowledge that the performance of this contract may result in the development of new proprietary concepts, methods, techniques, processes, adaptations, and ideas. XYZ Company will have unhindered right to use such processes and ideas for its own internal purposes. The vendor will have unrestricted right to use such processes and ideas for commercial purposes, including the right to obtain patents or copyrights.

12.2.10 Relocation

Sometimes it may become necessary to move equipment or whole systems from the original site to another site. The contract terms and conditions section of the RFP should include provisions for relocating purchased equipment without voiding vendor warranties or the terms of the agreement:

- In the event the equipment being maintained under the terms and conditions of this contract is moved to another location belonging to XYZ Company, the terms and conditions of this contract will continue to apply.
- Except in emergencies, XYZ Company will provide the vendor with at least thirty (30) days notice to move the equipment.
- Maintenance charges will be suspended on the date the dismantling of the

equipment in preparation for shipment is completed. Maintenance charges will be reinstated on the day the vendor completes equipment reassembly. XYZ Company will be charged for disassembly and reassembly at the vendor's then prevailing price for such services.

- Shipment to the new location will be by such means as are normally used by the vendor, by padded van or air freight, or any means specifically requested by XYZ Company. XYZ Company may ship the equipment *via* its own transportation or by commercial carrier or, at its option, provide the vendor with authorization to ship by commercial carrier on a prepaid basis, in which case XYZ Company will be invoiced for transportation, rigging, drayage, and insurance costs.

12.3 PROPOSAL SPECIFICATIONS

Section III of the RFP describes the format that vendors must follow in their responses. The purpose of mandating a particular format is to facilitate the review process. You do not want to waste time figuring out if the vendor has supplied the information you requested in the RFP; you want to be able to turn to the appropriate place in the proposal to find what you need. At the same time, you want to encourage vendors to include additional information that they may consider appropriate or helpful in evaluating their proposal.

12.3.1 Introduction

The following language is offered as the introduction to the proposal specifications part of the RFP:

- All documents submitted in response to XYZ Company's Request for Proposal must be clearly identified by title, volume, or document number with the pages numbered consecutively. Accessibility to the proper information is more likely to result in an accurate and complete assessment of the proposal during the evaluation process.
- All documents that comprise vendor proposals must be delivered to XYZ Company in sealed packages. Each package must be clearly labeled as follows:
  — Proposal for XYZ Company Data Communications Network;
  — Vendor's name;
  — Document name (Contractual Proposal, Technical Proposal, Financial Proposal, Reference Materials, *et cetera*);
  — Date of submission.
- The vendor's meeting is the appropriate forum for requesting clarification of any elements of the RFP which remain unclear. Written requests for clarification submitted prior to the vendor's meeting will be appreciated. XYZ

Company will treat such requests as confidential. Any delay in the schedule for receiving or evaluating proposals necessitated by a vendor's inquiry will be applied to all vendors.

12.3.2 Letter of Transmittal

To insure that there are no problems matching proposals with the proper vendors, you may want to specify the content of the cover letter that should accompany the proposal and each separate package that is considered a part of the proposal:

- Name and address of the vendor (or prime contractor);
- Name, title, and telephone number of the person authorized to commit the vendor to the contract;
- Name, title, and telephone number of the person to be contacted regarding the content of the vendor's proposal, if different from above;
- Name and address of any proposed subcontractors;
- Time validity of the offer stated in the proposal (you may specify that the offer be valid for 90 or 180 days, or anything in between — whatever seems appropriate to your situation);
- The cover letter must be signed by an officer of the company.

12.3.3 Proposal Format and Content

To facilitate the evaluation and comparison of proposals, you should plan a format and ask the vendors to adhere to it, possibly as a condition for acceptance. If you are doing an RFP for the first time and need help with specifying a proposal format, here are a few guidelines:

- Executive summary: This section of the proposal will provide a summary of the proposal to include a brief statement of the significant features of the proposal in its component parts. This section should include a statement of the vendor's capabilities and experience with projects of this nature and scope. Vendors may also include any additional information of a general nature which would aid the evaluation team in understanding the thrust of the proposal.
- Contract terms and conditions: Vendors must respond to the contract terms and conditions in Part II of this RFP, either by indicating verbatim acceptance or by including specific language for those provisions to which exception is taken. Failure to address the terms and conditions may result in the rejection of the proposal.
- Project workplan: The vendor must include with the proposal a detailed

description of the work to be done to fulfill the requirements of this RFP, including the target dates for the completion of each task. The workplan must include, but should not necessarily be limited to, the following items:
— A statement of the vendor's understanding of the objective and scope of the requested work;
— A detailed description of each major task associated with the project, including the total number of person-days and elapsed time. This description will identify any anticipated decision points that will involve participation by XYZ Company;
— A project organization chart that shows the involvement of XYZ Company and the vendor's staff;
— A list of the vendor's staff available for the project and a statement of their qualifications, including relevant education, technical level, and similar past experience. Upon selection, the vendor's staff cannot be changed without notifying XYZ Company in writing.

- Forms: All forms included with this RFP must be completed and returned as part of the vendor's proposal. The forms are designed to aid the evaluation process and to demonstrate compliance with this RFP. Failure to complete all of the specified forms may result in rejection of the proposal.
- Vendor qualifications: The qualifications of vendors are addressed throughout this RFP. Responses to the contractual, technical, and financial parts of the RFP will be used to determine the vendor's capabilities to provide a data-communications network to XYZ Company. In addition, the vendor must submit background statements to include:
— Financial statements for the last three (3) fiscal years;
— Three (3) references from financial institutions or creditors;
— A description of any litigation in which the vendor is currently involved;
— A list of subcontractors to whom the vendor intends to contract for purposes of completing the project described in this RFP. This list will include the name of each subcontractor, as well as their addresses, phone numbers, and points of contact. Upon selection, the vendor may not change subcontractors without notifying XYZ Company in writing.
— Three (3) references from customers for whom the vendor has performed similar work. This list will include the name of each customer, as well as their addresses, phone numbers, and points of contact.
- Technical proposal: This RFP describes the features and capabilities required for the data-communications system. Vendors must respond to each of the requirements. Failure to address each requirement may cause the proposal to be rejected from further evaluation. Since all evaluation team members are not technicians and do not necessarily have technical backgrounds, it will be in the best interest of the vendor to keep descriptions in nontechnical language, wherever possible.

- Alternate proposals: Alternate proposals may be submitted. Only those sections that are different from the original proposal need be submitted, provided all differences are clearly defined. Separate, sealed cost proposals clearly marked "Alternate Proposal" must also be submitted with each alternate proposal.
- Reference materials: Reference materials are those that are referred to in the proposal such as sales literature, technical manuals, training manuals, *et cetera*. Whatever materials are referenced in the proposal must be packaged separately and submitted as part of the proposal.
- Financial proposals: Vendors may submit separate pricing proposals which address one or more of the following options:
  — Rental price;
  — Straight purchase price;
  — Straight monthly long-term lease prices for five-year (60-month) and ten-year (120-month) periods.

In addition to the above, government agencies may want to consider the following option:

  — Tax-exempt installment purchase for five-year (60-month) and ten-year (120-month) periods.

Additional options or different time periods may be proposed at the vendor's discretion, provided the vendor responds to at least one of the four options described above.

12.4 TECHNICAL REQUIREMENTS

Part IV of the RFP describes the general aspects of your network that the vendor must address throughout the proposal, as well as specific equipment requirements. The phrasing of the introductory paragraph may be as simple as:

XYZ Company intends to purchase a data-communications system to replace existing equipment under lease from the ABC Leasing Company (see Appendix __). The new system will provide better diagnostic capabilities, greater configuration flexibility, and substantial cost savings over the system currently in use.

Do not forget to reference the diagram of your network in its current configuration. This diagram simply provides a bird's eye view of your entire network and is not intended to be too detailed.

Follow up the introductory paragraph with more detail on what you are trying to do with your network and how you want to go about accomplishing it. For each node of your network, provide a separate diagram showing the type and quantity of equipment in use. If you anticipate growth, provide separate diagrams showing the type and quantity of equipment you think you may require. If you anticipate growth, but do not know what your equipment requirements will be, supply enough data about current and projected traffic (data as well as voice), staffing levels, terminal stations, and type of transmissions (interactive or batch), as well as their breakdown by percentage, so that the vendor has enough information to propose a solution.

12.4.1 General Considerations

You may have some broad areas of concern that you expect vendors to address in their proposals, such as:

- Scope: The successful vendor will be required to furnish, install, interface to telephone company equipment, test, maintain, and provide training for the system and individual hardware components.
- Transmission speeds: The vendor must be able to provide equipment to allow transmission speeds of 1200, 2400, 4800, 9600, 14400, and 19200 b/s, which may differ from site to site.
- Transparency: The new data-communications system must be transparent to the user, with no alterations to data-terminal equipment or networking software required to implement transmissions.
- Data formats: At both the terminal sites and the CPU site, the system must support a combination of asynchronous and synchronous data-transmission formats.
- Data-terminal equipment: The data-terminal equipment currently supported
- by the XYZ Company network is listed in Appendix___.
- Cabling and wiring: The vendor must provide and install all cabling and wiring for the new data-communications network. This requirement applies to all XYZ Company locations and all user work stations, from telephone interface to user terminals. Where feasible, the vendor may use existing user site cabling and wiring. In any case, the vendor must provide detailed diagrams of all cabling and wiring, and provide appropriate labeling at termination points.
- Cabinets: Vendors must supply equipment cabinets when installing more than four modems or more than one multiplexer at a single site. It is preferred that the modems use a card type that is usable for both stand-alone and rack-mounted configurations, and that the card cage utilize a universal type back-plane to accommodate any mix of modem and multiplexer cards. Power distribution equipment must be included in the cabinets.
- Security: XYZ Company plans to maintain its present security system, which is implemented at the front end processor (FEP) under the following hardware and software constraints:

— From any given terminal, there is only one data path for transactions to reach the transaction processor.

— A terminal must be logged into the message-control program before any transaction will be passed to the transaction processor.

— All transactions have the log-in code appended as a prefix before the data is passed from the message-control program to the transaction processor.

— The transaction processor uses the log-in code in building the key to access any and all on-line data files.

The message-control program associates a hardware address (multiplexer subchannel) with a specific user terminal identification. The security file relates terminal identification with valid log-in codes. If a user at any location attempts to log in using a code that is not valid for that location, the user will be denied access to data files.

- Operating and maintenance procedures: The vendor will be responsible for developing, for XYZ Company's approval, the necessary operating and maintenance procedures for the network-management control system and the system in general. These procedures will be prepared prior to cutover of the first site, and revised as necessary during system implementation.

- Equipment labeling: The vendor will label all racks, cabinets, equipment, and boards, as well as connectors and cross cabling. Such labeling must be in plain view.

- The vendor will include the following information about the proposed solution:

  — Equipment requirements and costs by location;

  — Equipment configuration drawings by location;

  — Network configuration drawings showing all locations;

  — Space and power requirements for each location;

  — Environmental requirements for each location;

  — Technical documentation for all equipment;

  — Complete description of circuit requirements for each location.

12.4.2 Equipment Specifications

The technical requirements section of the RFP is your opportunity to request detailed information about vendor products. An example of a format that might be used to solicit vendor information about a centralized network-management control system follows.

- Centralized network-management control system: XYZ Company requires a centralized network-management control system to be installed at its present CPU location. This system must be of sufficient capacity to support and control the entire XYZ Company network. Minimum components required include:

— Central processing unit;
— Hard-disk storage;
— Network-management terminal with graphics capability;
— On-line printer;
— Local and remote monitoring devices.

The new network control system must support management activities at the operational and planning levels for:
— Failure management;
— Performance management;
— Configuration management;
— Inventory management.

Failure-management activities include problem determination and system restoral. Required operational-level functions will include positive alarms (audible, visual, and printed) for network-component failure or degradation. Alarm information will include nature of failure or degradation, and location.

Performance-management activities will include the optimization of response time and network-availability parameters. Response-time data must be available as both CPU response time and network response time. Long-term response-time data must be available for historical inquiry to aid in future planning and problem solving.

Configuration-management activities combine data from failure management and performance management to support the long-range planning of the network's topology.

Inventory-management activities require that an inventory data base be established which includes both active and spare parts. Inventory data combined with failure and performace data must provide the network manager with information to support critical network-management decisions.

Vendors must provide the following information on their network-management control system:
— System characteristics:
    Number of processors;
    Processor type;
    Main memory capacity;
    Operating system;
    Storage capacity;
    Storage-capacity expansion capability;
    Console display type.
— Technical-control features and functions:
    Alarm conditions;
    Number of alarm levels;
    Alarm types;

Fallback switching;
Switching method;
Monitoring;
Remote monitoring devices;
Type of monitoring signal.
— Network-management features and functions:
Data-base management system supported;
DBMS acquisition (bundled or separate);
Data recorded;
Reports available.
— Transmission specifications:
Maximum transmit and receive rates;
Transmission techniques supported;
Interfaces supported;
Maximum number of lines supported;
Expansion increments.

For each type of equipment, you would request from vendors appropriate information in similar detail. If you are concerned about possible future migration to ISDN, you should also ask vendors to address the compatibility or upgradability of their products with ISDN.

12.5 APPENDICES

Include various appendices that amplify key elements of the RFP. For example, you should include a summary diagram of your current network, with supplementary diagrams showing specific details that you expect vendors to address, such as present and planned locations for modems with drop-and-insert capabilities.

Use a separate appendix to list equipment currently in use on your network. Include the quantity of equipment by model and manufacturer. Any forms or questionnaires also merit appendices of their own, as do summary tables of voice-data traffic and any network-modeling studies, including their assumptions. A glossary of acronyms used in the RFP may even be warranted.

In general, any information that will assist vendors in assessing your needs and developing a proposal that addresses those needs is appropriate for an appendix. Do not forget to cite the appendices in the main body of the RFP.

12.6 SOME RFP DO'S AND DON'TS

Everyone has his or her own way of developing an RFP, but there are some things you should and should not do in preparing and issuing an RFP. First, the "don'ts":

- If you have a preferred vendor in mind, don't issue the RFP to vendors who cannot possibly meet the product specifications. Worse yet, don't specify the product and then issue the RFP to that vendor's competitors. Such attempts to "hot wire" the RFP are easily recognized as such in the vendor community. Some vendors feel they must respond to these proposals, if only to leave the possibility open for future business, but, as the issuer, be aware that you are only wasting the vendor's resources in what it knows to be a losing proposition. Underneath the thin veneer of cordiality, there lies a seething cauldron of ill will that may come back to haunt you, and if your organization is a government agency, you could be leaving yourself open to legal challenge by an unsuccessful vendor.

- Don't dictate delivery and installation schedules to vendors; more often than not your time frames are going to be unrealistic, which may cause the best vendor to "no-bid" the project. Let the vendors address these matters in their proposals; after all, they know best what their inventories look like, what their production output is, how long it takes to install their own products, and the man-power requirements of each job. Vendors know they are competing with other companies for your business; count on them to propose their best schedules. You can always nail down specific dates during the contract negotiations with the successful vendor.

- Don't specify features or capabilities that you know do not exist. Vendors tend to read too much into such requirements; either they think you are only on a fishing expedition and not really serious about the RFP, or they believe the RFP has been tailored to a specific vendor. Both interpretations may trigger a "no-bid" response to your RFP, or result in RFPs that simply ignore the requirement. There are two effective ways of handling this issue: one is to request that vendors address the requirement in terms of customizing existing products; the other is simply to describe the problem you have and let the vendors propose the possible solutions.

- If your RFP is for a large-scale network that involves multiple vendors, a high degree of equipment interfacing, and upward migration paths to features that may not yet exist, don't write the RFP as if this were a mere product procurement that only calls for the vendor to supply a price sheet and delivery schedule. You will leave yourself open to the unscrupulous vendor who low bids on the equipment and later overcharges you for integration services and ongoing support, once the project is too far along for you to waste time soliciting bids for that kind of work.

- Don't be fooled by vendors who offer "RFP support services." Vendors use this ploy to slant RFPs to their own products. Sometimes the slant is so subtle that only the vendor's close competitors will recognize what is really going on.

- Don't assume that you alone know the requirements of your entire organization. Perform a needs assessment among the various departments and major work-groups to insure that all the present and foreseeable requirements of the organization will be covered in the RFP.

There are also a number of things you can do to facilitate RFP development and insure vendor responses that address your organization's specific needs:

- Do allow enough time for planning — not just the kind of planning required for daily operations, but strategic planning. Stay updated on your organization's business plan, so you can anticipate future network requirements. Keep informed of new products and technologies. Read up on the latest merger and acquisition activity of current and potential vendors, and try to predict what effect this will have on your ability to expand your network. Learn to develop contingency plans that can be invoked virtually instantaneously if things do not go according to plan.
- Do take advantage of the creativity and problem-solving abilities of vendors. Although you may have a firm handle on your organization's network requirements, instead of imposing your own solutions on vendors, use the RFP as the means to request possible solutions. Vendors continually complain that they are not given the chance to provide this kind of input. Invariably, vendors will come up with solutions you had not considered, if only because they have the benefit of being able to draw upon more expertise over a number of specialized fields. The vendor's solution may even save you money over the life of the contract. In sum, build into your RFP enough flexibility to allow input from vendors.
- At the same time, have a technically knowledgeable staff person participate in developing the RFP. Allow that person to ride shotgun over all meetings with vendors, and let that person play a leading role in evaluating vendor proposals. If vendors know that you have an "ace in the hole," they will be discouraged at the outset from trying to pressure you into buying products you do not really need.
- If you do not have a technical whiz on staff, consider an outside consultant for the role, especially if the RFP requires an in-depth knowledge of available products, technologies, and architectures. The competitive marketplace has become saturated with a seemingly endless variety of products, and the pace of innovation boggles the mind. Unless you are continually involved with evaluating products, making a decision on a large capital purchase can be quite risky without some outside assistance. Consultants can bring objectivity to needs assessment, vendor evaluation, and product selection. Beyond that, consultants provide you with extra staff and lend credibility to your decision-

making. If you use a consultant to write your RFP, this chapter will make a good yardstick for evaluating the results.

- With a multisite network, issue a single RFP for the entire network, rather than separate RFPs for each site. Even though your time frame for completing the project may be as long as two or three years, you can still qualify for bigger volume discounts on equipment by lumping all of your requirements together under a single RFP.

- Finally, package the RFP in a professional manner by organizing it simply and logically, thereby making it easy for vendors to follow and helping them to develop a timely response that addresses the issues that are important to you. Make every effort to eliminate typographical errors and ambiguous language. When reprinting the RFP for distribution, make sure the pages are not spotty or streaked, reducing legibility. Make sure pages are properly numbered, diagrams properly labeled, and that acronyms are spelled out. Insure that the binding does not interfere with the text and that appropriate contact information is displayed in a prominent place at the beginning of the RFP.

12.7 REQUEST FOR INFORMATION

When a variety of possible solutions are available for your particular problem, or you are not quite up to date on the latest technology, you may want to solicit input from a multitude of vendors before issuing an RFP. In this case, the Request for Information (RFI) is the appropriate information gathering tool. The RFI is typically distinguished from the RFP by its brevity and informality.

You should be aware that many vendors dislike RFIs, because they believe the issuer is merely trying to do market research at their expense. Judging from the increasingly detailed and rigidly constructed RFIs being issued these days, vendors seem to have a legitimate gripe. Briefly, the RFI describes your problem and simply asks vendors to respond with an available or feasible solution, the degree of customization that may be required, and what is involved in implementing such a project. The RFI should not require a detailed response from vendors or specify a rigid response format. Even requesting financial statements and credit references constitutes an abuse of the RFI. After all, if you are not ready to commit to a purchase, you should not expect vendors to respond as if you are.

Some users justify using RFIs in place of RFPs because doing so provides them with more flexibility in selecting vendors. Unlike an RFP, the RFI does not convey a commitment to purchase, and there is no obligation to consider all responses with an evaluation process that is applied equally to all. Consequently, this lessens the chance of lawsuits from unsuccessful vendors. This kind of gameplaying, however, is recognized by vendors. With their resources stretched to the limit, vendors must give priority to RFPs. If your RFP is masquerading as an RFI, you risk not attracting the attention of the most qualified vendors.

12.8 CONCLUSION

The process of acquiring and accepting new products, and integrating them onto your network, can become quite a complex undertaking. If you allow time for planning, involve the operating groups within your organization in the needs assessment, put some effort into structuring the RFP, and stay flexible with regard to suggestions provided by vendors, you will not only minimize risk, but also insure selection of the best possible products and continuing vendor support for your network.

This is not to say that doing your homework is unnecessary and that you can achieve the best results by leaving everything to vendors, or even consultants. Keep in mind that vendors will only propose solutions that can be achieved with their available products, which may not be what you need. Consultants will only propose solutions that they are comfortable with; in doing so, they may exclude solutions that might be exactly what you need. At the same time, if you give the impression that you are giving vendors and consultants *carte blanche,* they tend to go off the deep end by giving you the best of *everything,* when you may actually be better off with considerably less. The RFP should give vendors an accurate picture of your requirements in as much detail as possible.

Finally, do not take anything in the vendors' proposals at face value. Use the risk-avoidance strategy to determine the organizational stability of vendors and the long-term reliability of their products.

REFERENCE

[1]   *Network World,* February 22, 1988, p. 23.

PART FOUR

Given a chance to bad-mouth vendors, blow the whistle or tell war stories
. . . most managers simply don't feel the need. There's no untapped vein of
anger and frustration . . . just resignation to the fact that you can get what
you want from vendors if you work at it. [1]

John Gantz
Executive Vice President
Technology Financial Services, Inc.
Chelmsford, Massachusetts

Chapter 13
Wrapping It Up

13.1 SOME PARTING TIPS FROM THE PROS

Consultants are a good source of information about vendors and specific products. Unlike most corporate managers, consultants meet with vendors every day and have, over the years, accumulated a vast storehouse of knowledge about the performance of numerous products. Through it all, they have become hardened against sales hype and vendor ploys. Since they are not connected with any single vendor or product, they also usually have no qualms about standing tough against the bully tactics of unscrupulous vendors, and otherwise looking out for the best interests of their clients. Among such consultants are Paul F. Kirvan, Judy Cherashore, Richard A. Kuehn, and James H. Morgan.

Paul F. Kirvan, an independent consultant in Turnersville, New Jersey, notes that insuring vendor and product reliability has become a critical activity for corporate managers. "Before making a major equipment purchase, especially if the manufacturer, distributor, and product are not well known, managers should make sure they pass several stringent tests." Kirvan offers this back-pocket checklist for vendor reliability:

- Company's financial position;
- Number of years in business;
- Number of years manufacturing or selling product of interest;
- Channels of distribution; specifically, how many, where located, exclusivity of product, years selling product;
- Overall market position;
- Growth (or decline) over the past two to three years;
- Number of employees; increase or decrease over the past two to three years;

- Number of customers; increase or decrease over the past two to three years;
- Reasons for increase or decrease;
- Demographics of customer base;
- Number of employees in R&D, product development;
- Number of employees in customer service;
- Number of employees in technical support and maintenance;
- Number of employees in software support;
- Activities in R&D, number of new product announcements, patents, product recalls;
- Presence or absence of quality control–quality assurance programs;
- Current and typical inventory levels.

Kirvan counsels buyers to confirm the vendor's financial position and market strength by examining annual reports and Form 10-Ks, as well as by discussing the vendor's performance with leading brokerage houses, evaluating current stock price-earnings ratios and overall stock performance for the last 12 to 18 months, and noting evidence of any merger or takeover activities.

"Once company research has been completed," says Kirvan, "product reliability should be evaluated":

- Determine mean time between failures (MTBF);
- Determine mean time to repair (MTTR);
- Identify and speak with at least three customers who have the product installed and working;
- Make site visits to at least two users of the product to see typical installation, vendor workmanship, and product operation;
- From these customers, identify at least two other customers whose names were not provided by the vendor; these can provide more objective analysis, even reveal negative aspects of product or vendor;
- Examine product warranties;
- Examine service contracts available for product;
- Determine availability of spare and replacement parts;
- Examine results of independent testing firms, if available.

"After researching the above areas," advises Kirvan, "managers should obtain opinions and recommendations from reliable consultants who have proven objectivity and no recognized allegiance to a specific product or vendor." Kirvan also notes that if you are a member of an association of end users, or of another similar organization, you should discuss your situation with fellow members at the next regularly scheduled meeting, or simply call a few members you know for their observations.

Judy Cherashore, a consultant with the telecommunications consulting services group of Richard A. Eisner and Company (New York, N.Y.), warns that buyers usually face a broad range of vendors and personalities when comparison shopping, and each one is intent on selling his product as the best and only product. Cherashore suggests interviewing at least four to six vendors and obtaining an understanding of each vendor's product before attempting to make a selection. This process, although important, can overwhelm even the most experienced buyers. Therefore, Cherashore advises seeking a telecommunications consultant who knows the industry and can match your requirements with the proper vendors.

"To avoid unnecessary risks in the purchase of high technology, the first step is to find a good consultant. Vendors see the value of working with consultants and making their clients happy because it means more business down the road," observes Cherashore. "A vendor working with a single customer may be tempted to make the sale and run."

Where are consultants to be found? Cherashore says you can start with the Yellow Pages under "Telecommunications Consultant" or "Communications Consultant." You can also contact the Society of Telecommunications Consultants (New York, N.Y.) for their directory, which lists the experience and qualifications of its more than 200 members. "Another good source is to ask business friends for the names of consultants they have used," she says. "Then call at least four references for each consultant."

According to Cherashore, upon finding a good consultant, these are the steps you can expect that person to go through with you, when looking for a telephone system, for example:

- Discuss your present telephone system, as well as your current and future needs.
- Set up a meeting with several reliable vendors who have a base in your geographic area. When an emergency occurs with your system, you want the vendor chosen to be close enough to respond quickly.
- Go on field trips with the vendors to see their systems in operation.
- Discuss the Request for Proposal (RFP) designed by your consultant for your company's needs. The RFP should be sent out to at least three of the vendors you have visited during your field trips.
- Review the completed RFP, which will include requests for references on the vendor and product, as well as such issues as union affiliation and response time for minor and major outages.
- In reviewing vendor proposals, decide which references to check and call those in your geographic area. Some vendors have several offices, one of which may be better run than the others. You want the reference on the office that would be servicing your company.
- Ask pointed questions when checking references: How long have you had the system? Was the installation reasonably smooth? Were the technicians neat, clean, polite, and cooperative? Has the system ever gone down? Did the vendor respond quickly? Was the problem resolved within a reasonable time?

Cherashore cites another benefit of having a consultant assist you with the purchase: "Consultants know how to ask for the things that are in your best interest — and they're wise enough to get it in writing. They are not afraid to negotiate the price and terms of the contract, making sure that your last payment isn't due until 30 days after the successful cutover of the new telephone system."

"When you add it up," says Cherashore, "choosing a good consultant will save you time, money, and aggravation."

Richard A. Kuehn, principal of RAK Associates (Cleveland, Ohio), notes that long after the sales hype and price are forgotten, any technical system settles into the requirement for reliable day-to-day operation. "Ultimately, customer and user satisfaction can be distilled into a combination of product reliability which is dependent upon, to a great degree, vendor stability and/or support."

According to Kuehn, "never has any sales presentation done other than promise the highest levels of reliability and support. If the buyer is to avoid the relative high risk of simply believing sales presentations, there must be methods to further verify the key factors that contribute to a project's success."

Kuehn acknowledges that a product's reliability may, in fact, be the most difficult thing to deal with. He notes that while there are methods of calculating and verifying MTBF and MTTR, there is still always the possibility of the occasional "lemon." He recommends, therefore, that as a standard portion of the RFP process, the manufacturer's estimate of MTBF be requested. "While this is not a guarantee of failure rate, it is a benchmark on which to judge future performance." Even before MTBF is discussed with a vendor, Kuehn advises coming up with a simple assessment of a failure rate that you can live with, being mindful, of course, that a 99.9% availability rate could result in a single three-hour failure during the course of the year.

"While the manufacturer's statement will not make it so, the ability to have a contracted MTBF reference certainly is helpful in the event a 'lemon' is delivered. From a reliability perspective, an understanding of the product and ability to secure — generally at additional cost — further safeguards in the system and/or equipment through redundancy of processors, critical electronic parts, and universal power systems must be further weighed on a 'risk *versus* reward' basis."

In many cases, according to Kuehn, it is much more difficult to deal with the issue of vendor stability. Some level of stability can be measured by the market share of the particular item involved. However, notes Kuehn, it is necessary to view that market share from a national perspective. Even then, vendor stability and support differ according to the individual office supplying equipment at the local level. Service and coverage can vary greatly across the country. Nevertheless, Kuehn cites several items which can generally be verified and provide relatively good indicators of the support that can be expected.

"First is the obvious inquiry of references provided by the vendor. If at all possible, attempt to go several levels into this referencing process by asking those references who they themselves talked to when contacting references for their purchase. It is surprising how many times that references provided by the vendor turn out to be not very good."

Next, advises Kuehn, determine the number of factory trained and certified technicians available at the local-office level to supply support. Make sure you visually check the depth of stocked service parts. Those parts should be reserved as "service spares" and not be available for expansion or additions to systems. Once the quantity of trained and certified technicians is known, determine the quantity of systems or, in the case of a telephone system, station lines, being serviced. In the telephone-system instance, the ratio should be approximately one technician per 800 to 1,200 instruments. According to Kuehn, deviations from this ratio would indicate either a highly unprofitable service department or a lack of the depth necessary to provide the desired rapid service response.

Kuehn also suggests that visits to other installations serviced and maintained by the local vendor's office can prove invaluable. "Viewing terminal strips in the main equipment room and the wiring on the distribution frames on various floors will provide a great deal of insight relative to the vendor's pride in a particular installation. Clean equipment rooms with all cable neatly tied, laced and labeled on terminal strips, as well as up-to-date site documentation, indicates a high level of pride in the installation. The opposite is usually indicative of the type of support attitude that could be expected."

While attention to these factors will neither be all-encompassing nor a substitute for other areas of equipment or application review, Kuehn notes that satisfactory responses to all of these concerns will materially assist in selecting an ideal vendor in potentially risky purchase situations.

Consultants working on the leading edge of technology are more exposed to risk, because there is less precedent to draw upon. Over the years, such consultants have developed their own guidelines for avoiding risk — or at least minimizing it. One of these is James H. Morgan, the principal of J. H. Morgan Consultants (Morristown, N.J.).

Morgan notes that high-technology purchasing is hamstrung by the continued presence of throwbacks to older days, which, he claims, still govern the majority of high-technology purchases. Some of these outmoded ideas include:

- Purchasers still go with the lowest price;
- Purchasers still ignore the lifetime costs of systems;
- Specs are still poorly written;
- Purchasing still meets Telecom at the end of the line;
- Project teams are still seldom formed;
- Nontechnical factors are still underplayed in the evaluation process;
- Contracts are still written in a last-minute rush.

To correct these outdated concepts, Morgan offers some practical advice which has proven useful many times over in his experience with high-technology purchasing.

"Purchasers must go with best value — not lowest price."

Many purchasing departments just do not know how to purchase high-technology systems. They continue to apply the rules used for buying pencils, vehicles, sprinkler systems, and the like — they go with the lowest bid.

Purchasing Departments must establish a separate set of rules called "High-Technology Procurement." Coming under this heading would be data-processing or management-information systems, telecommunications systems, and other computer-oriented systems. The common denominator is highly sophisticated, intelligent, fast-moving technology — usually heavily software oriented. Examples are computers, PABXs, LANs, fiber optics, radio systems, network-management systems, wiring systems, and related components and subsystems.

"Lowest price" must be replaced with "best value" as the main purchasing guideline. Allowance must be made for the evaluator's judgement, because high technology does not lend itself to the simple "black and white" comparisons that were so common in the past, when all wiring systems or all phone systems looked alike.

Tip No. 2
"Purchasers must look at systems' lifetime cost."

This should have been done all along, but it becomes even more important for high-tech systems because of high tech's complexity and boundary crossing. Purchase price certainly is important, but over the system's lifetime, typically four to seven years, it can be swamped out by other, recurring "hidden" costs. The culprits are people saying, "Oh, that goes on someone else's budget." Examples of these extra costs, too often ignored, are: maintenance; staff changes, especially increases or decreases of numbers and skill levels; utilities (power, UPS, HVAC); floor space; and cost-saving features.

As applied to a PABX, for example, cost-saving features include LCR, SMDR, and network optimization. According to George Durar, President of ATM Consultants (Denville, N.J.), a 1,000-line PABX initially costs $1 million, which incurs $800,000 a year in toll charges. With appropriate cost-saving features, typically 25% can be saved on tolls which, over a five-year period, can pay for the original equipment! The purchasing department must insure that this lifetime cost analysis is always made.

Tip No. 3
"Write detailed, but functional, specifications."

Experience has shown that the smoothest-running projects started off with specs that were detailed, yet still only functional. This is the best time and money spent on a project. Today, even the spec for building wiring must be 50 pages or more.

It is no longer possible, however, to go beyond a functional spec into a design spec for high-tech purchases. This would mean doing a very detailed design, which may cost vendors many millions of dollars, which purchasers obviously cannot afford. Besides, all vendors would "no bid" the RFP, unless, by sheer coincidence, or by questionable means, the purchaser's design fit a particular vendor's offering. Even then, competitive bidding is out.

Detailed, functional specs encourage competent vendors. Incompetent vendors drop out, because they rely on weak specs — the so-called three-pager. Yet, this is exactly what most purchasing departments seek.

Other solid documentation that is common to successful high-tech procurements include (in order): Strategic Plan, Spec-RFP, Evaluation Criteria, User-Written Contracts (one each for the system and for maintenance), Project Management, Acceptance Test Procedure, and User Manuals. The Purchasing-Telecom team must assure that such documentation is accomplished or, in the case of user manuals, provided by the vendor.

Tip No. 4
"Purchasing and Telecom must work together from the start."

The likelihood of high-tech project success greatly increases when Telecom or other technical departments bring in Purchasing immediately upon identification of the project's need. In the past, it was common to invite Purchasing's involvement after the spec was written.

This now requires Purchasing to acquire some technical savvy — not a great deal, but some. Purchasing should attend vendors' preliminary presentations and site visits, cosponsor the bidders' conference, and coauthor the RFP with Telecom.

This experience enables Purchasing to see first hand that going for the lowest-cost award is wrong. They can see, perhaps for the first time, that apples-to-apples comparisons are near impossible with high-tech purchases, although they may very well be possible with purchases of pencils — or even apples. The "best-value" approach will soon become obvious to Purchasing when they can see for themselves the hundreds of variables that go into the decision to buy today's sophisticated, high-tech systems.

Tip No. 5
"Form a project team for every major purchase."

Today's high-tech telecom projects must have a project team consisting not only of Telecom and Purchasing, but MIS-dp, Finance, Facilities, Legal, and representatives of the various user departments. The wrong, or at least less desirable, telecom system will almost certainly be purchased if inputs from other departments are not solicited and taken into consideration. This is because today's telecom systems typically carry MIS-dp's data, run through Facilities' wiring, and greatly affect user productivity in every other department. Telecom people can no longer write today's spec alone, although many still try.

Tip No. 6
"Understand that hardly half the evaluation is technical."

Nontechnical factors have risen up to claim half the weight or more in a high-tech evaluation. People with technical blinders do not always see this, so Purchasing must be very sensitive to this issue and actively participate in vendor selection — preferably side by side with Telecom.

Will the manufacturer still be in business in five years, or even in two years? How good is the vendor's track record on service and upgrades? Does the vendor have enough capital for continual upgrades and to address unanticipated field problems? Is the vendor ripe for a takeover? How strongly committed is the vendor's upper management? There are literally dozens of nontechnical issues like these that merit consideration.

Technical people are disappointed because the decision to go with one vendor over another may be decided on nontechnical factors, before they get a chance to delve into such technical details as backplane speed, modulation scheme, bit *versus* character orientation, protocol, software architecture, and so on.

The high-tech field is already littered with numerous carcasses of vendors who had clever, wonderful technology, but, unfortunately, bit the dust in today's fast-track arena.

Tip No. 7
"Write a prototype contract."

Here is a unique concept that is seldom used, but can greatly assist you in purchasing and installing the right system. Under the "prototype contract" concept, the Purchasing-Telecom team begins to write the contract on the very first day the project is established. It becomes another up-front effort *versus* the traditional method of trying to write the contract at the last minute. The advantages to writing a prototype contract include:

- It provides a convenient place to write down desires immediately; waiting too long to start writing could result in a 70% loss of ideas.
- It provides a collection point for vendor literature, which may become official attachments to the final contract.
- If shown to vendors early, say at the bidders' conference, the prototype contract may cause incompetent vendors to drop out.
- There is no scramble to write the contract at the last minute, and less chance of getting careless. This eliminates the need to rush everything through Legal and other departments for approval.

All of this assumes, of course, that the purchaser is smart enough to realize that writing one's own contract results in less risk than merely accepting the vendor's standard two-pager. Morgan states that these concepts have demonstrated their effectiveness in high-tech purchasing again and again. He predicts that all purchasers will eventually use them — it is just a matter of time. "Old habits can be hard to shake. Sometimes the Purchasing Department will resist. Sometimes it is upper management, which enlightened Purchasing and Telecom must convince. And the sooner the better for all concerned except, of course, the weak vendor."

13.2 PROTECT YOUR INVESTMENT

Risk is endemic to the purchase of high-technology products and services, mainly because the pace of innovation is such that no vendor can maintain a commanding lead for long. Moreover, the market is a moving target; it is difficult for well-intentioned vendors to predict what users will want a year from now and then to be right on target after a twelve-month development effort, and that is if users do not change their minds, or regulatory changes do not pull the rug out from under the vendor.

As a purchaser, your strategy for risk avoidance boils down to protecting the investment. This means taking appropriate steps to insure that vendors are financially stable, competitively positioned, and properly staffed and managed. It also means taking precautionary measures to insure that you do not become locked into proprietary products, which could result in premature obsolescence, limited connectivity potential, or cause you to become overly dependent on a single vendor. In such cases, the result is inevitably higher operating costs, delayed implementation, and, of course, lost credibility. By way of review, make sure you evaluate the vendor in terms of the twelve criteria discussed in Chapter 1:

- Research and Development
- Applications Engineering

- Quality Assurance
- Repair and Return
- Customer Service
- Technical Documentation
- Training
- Line of Business
- References
- Escrow Protection
- Quality of Salespeople
- Human Resources

This investigative procedure requires visits to vendor and customer sites, detailed discussions with vendor personnel, as well as verification procedures that entail background checks, talks with third parties, and a thorough reading of product documentation. To insure a sound purchasing decision, you will have to develop a plan that takes into account the needs of other groups in the organization, preferably with their active participation during the planning stage, and, if you are not staffed to do the necessary legwork, you should seek the help of an experienced consultant who can augment your staff with one of his or her own. After all, the future competitive position of your company may be determined by the decisions you make today.

13.3 PROTECT YOUR NETWORK

Networks are not something that you can set up and just walk away from. They require constant attention. In fact, the only thing that you can be sure of about your network is that it will continue to change. Some of the factors that contribute to change include:

- Company growth, including the addition or relocation of personnel and physical plant;
- Merger or acquisition activities;
- Entrance into new markets or expansion into new territories;
- Competitive pressures, which may force the company into new ways of reaching potential customers and of improving linkages to present customers;
- Economic climate, which may either limit or facilitate expansion, according to such factors as the money supply, condition of the trade balance, and the mood of Wall Street;
- Markets, the projection of which may force production cuts, reductions in force, plant closings, and a scaling back in the number and types of services and products offered.

So many other variables enter into the rate at which your network will change;

such things as tariffs, new products and services from vendors, and the pace of technological innovation in the computer and communications industries. What does all this mean within the context of protecting your network? It means that product selection must be governed by such criteria as:

- Modularity
- Reliability
- Configuration Flexibility
- Upgradeability
- Compatibility (Standards)
- Expandability
- Maintainability
- Affordability

In choosing products with these attributes, you will not be locked into single-vendor dependence. Your ability to expand and enhance your network will not be hampered. Your exposure to a variety of high-risk situations will be minimized. Thus, your network will be protected.

13.4 PROTECT YOUR CREDIBILITY

Consultants and users can regale each other for hours on end with stories of vendor duplicity, but when you get right down to it, many problems with vendor performance are the result of poor planning on the part of buyers; specifically, ill-defined needs, ambiguous RFPs, or just plain incompetence. When things go wrong, and you are finally called on the carpet by top management for the bad purchase decision, the words "I just assumed . . . " will fall on deaf ears. While you are making excuses, others are wondering why you did not do your homework. Why you did not ask the right questions. Why you did not think. In the real world of business, it is difficult, if not impossible, to recover from lost credibility. Even if you are around long enough to be given another chance, others will be looking over your shoulder at your every move, an indication that you really have not overcome the credibility gap.

Aside from heeding the strategy for risk avoidance described in this book — when it is appropriate for your situation — there are some things you can do to build credibility within the organization and, in the process, prepare yourself for making the tough decisions that lie ahead.

- Stay informed of trends in technologies, especially those facilitating the convergence of computers and communications. Also track developments in connectivity, especially those that bridge local-area networking with wide-area networking.
- Do not wait to get into the purchasing mode before you talk to hardware

and software vendors about their product development plans; make this an ongoing activity. Since your time is limited, you can spend it wisely by attending trade shows and seminars, where information-gathering opportunities abound.

- At the same time, there is no need to study every morsel of information that comes your way. Often, information that seems useful has no immediate relevance to your organization's present needs. You should attempt to categorize information gathered from trade shows and periodicals, and file it for future reference — even if it means ripping your favorite magazines to shreds. The types of information worth saving include: technical tutorials, news items on the vendors you do business with, market analyses, trends in technology, and user case studies. When the need arises for this information, you will know exactly where to find it, and it will already be categorized, and being able to pull out facts and figures quickly will enhance your image as an expert.

- One of the best things you can do is get to know your company's business inside and out. This means getting your hands on the business plan and devoting the time necessary to study it thoroughly. In reading it, you should determine how computer resources and the communications network can be used to help achieve the organization's objectives.

- Learn the language of accounting. This will help you sell projects to senior management, who typically evaluate large purchases on the basis of their return on investment (ROI).

- Learn the language of marketing. This will help you sell projects to senior management on the basis of improving the competitive position of the company, a situation that does not lend itself to strict reliance on the numbers.

- You can get to know the business by getting some hands-on experience, such as by volunteering for special projects, study groups, or steering committees. These activities will provide you with detailed knowledge about the special problems different operating groups face. This information will help you be more responsive to the communications needs of various users when it comes time to upgrade or expand the information systems and communications facilities.

In expanding your perspective beyond a particular technical discipline, you accumulate the knowledge base with which to make sound judgements about products, technologies, and applications, and through it all, you are protecting your greatest asset — your credibility.

REFERENCE

[1]   Gantz, John, "Do vendors deliver what they promise?" *Telecommunications Products and Technology,* May 1988, pp. 41–55.

Glossary

ABATS	Automatic Bit Access Test System
ACD	Automatic Call Distribution
ACK	Positive Acknowledgement
A-D-A	Analog-Digital-Analog
ADPCM	Adaptive Differential Pulse Code Modulation
AIOD	Automatic Identified Outward Dialing
AIS	Alarm Indication Signal
ALI	Automatic Location Information
AM	Amplitude Modulation
ANI	Automatic Number Identification
ANSI	American National Standards Institute
APA	Adaptive Packet Assembly
APPC	Advanced Program-to-Program Communications
Arpanet	Advanced Research Projects Agency Network
ASCII	American Standard Code for Information Interchange
AT&T	American Telephone and Telegraph
ATM	Automated-Teller Machine
AWG	American Wire Gauge
B8ZS	Binary 8-Zero Suppression
BER	Bit-Error Rate
BERT	Bit-Error Rate Tester
BIOS	Basic Input-Output System
BMS	Bandwidth Management System
BOC	Bell Operating Company
BSC	Binary Synchronous Communications

CAD	Computer-Aided Design
CAM	Computer-Aided Manufacture
CAMA	Centralized Automatic Message Accounting
CASE	Computer-Aided Software Engineering
CCITT	Consultative Committee for International Telegraph and Telephone
CCR	Customer Controlled Reconfiguration
CD-ROM	Compact Disc–Read Only Memory
CDR	Call Detail Recording
CO-LAN	Central Office Based Local Area Network
COD	Cash on Delivery
CPE	Customer-Premises Equipment
CP/M	An Operating System
CPU	Central Processing Unit
CRC	Cyclic Redundancy Checking
CRT	Cathode-Ray Tube
CSA	Corporate Service Amendment
CSMA	Carrier Sense Multiple Access
CSMA-CA	Carrier Sense Multiple Access–Collision Avoidance
CSMA-CD	Carrier Sense Multiple Access–Collision Detect
CSR	CENTREX Station Rearrangement
CSU	Channel Service Unit
CTS	Customer Test Service
CVSD	Continuously Variable Slope Delta
DAA	Data Access Arrangement
DACS	Digital Access and Cross-Connect System
DAT	Digital Audio Tape
DDD	Direct Distance Dialing
DDN	Defense Data Network
DDS	Dataphone Digital Service
DDS-SC	Dataphone Digital Service with Secondary Channel
DES	Digital Encryption Standard
DFT	Distributed Function Terminal
DID	Direct Inward Dialing
DIP	Dual In-Line Package
DIU	Digital Interface Unit
DOA	Dead on Arrival
DOD	Department of Defense
DOD	Direct Outward Dialing
DOS	Disk Operating System

DOV	Data Over Voice
DPSK	Differential Phase-Shift Keying
DRAM	Dynamic Random Access Memory
DRINET	Defense Research Internet
DRP	Disaster Recovery Plan
DSI	Digital Speech Interpolation
DSU	Digital Service Unit
DTE	Data Terminal Equipment
DTR	Data Terminal Ready
E&M	Receive and Transmit Leads of a Trunk
EAROM	Electronically Alterable Read Only Memory
EAS	Extended Areas of Service
EBCDIC	Extended Binary Coded Decimal Interchange Code
EDI	Electronic Data Interchange
EEPROM	Electronically Erasable Programmable Read Only Memory
EGA	Enhanced Graphics Adaptor
EIA	Electronics Industry Association
EMS	Expanded Memory Specification
EOT	End of Transmission
ESF	Extended Superframe Format
ETN	Electronic Tandem Network
FCC	Federal Communications Commission
FDDI	Fiber Distributed Data Interface
FDM	Frequency-Division Multiplexing
FEP	Front End Processor
FPU	Floating-Point Unit
FSK	Frequency-Shift Keying
FT1	Fractional T1
FTC	Federal Trade Commission
FX	Foreign Exchange
HASP	Houston Automatic Spooling Program
HDLC	High-Level Data Link Control
HSR	Helical Scan Recording
HVAC	Heating, Ventilation, and Air Conditioning
I/O	Input/Output
IBM	International Business Machines Corporation
IDF	Intermediate Distribution Frame

IEEE	Institute of Electrical and Electronics Engineers
In-WATS	Inward-Wide Area Telecommunications Service
IP	Internet Program
ISDN	Integrated Services Digital Network
ISN	Information Systems Network
JIT	Just in Time
KFLOPS	Thousands of Floating-Point Operations per Second
KTS-PBX	Key Telephone System–Private Branch Exchange
LAN	Local Area Network
LAP-M	Link Access Protocol–Modem
LAPD	Link Access Procedure D
LAT	Local Area Transport
LATA	Local Access and Transport Area
LCR	Least-Cost Routing
LED	Light-Emitting Diode
LPC	Linear Predictive Coding
LSI	Large-Scale Integration
LT	Link Termination
MAC	Media Access Control
MAP	Manufacturing Automation Protocol
MDF	Main Distribution Frame
MES	Master Earth Station
MFLOPS	Millions of Floating-Point Operations per Second
Milnet	Military Network
MIPS	Millions of Instructions per Second
MIS-dp	Management Information System–data processing
MMS	Manufacturing Message Standard
MNM	Multidrop Network Management
MNP	Microcom Network Protocol
MRP	Material Requirements Planning
MRSA	Midrange Service Agreement
MTBF	Mean Time Between Failure
MTTR	Mean Time to Repair
NAK	Negative Acknowledgement
NCP	NetWare Core Protocol
NCS	Network Computing System
NetBIOS	Network Basic Input-Output System

NeWS	Network-Extensible Windowing System
NFS	Network File System
NIU	Network Interface Unit
NSA	National Security Agency
NSF	National Science Foundation
NT	Network Termination
NTT	Nippon Telephone Telegraph
OCR	Optical Character Reader
OEM	Original Equipment Manufacturer
ONA	Open Network Architecture
OPX	Off-Premises Extension
OS/2	Operating System/2
OSI	Open Systems Interconnection
PABX	Private Automatic Branch Exchange
PAD	Packet Assembler-Disassembler
PARC	Palo Alto Research Center
PBX	Private Branch Exchange
PCM	Pulse Code Modulation
PDS	Premises Distribution System
PNB	Pacific Northwest Bell
PNX	Private Network Exchange
POP	Point of Presence
POS	Point of Sale
POTS	Plain Old Telephone Service
PSAP	Public Service Answering Point
PSK	Phase-Shift Keying
PTE	Packet Transport Equipment
PTT	Post, Telegraph, and Telephone Administration
PUC	Public Utilities Commission
QA	Quality Assurance
QAM	Quadrature Amplitude Modulation
QIC	Quarter-Inch Cartridge
R&D	Research and Development
RAM	Random Access Memory
RF	Radio Frequency
RFI	Request for Information
RFP	Request for Proposal
RFS	Remote File Sharing

RISC	Reduced Instruction Set Computing
RLL	Run Length Limited
ROI	Return on Investment
RPC	Remote Procedure Call
SCS	Scientific Computer Systems Inc.
SCSI	Small Computer System Interface
SDLC	Synchronous Data Link Control
SDN	Software Defined Network
SMB	Server Message Block
SMDR	Station Message Detail Recorder
SNA	System Network Architecture
SONET	Synchronous Optical Network
SPARC	Scalable Processor Architecture
SQL	Structured Query Language
SS#7	Signaling System No.7
STC	Society of Telecommunications Consultants
STDM	Statistical Time-Division Multiplexer
STP	Signal Transfer Point
STS	Shared Tenant Services
STX	Start of Transmission
TA	Terminal Adaptor
TASI	Time Assignment Speech Interpolation
TCM	Trellis-Coded Modulation
TCP	Transmission Control Program
TCP-IP	Transmission Control Protocol–Internet Protocol
TDM	Time-Division Multiplexing
TDMA	Time-Division Multiple Access
TDR	Time Domain Reflectometer
TE	Terminal Equipment
TOP	Technical and Office Protocol
TV	Television
TWX	Teletypewriter Exchange
UL	Underwriter's Laboratories
UM	Universal Module
UNIX	An Operating System
UNMA	Unified Network Management Architecture
UOS	Universal Operations Systems
UPI	United Press International
UPS	Uninterruptible Power Supply

V&H	Vertical and Horizontal
VAR	Value-Added Reseller
VCR	Video Cassette Recorder
VCS	Virtual Circuit Switch
VF	Voice Frequency
VGA	Video Graphics Array
VLSI	Very Large-Scale Integration
VMS	Visual Memory System
VOM	Volt-Ohm Meter
VSAT	Very Small Aperture Terminal
WAN	Wide Area Network
WATS	Wide Area Telecommunications Service
WORM	Write Once–Read Many
WYSIWYG	What You See Is What You Get
XNS	Xerox Network Systems

Bibliography

Beizer, Boris. *Personal Computer Quality: A Guide for Victims and Vendors,* New York: Van Nostrand Reinhold, 1986.

Connell, John L., and Linda Shafer. *The Professional User's Guide to Acquiring Software,* New York: Van Nostrand Reinhold, 1987.

Ditto, Steve. *Buying Short-Haul Microwave: The Official Guide to Choosing, Acquiring and Using a Short-Haul Microwave System in North America,* New York: Telecom Library, 1988.

Flanagan, William. *The Teleconnect Guide to T-1 Networking: How to Buy, Install, and Use T-1 Circuits,* New York: Telecom Library, 1987.

Newton, Harry. *Which Phone System Should I Buy?,* New York: Telecom Library, 1987.

Schank, Roger. *The Creative Attitude: Learning to Ask and Answer the Right Questions,* New York: Macmillan, 1988.

Self, Robert. *Long Distance for Less: The Official Guide to Long Distance Telephone Services in the United States,* New York: Market Dynamics, 1988.

T-1 Multiplexer Industry Analysis, Dedham, MA: Vertical Systems Group, 1988.

Thaker, Chet. *Negotiating Telecommunications Contracts: The Official Guide to Buying Telecommunications Products and Services,* New York: Telecom Library, 1988.

Young, Jeffrey S. *Steve Jobs: The Journey Is the Reward,* Glenview, IL: Scott, Foresman and Company, 1988.

Index

The Author

Until recently, Nathan Muller was an independent communications consultant in Huntsville, Alabama. Since writing this book, he has joined the staff of General DataComm, Inc., in Middlebury, Connecticut, where he heads the company's Consultant Relations Program.

Nathan Muller's eighteen-year career in the computer and communication industries includes a four-year stint in the United States Marine Corps as a communications and electronics instructor at MCRD, San Diego, from 1971 to 1975. He has a B.A. in Political Science from San Diego State University and an M.A. in Social and Organizational Behavior from George Washington University.

He has held numerous technical and marketing positions with such companies as Planning Research Corporation, Cable & Wireless Communications (formerly TDX systems, Inc.), and ITT Telecom. He codesigned the first telecommunication courses offered at Northern Virginia Community College as part of that school's evening program for local professionals. He has written numerous articles on marketing, technical, management, and regulatory topics for such publications as *Data Communications, Telephony, Telephone Engineer & Management, Telecommunications, Network World,* and *Communications Week.*

The opinions expressed in this book are entirely the author's own and do not necessarily reflect the views of General DataComm, Inc.